Fundamental
Mathematics
for Health Careers

3rd Edition

Delmar Publishers Online Services

Fundamental Mathematics for Health Careers

3rd Edition

Jerome D. Hayden
Mathematics Department Head K-12
McLean County Unit District No. 5

Howard T. Davis
Educational Consultant
Retired Assistant Superintendent
Curriculum & Instruction
McLean County Unit District No. 5

Delmar Publishers, Inc.™

1945 - 1995
50 years

I(T)P An International Thomson Publishing Company

Albany • Bonn • Boston • Cincinnati • Detroit • London • Madrid
Melbourne • Mexico City • New York • Pacific Grove • Paris • San Francisco
Singapore • Tokyo • Toronto • Washington

NOTICE TO THE READER

Cover photo courtesy of Gary Conner
Cover Design: Highwire Digital Arts

Delmar Staff
Publisher: Robert D. Lynch
Developmental Editor: Mary Clyne
Production Coordinator: Larry Main
Art and Design Coordinator: Nicole Reamer

COPYRIGHT ©1996
By Delmar Publishers Inc.
an International Thomson Publishing Company

The ITP logo is a trademark under license.

Printed in the United States of America

For more information, contact:

Delmar Publishers
3 Columbia Circle, Box 15015
Albany, New York 12212-5015

International Thompson Publishing Europe
Berkshire House 168 - 173
High Holborn
London WC1V 7AA
England

Thomas Nelson Austrailia
102 Dodds Street
South Melbourne, 3205
Victoria, Australia

Nelson Canada
1120 Birchmount Road
Scarborough, Ontario
Canada M1K 5G4

International Thomson Editores
Campos Eliseos 385, Piso 7
Col Polanco
11560 Mexico D F Mexico

International Thomson Publishing GmbH
Königswinterer Strasse 418
53227 Bonn
Germany

International Thomson Publishing Asia
221 Henderson Road
#05-10 Henderson Building
Singapore 0315

International Thomson Publishing — Japan
Hirakawacho Kyowa Building, 3F
2-2-1 Hirakawacho
Chiyoda-ku, tokyo 102
Japan

1 2 3 4 5 6 7 8 9 10 XXX 01 00 99 98 97 96 95

Library of Congress Cataloging-in-Publication Data

Hayden, Jerome D.
    Fundamental mathematics for health careers / Jerome D. Hayden, Howard T. Davis. — 3rd ed.
    p.   cm.
    Includes index.
    Rev. ed. of: Mathematics for health careers. © 1980.
    ISBN: 0-8273-6689-2. — ISBN 0-8273-6689-2
    1. Mathematics.   I. Davis, Howart T.   II. Hayden, Jerome D.
    Mathematics for health careers.   III. Title.
    IV. Title: Mathematics for health careers.
        (DNLM: 1. Mathematics.   QA 39.2 H414f 1996)
    QA39.2.H37 1996
    613'.13'02461—dc20
    DNLM/DLC   95-14206
    for Library of Congress   CIP

Contents

Preface

Mathematics is an important foundation for all that we do. Even primitive civilizations developed methods of counting and measuring. Modern technology has carried the use and application of mathematics to new levels. Healthcare workers will find many opportunities to apply the math skills developed through the use of *Fundamental Mathematics for Health Careers*.

This is a comprehensive text designed to cover a a wide range of healthcare services. Space does not permit extensive development in any one, specialized area. Applications are taken from real-life situations whenever possible. The user of this text must recognize that healthcare service is continually changing and improving. While particular applications may become dated, the methods used to solve related problems remain the same.

Students, teachers and reviewers of previous editions have provided valuable suggestions for continued improvement and revision of this current edition. We thank these many individuals for sharing their thoughts. This input has led to reorganization, expansion and additions to the previous edition. Students and teachers have been provided with learning objectives, vocabulary words, carefully developed examples, applications, solutions, tests and summary overviews.

Calculators and computers play important roles as tools for solving problems in healthcare service. Tips are introduced throughout the book to indicated uses of the calculator. Selecting the most appropriate calculator can be a challenge. Good solar or battery powered scientific calculators may be purchased for under \$25. Scientific calculators possess an algebraic logic system, which lets users enter problems from left to right. A typical key configuration will contain at least the following keys:

$$y^x \text{ or } x^y, \ (\ ), \ \log, \ \ln, \ \sin, \ \cos, \ \tan, \ +/-, \ \sqrt{}, \ \text{and } x^2$$

If you are not certain the calculator, is scientific, try entering the following problem:

$$3 + 2 \times 5 =$$

The keying sequence directs the user to press the keys labeled 3, +, 2, ×, 5 and = in that order. A scientific calculator with algebraic logic should display an answer of 13.

$$3 + 2 \times 5 = 13$$

Key the same sequence into a non-scientific calculator and the displayed answer is 25!

Computer use has rapidly expanded in the past twenty years. Even in their sophisticated design, computers still let the user store, process, retrieve and apply data. Numerous programs have been developed for specialized applications. Some use additional data entered on-site by a health service worker; some collect data directly, whether it comes from a patient across the street or around the world by means of satellite transmission. This text only introduces some of the most basic applications. Your real understanding will begin when you start using specific programs relating to your work.

Expanding your awareness of math as it relates to healthcare service can be accomplished in many ways. Common publications such as newspapers and magazines frequently include graphs, charts and statistical information relating to the health of humans. Professional journals deal more intensively with topics and often include more advanced applications of mathematical expressions. Your life is touched by mathematics each day, even if you don't notice it. The skills you develop upon successful completion of this text will greatly assist you both personally and professionally.

Acknowledgments

The development of this text began with questions from students about the applications of mathematics in the field of health care. From its beginning in 1980, each revision has continued to reflect the thoughts of students, teachers and health care workers. This remains the strength of this text. The first edition of this text was more than twice the size of the present edition. Every effort has been made to focus on key mathematical concepts, carefully supplemented examples, step-by-step solutions, sample problems, and selected applications.

The support of Tina Starkey has been greatly appreciated with her review of all problems and the development of worked-out solutions.

PROFESSIONAL RESOURCE

Patricia A. Payne
Family Nurse Practitioner
Veterans Administration
VA Medical Center
Seattle, WA

Steve Penneke
Assistant Director of Pharmacy Services
BroMenn Healthcare Pharmacy
Normal, IL

Cindy Maurer
Radiographic Technologist
Family Practice Associates of Central Illinois
Bloomington, IL

Laurie Benjamin
Medical Technologist
St. Joseph Medical Center
Bloomington, IL

Edith Bare
Chemistry Section Manager
Boswell Hospital
Sun City, AZ

Section One

Common Fractions

Unit 1

Introduction to the Mathematical System

Objectives

After studying this unit the student should be able to:
- **Correctly use basic mathematics terminology.**
- **Use divisibility rules for 2, 3, 4, 5, 6, 8, 9, 10.**
- **Compare whole numbers.**
- **Find the common multiple of two or more numbers.**
- **Give the prime factors of a number.**

Whole numbers are the basis of mathematics. Numerals, together with the basic operations of addition, subtraction, multiplication, and division, comprise the mathematical system. Understanding the proper mathematical terminology is essential for using this mathematical system.

DEFINITIONS

- *Addition* is a fast way of counting. The result is the *sum*.
- *Subtraction* is the opposite of addition. The result is the *difference*.
- *Multiplication* is repeated addition. The numbers that are multiplied are the *factors*. The result is the *product*.
- *Division* is the opposite of multiplication. The result is the *quotient*.

1.1 Exercises

In 1–12, find each result using paper and a pencil. Check your results using a calculator.

1. 3,254
    2,730
    3,149
    + 5,682

2. 531
    − 207

3. 243
    × 17

4. $46\overline{)2,392}$

5. 79,413
    65,042
    + 73,294

6. 9,028
    − 6,139

7. 256
    × 904

8. $104\overline{)25,584}$

9. 643
    805
    1,307
    94,315
    +   61

10. 49,271
    − 1,046

11. 4,361
    × 7,831

12. $471\overline{)256,224}$

Numbers may be considered in many ways. Numbers may be compared to find the greater or lesser quantity, and factored to determine the least common multiple (LCM) or the greatest common factor (GCF).

DIVISIBILITY

A number is *divisible* by another if the quotient is a whole number and the remainder is zero. Divisibility tests are a quick way of finding out if one number is divisible by another number without exactly dividing.

A number is divisible by:

- **2** if the ones digit is even (0, 2, 4, 6 or 8).
- **3** if the sum of the digits is divisible by 3.
- **4** if the number formed by the last two digits is divisible by 4.
- **5** if the ones digit is 0 or 5.
- **6** if the number is divisible by both 2 and 3.
- **8** if the number formed by the last three digits is divisible by 8.
- **9** if the sum of the digits is divisible by 9.
- **10** if the ones digit is 0.

Note: The number 7 has no divisibility pattern.

EXAMPLE 1

Determine whether 216 is divisible by 2, 3, 4, 5, 6, 8, 9 or 10.

2: The ones digit (6) is even. Thus, 216 is divisible by 2.
3: The sum of the digits, 2 + 1 + 6 = 9, is divisible by 3. Thus, 216 is divisible by 3.
4: The number formed by the last two digits (16) is divisible by 4. Thus, 216 is divisible by 4.
5: The ones digits is not 0 or 5. Thus, 216 is not divisible by 5.
6: 216 is divisible by both 2 and 3. Thus, 216 is divisible by 6.
8: The number formed by the last three digits (216) is divisible by 8. Thus, 216 is divisible by 8.
9: The sum of the digits, 2 + 1 + 6 = 9, is divisible by 9. Thus, 216 is divisible by 9.
10: The ones digit is not 0. Thus, 216 is not divisible by 10.

1.2 Exercises

Group these numbers according to divisibility.

1,764	11,130	296,048
3,278	49,000	446,075
65	6,775	786,312
365	82,722	954,560
7,420	24,835	1,526,200

1. by 1   2. by 2   3. by 3   4. by 4

5. by 5   6. by 6   7. by 8   8. by 9

9. by 10   10. Explain what the display on your calculator shows when a number is divisible by another number.

COMPARING WHOLE NUMBERS

Numbers that represent the same quantity are *equal*. The symbol used to show equality is =.

EXAMPLE 1

$3 + 5 = 18 - 10$
$84 \div 4 = 7 \cdot 3$ ◄ *Note:* $\times$ and $\cdot$ will each be used to show multiplication.

Numbers that do not represent the same quantity are *unequal*. The symbol used to show that two numbers are not equal is $\neq$.

EXAMPLE 2

$88 - 13 \neq 75 + 13$
$36 \div 3 \neq 3 \cdot 3$

Numbers that do not represent the same quantity are *greater than* the quantity or *less than* the quantity. The symbol for greater than is >. The symbol for less than is <. To help remember which symbol is which, do the following:

1. Place one dot by the smaller quantity.
2. Place two dots by the larger quantity.
3. Connect the dots.

   ex.: $8 < 12$ $9 > 5$

EXAMPLE 3

$88 - 13 < 75 + 13$ since $75 < 86$.
$36 \div 3 > 3 \cdot 3$ since $12 > 9$.

DIVISION INVOLVING ZERO

In a division problem involving zero, there are three possibilities:
- Zero is the dividend.
- Zero is the divisor.
- Zero is the dividend and the divisor.

EXAMPLE 1

Zero is the dividend.
$0 \div 5 = ?$
$0 \div 5 = ?$ means $? \times 5 = 0$
Since any number times zero is zero, $0 \times 5 = 0$ and $0 \div 5 = 0$.
$0 \div 5 = 0$ $0 \times 5 = 0$

- Zero divided by any number equals zero.

EXAMPLE 2

CALCULATOR DISPLAY

Zero is the divisor.

E r r o r $4 \div 0 = $ Error

$4 \div 0 = ?$
$4 \div 0 = ?$ means $? \times 0 = 4$
Since there is no number times zero that equals 4, $? \times 0 = 4$ is not possible and $4 \div 0 = ?$ is not possible. It is said that $4 \div 0$ is not defined.

4 ÷ 0 = not defined   no number × 0 = 4

- Any number divided by zero is not defined.

EXAMPLE 3   CALCULATOR DISPLAY

Zero is the dividend and the divisor.   E r r o r   0÷0=Error
$0 \div 0 = ?$
$0 \div 0 = ?$ means $? \times 0 = 0$
Since any number times zero equals zero, it is said that $0 \div 0$ is not defined.
$0 \div 0 =$ not defined   any number × 0 = 0

- Zero divided by zero is not defined.

1.3 Exercises

In 1–18, compare each pair of numbers. Use the symbols <, >, =.

1. $5 \underline{\ ?\ } 7$
2. $7 \underline{\ ?\ } 5$
3. $13 \cdot 3 \underline{\ ?\ } 3 \cdot 13$
4. $42 \cdot 2 \underline{\ ?\ } 4 \cdot 21$
5. $7 \cdot 2 \cdot 3 \underline{\ ?\ } 7 \cdot 6$
6. $5 \cdot 8 \cdot 2 \underline{\ ?\ } 40 \cdot 0$
7. $36 + 2 \underline{\ ?\ } 40 - 8$
8. $21 - 7 \underline{\ ?\ } 28 \div 2$
9. $99 - 9 \underline{\ ?\ } 90 - 9$

10. $100 \div 10 \underline{\ ?\ } 19 + 9$
11. $0 \cdot 8 \underline{\ ?\ } 0 \div 8$
12. $1 \cdot 9 \underline{\ ?\ } 9 \div 1$
13. $16 \cdot 3 \underline{\ ?\ } 42 \div 6$
14. $55 \div 5 \underline{\ ?\ } 3 \cdot 4$
15. $23 \cdot 5 \underline{\ ?\ } 15 + 25 + 75$
16. $10 + 20 + 50 \underline{\ ?\ } 160 \div 2$
17. $1,000 - 72 \underline{\ ?\ } 459 \cdot 2$
18. $890 - 90 \underline{\ ?\ } 90 + 800$

FACTORS, MULTIPLES, PRIME NUMBERS

Factors and multiples are a valuable tool in using the mathematical system. Prime numbers help to simplify the process of finding factors and multiples.

DEFINITIONS

- *Factors* are the numbers being multiplied to find a product.
- *Common factors* of two or more numbers are the factors that are common to both numbers.
- *Factorization* is the process of finding the factors of a number.
- A *multiple* is the product of a given number and another factor.
- A *common multiple* is a number which is a multiple of each of two or more numbers.
- A *prime number* is a natural number, other than 1, having only 1 and itself as factors. The first five prime numbers are 2, 3, 5, 7 and 11.
- A *composite number* is any number with more than two factors.
- A *prime factor* is a prime number that is a factor of a number.
- *Prime factorization* is the process of finding the factors of a number using only prime numbers.

Counting by 3 gives the numbers
3, 6, 9, 12, 15, 18, 21, 24, 27, 30, . . .
The numbers are <u>multiples</u> of 3.
The number 12 is a multiple of 3.

   $12 = 4 \cdot 3$

Three is a <u>factor</u> of 12. Four is a <u>factor</u> of 12. Four is not a <u>prime factor</u> of 12 because four is not a <u>prime number</u>.

Counting by 5 gives the numbers
5, 10, 15, 20, 25, 30, 35, 40, 45, 50, . . .
The numbers are <u>multiples</u> of 5.
The number 35 is a multiple of 5.

   $35 = 7 \cdot 5$

Five is a <u>factor</u> of 35. Seven is a <u>factor</u> of 35.

The numbers 15 and 30 are multiples of both 3 and 5. Thus, the numbers 15 and 30 are <u>common multiples</u>.

Any number can be expressed using factors.

EXAMPLE 4

Find two factors of 28.
$28 = 7 \cdot 4$
Four is a factor of 28. Seven is a factor of 28.
Since 7 is also a prime number, it is called a <u>prime factor</u>.

EXAMPLE 5

Find the factors of 33.
$33 = 11 \cdot 3$ *or* $1 \cdot 33$
Eleven is a factor of 33. Three is a factor of 33.
Since both 11 and 3 are prime numbers, both numbers are <u>prime factors</u> of 33.

EXAMPLE 6

Use prime factorization to find the prime factors of 60. This can be done using either a factor tree or successive division on your calculator.

$$
\begin{array}{c}
60 \\
2 \cdot 30 \\
2 \cdot 2 \cdot 15 \\
2 \cdot 2 \cdot 3 \cdot 5
\end{array}
\qquad\qquad
\begin{array}{r}
5 \\
3\overline{)15} \\
2\overline{)30} \\
2\overline{)60}
\end{array}
$$

    $60 = 2 \cdot 2 \cdot 3 \cdot 5$ $60 = 2 \cdot 2 \cdot 3 \cdot 5$

a. A Factor Tree   b. Successive Division on Your Calculator

The numbers 2, 3 and 5 are prime factors.

PROCEDURE FOR FINDING THE PRIME FACTORIZATION OF A NUMBER

Step 1	Step 2	Step 3
Write down the number to be prime factored.	Try to divide the number to be factored by 2, the first prime number. If 2 will not divide the number evenly, go to the next prime number 3, etc. Use the tests for divisibility to quicken the process. Continue until you find a prime factor.	Write down both the first prime factor (divisor) that works along with the quotient (answer).

Step 4	Step 5	Step 6
Repeat Step 2 on the quotient found in Step 3.	Continue steps 2, 3, and 4 until all the factors are prime.	Check your work by multiplying all the factors together. Their product should be the original number.

1.4 Exercises

In 1–6, find two common multiples of each group of numbers.

1. 2, 5
2. 9, 4

3. 7, 6, 8
4. 5, 9, 2, 7

5. 3, 4, 5
6. 3, 5, 7, 2

In 7–12, find the prime factorization of each number.

7. 42
8. 68

9. 99
10. 60

11. 125
12. 132

13. List the first dozen prime numbers.

14. Explain why one (1) is not prime.

For problems 15–24:

   a. Enter into your calculator each symbol, in order from left to right. Press $\boxed{=}$ to get the answer to each expression.
   b. Circle YES if the answer is prime or NO if the answer is not prime.

Expression	Answer	Prime
3 • 4 + 1	15. _____	16. Y or N
42 ÷ 6 + 9	17. _____	18. Y or N
4120 − (14 • 130)	19. _____	20. Y or N
(52 • 8) + (832 ÷ 416)	21. _____	22. Y or N
1576 + 24 − (3000 ÷ 2)	23. _____	24. Y or N

APPLICATIONS

Almost four million people work in health-related occupations. Of this number, hospitals employ about one-half of the workers in the health field. Clinics, laboratories, pharmacies, nursing homes, public health agencies, mental health centers, private offices, and patients' homes are places of employment of the other one-half of the workers.

1.5 Exercises

1. Using a city directory, determine the number of hospitals, clinics, laboratories, pharmacies, nursing homes and other health care centers that are located in the community.

2. Tour one or more of these health care facilities and list the type of health care positions that are available. Group the health care workers by occupation and observe the distribution. Note the differences between smaller and larger facilities.

3. Inquire about one or more of the health care positions. Find out the nature of the job, training and other qualifications, and employment outlook. Compare the finding with smaller and larger facilities.

4. Develop a list of additional sources that may be used to obtain further information about health-related occupations.

5. Consult with a county hospital or clinic supervisor to determine how they estimate the need for health care services in their community.

6. Consult the "Help Wanted" advertisements in at least one major city newspaper to determine the opportunities for health care related employment in that community.

7. Through consultation with your instructor, medical supervisor, or others employed in the health care area, develop a list of courses and degrees you must complete to advance to the level of health care service you want to provide during your career.

8. Data revealed an average of 2 hospitals in each of four cities having populations from 50,000 to 75,000. Using this information, how many total hospitals would be found in 12 cities?

9. An area vocational center (AVC) considered starting a health career course for high school students. A survey found that 7 students out of 100 at grade 11 were interested in taking the course.

   a. How many total students would be expected to show this interest if the total 11th grade population served by the AVC was 1400?

   b. Each section of class could serve only 20 students. How many sections must be created to serve all interested students?

10. Mortality rates were studied for a population where the number of deaths for 25- to 35-year-old males was 8 per 1,000 each year.

   a. If there was a total of 27,000 males in this age group, how many deaths would be expected to occur for this group during the next 12 months?

   b. One year the number of deaths increased to 127 deaths per 1,000 for the males in this population. Give some reasons why this increase may have resulted.

Unit *2*

Introduction to Common Fractions and Mixed Numbers

Objectives

After studying this unit the student should be able to:
- **Express fractions in lowest terms.**
- **Express fractions as equivalent fractions.**
- **Compare fractional values.**

FRACTIONAL PARTS

A fraction is a comparison between a part of a whole and the whole.

EXAMPLE 1

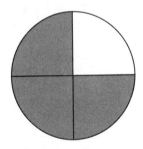

In this figure 3 of the 4 regions are shaded.
The fraction $\frac{3}{4}$ represents this comparison.

A fraction is also a comparison of a part of a group to the whole group.

EXAMPLE 2

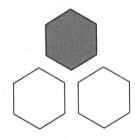

In this figure, 1 hexagon of the 3 hexagons is shaded.
The fraction $\frac{1}{3}$ represents this comparison.

DEFINITIONS

- The *numerator* is the number of parts being used. It is the number above the fraction bar. $\longrightarrow$
- The *denominator* is the number of equal parts needed to make a whole object. It is the number below the fraction bar. $\longrightarrow$

$$\frac{3}{4}$$

- *Equivalent fractions* are two or more fractions which represent the same part of the whole or the same part of a group.
- The *greatest common factor* (GCF) of two or more numbers is the largest factor common to all numbers. The GCF is used when reducing a fraction to lowest terms.
- A fraction is in *lowest terms* if the greatest common factor (GCF) of both the numerator and the denominator is 1.
- The *lowest-term fraction* is the simplest fraction in a group of equivalent fractions.
- A *mixed number* is a number having a whole number part and a fractional part.

EQUIVALENT FRACTIONS

These fractions name the shaded part of each region.

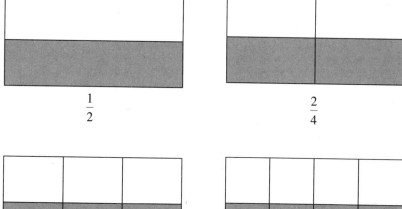

$$\frac{1}{2} \qquad\qquad \frac{2}{4}$$

$$\frac{3}{6} \qquad\qquad \frac{4}{8}$$

The same part is named by each of these fractions. The fractions $\frac{1}{2}, \frac{2}{4}, \frac{3}{6}$, and $\frac{4}{8}$ are <u>equivalent fractions</u>.

Multiplication is used to find equivalent fractions. The fraction is multiplied by a fractional form of 1. This means that the numerator and the denominator are multiplied by the same number.

EXAMPLE 1

Find four fractions that are equivalent to $\frac{1}{2}$.

Fraction	Equivalent Fractions			
$\frac{1}{2}$	$\frac{1 \times \boxed{2}}{2 \times \boxed{2}} = \frac{2}{4}$	$\frac{1 \times \boxed{3}}{2 \times \boxed{3}} = \frac{3}{6}$	$\frac{1 \times \boxed{4}}{2 \times \boxed{4}} = \frac{4}{8}$	$\frac{1 \times \boxed{5}}{2 \times \boxed{5}} = \frac{5}{10}$

The fractions $\frac{2}{4}, \frac{3}{6}, \frac{4}{8}$, and $\frac{5}{10}$ are equivalent to $\frac{1}{2}$.

Note: The preceding multiplication could have been done using parentheses to show multiplication.

$$\left(\frac{1}{2}\right)\left(\frac{2}{2}\right) = \frac{2}{4} \quad \left(\frac{1}{2}\right)\left(\frac{3}{3}\right) = \frac{3}{6} \quad \left(\frac{1}{2}\right)\left(\frac{4}{4}\right) = \frac{4}{8} \quad \left(\frac{1}{2}\right)\left(\frac{5}{5}\right) = \frac{5}{10}$$

2.1 Exercises

In 1–4, each figure suggests a pair of equivalent fractions. Name the pair of equivalent fractions for each.

1.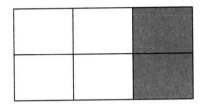

3.

2.

4.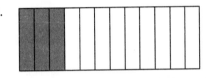

In 5–12, use multiplication to find equivalent fractions for each fraction in this chart.

	Fraction	Equivalent Fractions				
5.	two fifths	$\dfrac{2}{5}$	$\dfrac{4}{10}$	$\dfrac{6}{15}$		
6.	one fourth	$\dfrac{1}{4}$	$\dfrac{2}{8}$			
7.	three halves	$\dfrac{3}{2}$	$\dfrac{6}{4}$	$\dfrac{9}{6}$		
8.	one third	$\dfrac{1}{3}$				
9.	nine tenths	$\dfrac{9}{10}$				
10.	three fourths					
11.	four fifths					
12.	seven tenths					

Fractions can be shown to be equivalent by using cross products. Cross products are found this way.

$$\frac{2}{5} \searrow \frac{4}{10} \qquad \frac{2}{5} \nearrow \frac{4}{10}$$

$$10 \cdot 2 = 20 \qquad\qquad 5 \cdot 4 = 20$$

Since 20 = 20, the fractions are equal. This equivalence is written $\dfrac{2}{5} = \dfrac{4}{10}$.

EXAMPLE 2

Using cross products determine if $\frac{1}{3}$ is equivalent to $\frac{2}{6}$.

$$\frac{1}{3} \searrow \frac{2}{6} \qquad\qquad \frac{1}{3} \nearrow \frac{2}{6}$$

$6 \cdot 1 = 6$ $\qquad\qquad$ $3 \cdot 2 = 6$

Since $6 = 6$, the fractions are equivalent. $\frac{1}{3} = \frac{2}{6}$.

EXAMPLE 3

Is $\frac{3}{5}$ equivalent to $\frac{5}{10}$?

$$\frac{3}{5} \searrow \frac{5}{10} \qquad\qquad \frac{3}{5} \nearrow \frac{5}{10}$$

$10 \cdot 3 = 30$ $\qquad\qquad$ $5 \cdot 5 = 25$

Since $30 \neq 25$, the fractions are not equivalent. $\frac{3}{5} \neq \frac{5}{10}$.

- Two fractions are equivalent if the cross products are equal.
  Cross products can also be used to find equivalent fractions.

EXAMPLE 4

Find the denominator which makes $\frac{3}{4}$ and $\frac{6}{?}$ equivalent fractions.

$\dfrac{3}{4} = \dfrac{6}{?}$ $\qquad$ Check using your calculator.

$3 \cdot ? = 4 \cdot 6$

$3 \cdot ? = 24$

$3 \cdot 8 = 24$, so

$\dfrac{3}{4} = \dfrac{6}{8}$

CHECK

$\dfrac{3}{4} \overset{?}{=} \dfrac{6}{8}$

$8 \cdot 3 \overset{?}{=} 4 \cdot 6$

$24 = 24$, so

$\dfrac{3}{4} = \dfrac{6}{8}$

CHECK

$\dfrac{3}{4} \overset{?}{=} \dfrac{6}{8}$

$3 \div 4 \overset{?}{=} 6 \div 8$

$0.75 = 0.75$, so

$\dfrac{3}{4} = \dfrac{6}{8}$

EXAMPLE 5

Find the numerator which makes $\frac{1}{2}$ and $\frac{?}{12}$ equivalent fractions.

$\dfrac{1}{2} = \dfrac{?}{12}$ $\qquad$ Check using your calculator.

$2 \cdot ? = 12 \cdot 1$

$2 \cdot ? = 12$

$2 \cdot 6 = 12$, so

$\dfrac{1}{2} = \dfrac{6}{12}$

CHECK

$\dfrac{1}{2} \overset{?}{=} \dfrac{6}{12}$

$12 \cdot 1 \overset{?}{=} 2 \cdot 6$

$12 = 12$, so

$\dfrac{1}{2} = \dfrac{6}{12}$

CHECK

$\dfrac{1}{2} \overset{?}{=} \dfrac{6}{12}$

$1 \div 12 \overset{?}{=} 6 \div 12$

$0.5 = 0.5$, so

$\dfrac{1}{2} = \dfrac{6}{12}$

EXAMPLE 6

Using your calculator, find the decimal equivalent for each fraction (numerator divided by denominator) to determine if these fractions are equivalent or not equivalent.

Is $\frac{3}{4}$ equivalent to $\frac{7}{8}$?

$3 \div 4 \underline{\quad ? \quad} 7 \div 8$

$0.75 \neq 0.875$, so

$\frac{3}{4}$ is not equivalent to $\frac{7}{8}$

Is $\frac{1}{5}$ equivalent to $\frac{2}{10}$?

$1 \div 5 \underline{\quad ? \quad} 2 \div 10$

$0.2 = 0.2$, so

$\frac{1}{5}$ is equivalent to $\frac{2}{10}$

2.2 Exercises

In 1–12, use cross products to determine if these fractions are equivalent or nonequivalent. Use the symbol = for equivalent fractions and the symbol ? for nonequivalent fractions.

1. $\frac{1}{2} \underline{\quad ? \quad} \frac{5}{10}$

2. $\frac{1}{3} \underline{\quad ? \quad} \frac{3}{6}$

3. $\frac{4}{8} \underline{\quad ? \quad} \frac{8}{16}$

4. $\frac{4}{6} \underline{\quad ? \quad} \frac{6}{9}$

5. $\frac{10}{12} \underline{\quad ? \quad} \frac{12}{14}$

6. $\frac{1}{3} \underline{\quad ? \quad} \frac{2}{9}$

7. $\frac{80}{100} \underline{\quad ? \quad} \frac{8}{10}$

8. $\frac{3}{6} \underline{\quad ? \quad} \frac{6}{3}$

9. $\frac{0}{3} \underline{\quad ? \quad} \frac{2}{4}$

10. $\frac{3}{1} \underline{\quad ? \quad} \frac{9}{3}$

11. $\frac{2}{4} \underline{\quad ? \quad} \frac{4}{8}$

12. $\frac{2}{2} \underline{\quad ? \quad} \frac{4}{4}$

 In 13–24, use your calculator to find the decimal equivalent for each fraction (numerator divided by denominator) to determine if these fractions are equivalent or not equivalent. Circle the correct symbol.

13. Is $\frac{6}{100}$ equivalent to $\frac{2}{50}$?

$\underline{\qquad} \overset{=}{\neq} \underline{\qquad}$

14. Is $\frac{8}{32}$ equivalent to $\frac{4}{16}$?

$\underline{\qquad} \overset{=}{\neq} \underline{\qquad}$

15. Is $\frac{8}{2}$ equivalent to $\frac{1}{4}$?

$\underline{\qquad} \overset{=}{\neq} \underline{\qquad}$

16. Is $\frac{3}{3}$ equivalent to $\frac{5}{5}$?

$\underline{\qquad} \overset{=}{\neq} \underline{\qquad}$

17. Is $\frac{4}{5}$ equivalent to $\frac{16}{20}$?

$\underline{\qquad} \overset{=}{\neq} \underline{\qquad}$

18. Is $\frac{5}{2}$ equivalent to $\frac{2}{5}$?

$\underline{\qquad} \overset{=}{\neq} \underline{\qquad}$

19. Is $\frac{10}{3}$ equivalent to $\frac{20}{6}$?

$\underline{\qquad} \overset{=}{\neq} \underline{\qquad}$

20. Is $\frac{9}{4}$ equivalent to $\frac{6}{3}$?

$\underline{\qquad} \overset{=}{\neq} \underline{\qquad}$

21. Is $\frac{1}{2}$ equivalent to $\frac{3}{8}$?

$\underline{\qquad} \overset{=}{\neq} \underline{\qquad}$

22. Is $\frac{0}{5}$ equivalent to $\frac{5}{0}$?

$\underline{\qquad} \overset{=}{\neq} \underline{\qquad}$

23. Is $\dfrac{1}{10}$ equivalent to $\dfrac{0}{5}$?

———— $\overset{=}{\neq}$ ————

24. Is $\dfrac{10}{0}$ equivalent to $\dfrac{0}{10}$?

———— $\overset{=}{\neq}$ ————

In 25–30, use cross products to find the numerator or denominator that makes each pair of fractions equivalent. Check using a calculator.

25. $\dfrac{5}{10} = \dfrac{?}{12}$

26. $\dfrac{30}{?} = \dfrac{30}{100}$

27. $\dfrac{?}{3} = \dfrac{2}{6}$

28. $\dfrac{40}{50} = \dfrac{?}{10}$

29. $\dfrac{10}{?} = \dfrac{15}{30}$

30. $\dfrac{25}{100} = \dfrac{50}{?}$

In 31-36, check by finding the decimal equivalent for each fraction.

31. $\dfrac{4}{10} = \dfrac{?}{20}$

32. $\dfrac{?}{15} = \dfrac{2}{3}$

33. $\dfrac{2}{5} = \dfrac{4}{?}$

34. $\dfrac{18}{?} = \dfrac{24}{4}$

35. $\dfrac{?}{7} = \dfrac{6}{14}$

36. $\dfrac{25}{5} = \dfrac{15}{?}$

EXPRESSING FRACTIONS IN LOWEST TERMS

Numbers being multiplied to find a product are called <u>factors</u>.

EXAMPLE 1

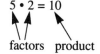

factors product   product factors

The <u>common factors</u> of two or more numbers are the factors that are common to both numbers.

EXAMPLE 2

Find the common factors of 20 and 12.

The factors of 20 are 1, 2, 4, 5, 10, 20.
The factors of 12 are 1, 2, 3, 4, 6, 12.
The common factors are 1, 2, and 4.

The number 4 is the greatest number that is common to both 20 and 12. It is called the <u>greatest common factor</u>.

A fraction is in lowest terms if the greatest common factor (GCF) of both the numerator and the denominator is 1. To find a lowest-term fraction, the numerator and denominator can be divided by the greatest common factor. This means dividing by a fractional form of 1.

EXAMPLE 3

Express $\dfrac{12}{20}$ in lowest terms.

Fraction	**Factors**	**GCF**
$\dfrac{12}{20}$	1, 2, 3, 4, 6, 12 → 1, 2, 4, 5, 10, 20 →	→ 4

$$\dfrac{\text{Numerator} \div \text{GCF}}{\text{Denominator} \div \text{GCF}} \qquad \text{Equals} \qquad \begin{array}{c}\text{Lowest-term}\\ \text{Fraction}\end{array}$$

$$\dfrac{12 \div 4}{20 \div 4} \qquad = \qquad \dfrac{3}{5}$$

The fraction $\dfrac{3}{5}$ is in lowest terms since the greatest common factor of 3 and 5 is 1.

A lowest-term fraction can also be found by <u>prime factorization</u>.

EXAMPLE 4

Using prime factorization, find the lowest-term fraction for $\dfrac{42}{105}$.

The prime factors of 42 are 2, 3, and 7.

The prime factors of 105 are 5, 3, and 7.

$$\begin{aligned}42 &= 2 \cdot 3 \cdot 7\\ 105 &= 5 \cdot 3 \cdot 7\end{aligned}$$

The greatest common factor is $3 \cdot 7$, *or* 21.

Divide both the numerator and denominator by 21. $\dfrac{42}{105} = \dfrac{2}{5}$

Note: The division can be shown as $\dfrac{42}{105} = \dfrac{2 \cdot \cancel{3}^{1} \cdot \cancel{7}^{1}}{5 \cdot \cancel{3}_{1} \cdot \cancel{7}_{1}} = \dfrac{2}{5}$

2.3 Exercises

In 1–9, use these prime factorizations to find each lowest-term fraction.

1. $\dfrac{2 \cdot 3 \cdot 5}{2 \cdot 3 \cdot 7}$

2. $\dfrac{11 \cdot 7 \cdot 5}{13 \cdot 5 \cdot 7}$

3. $\dfrac{3 \cdot 7 \cdot 2}{3 \cdot 7 \cdot 2 \cdot 5}$

4. $\dfrac{3 \cdot 2 \cdot 5 \cdot 2}{5 \cdot 2 \cdot 2 \cdot 7}$

5. $\dfrac{7 \cdot 3 \cdot 5 \cdot 31}{31 \cdot 7 \cdot 5 \cdot 3}$

6. $\dfrac{5 \cdot 7 \cdot 2 \cdot 5}{2 \cdot 5 \cdot 5 \cdot 7 \cdot 3}$

7. $\dfrac{3 \cdot 3 \cdot 2 \cdot 3 \cdot 2}{2 \cdot 2 \cdot 7 \cdot 3}$

8. $\dfrac{2 \cdot 5 \cdot 11 \cdot 3}{3 \cdot 11 \cdot 5 \cdot 2}$

9. $\dfrac{3 \cdot 3 \cdot 11 \cdot 11}{3 \cdot 11 \cdot 3}$

In 10–18, reduce each fraction to lowest terms.

10. $\dfrac{6}{12}$

11. $\dfrac{4}{10}$

12. $\dfrac{5}{20}$

13. $\dfrac{15}{15}$

14. $\dfrac{25}{50}$

15. $\dfrac{2}{6}$

16. $\dfrac{3}{9}$ 17. $\dfrac{2}{14}$ 18. $\dfrac{4}{16}$

In 19–24, name the lowest-term fraction for each fraction or each group of fractions.

19. $\dfrac{2}{6}, \dfrac{3}{9}, \dfrac{4}{12}$ 20. $\dfrac{2}{8}, \dfrac{3}{12}, \dfrac{4}{16}$ 21. $\dfrac{6}{20}, \dfrac{9}{30}, \dfrac{12}{40}$

22. $\dfrac{4}{10}, \dfrac{6}{15}, \dfrac{8}{20}$ 23. $\dfrac{8}{4}, \dfrac{4}{2}, \dfrac{6}{3}$ 24. $\dfrac{10}{2}, \dfrac{20}{4}, \dfrac{15}{3}$

FRACTIONS AND THE NUMBER LINE

In the customary system of measure, ruler scales are typically marked off in one-fourths or in one-eighths.

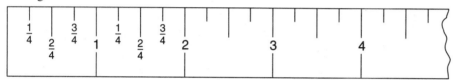

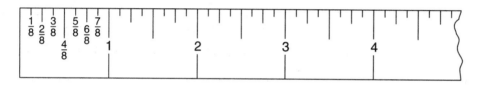

Think of the ruler divisions as points on a number line. For each fraction, there corresponds exactly one point on the number line with those points being consecutively numbered as if being counted.

Fractions can be represented on a number line. For each fraction, there corresponds exactly one point on the number line.

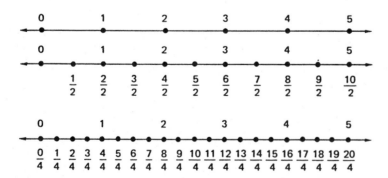

Fractions which name the same point on the number line are called *equivalent fractions*. The fractions $\dfrac{1}{2}$ and $\dfrac{2}{4}$ are equivalent fractions. The fraction $\dfrac{1}{2}$ is the lowest-term fraction for the fraction $\dfrac{2}{4}$.

EXAMPLE 1

Using the number line, find fractions which are equivalent to 3.

$\frac{6}{2}, \frac{12}{4}$, and 3 name the same point on the number line. The fractions $\frac{6}{2}$ and $\frac{12}{4}$ are equivalent to 3.

EXAMPLE 2

Find the whole number equivalent to $\frac{20}{4}$.

The number 5 and $\frac{20}{4}$ name the same point on the number line.

The number 5 is equivalent to $\frac{20}{4}$.

2.4 Exercises

1. In the metric system, scales are marked off in one-tenths. Name the fraction that corresponds to the point indicated by each letter.

A. _____   B. _____   C. _____
D. _____   E. _____   F. _____

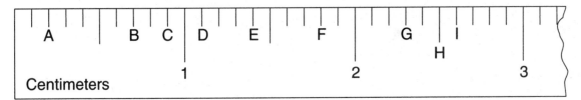

Centimeters

G. _____   H. _____   I. _____

In 2 and 3, name the fraction that corresponds to the point indicated by each letter.

2.

A. _____   B. _____   C. _____
D. _____   E. _____   F. _____
G. _____   H. _____   I. _____

3.

A. _____   B. _____   C. _____
D. _____   E. _____   F. _____

G. _____   H. _____   I. _____

In 4–21, draw three number lines. Label on each one the points for the whole numbers 0, 1, 2, 3, 4, and 5. Use one number line for problems 4–9, the second number line for problems 10–15, and the third number line for problems16–21. On each number line, mark the appropriate equal divisions and locate those points. Use equivalent fractions and lowest-term fractions as an aid.

4. $\dfrac{7}{2}$ 5. $\dfrac{4}{2}$ 6. $\dfrac{9}{2}$

7. $\dfrac{3}{2}$ 8. $\dfrac{2}{2}$ 9. $\dfrac{8}{2}$

10. $\dfrac{9}{3}$ 11. $\dfrac{2}{3}$ 12. $\dfrac{4}{3}$

13. $\dfrac{14}{3}$ 14. $\dfrac{10}{3}$ 15. $\dfrac{12}{3}$

16. $\dfrac{11}{4}$ 17. $\dfrac{2}{4}$ 18. $\dfrac{16}{4}$

19. $\dfrac{19}{4}$ 20. $\dfrac{15}{4}$ 21. $\dfrac{22}{4}$

COMPARING FRACTIONS

Different methods can be used to compare fractions. One method is by comparing shaded regions.

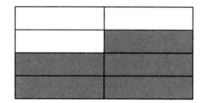 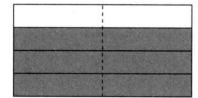

$$\frac{5}{8} < \frac{6}{8}$$

A second method is by using the number line.

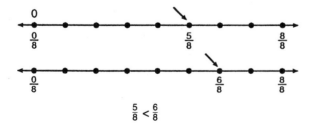

$$\frac{5}{8} < \frac{6}{8}$$

EXAMPLE 1

Using shaded regions, compare $\dfrac{2}{4}$ and $\dfrac{3}{4}$.

$\dfrac{2}{4}$ 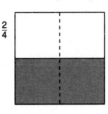 $\dfrac{3}{4}$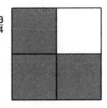

$$\dfrac{2}{4} < \dfrac{3}{4}$$

EXAMPLE 2

Using the number line, compare $\dfrac{3}{4}$ and $\dfrac{5}{4}$.

```
      0                      1
  ←———•——•——•——•——•——•———→
      0    1    2    3    4    5
      4    4    4    4    4    4
```

Since $\dfrac{3}{4}$ is to the left of $\dfrac{5}{4}$, $\dfrac{3}{4} < \dfrac{5}{4}$.

Note: In both Example 1 and 2 the fractions being compared had a common denominator. How can fractions be compared that do not have a common denominator?

A third method that uses the idea of cross products can be used to compare fractions with different denominators.

EXAMPLE 3

Compare $\dfrac{5}{8}$ with $\dfrac{3}{4}$.

$$\dfrac{5}{8} \;\boxed{?}\; \dfrac{3}{4}$$

$$\dfrac{5}{8} \times \dfrac{3}{4}$$

$$20 < 24$$

So, $\dfrac{5}{8} < \dfrac{3}{4}$.

2.5 Exercises

In 1–6, compare each pair of fractions by using shaded regions. Use the symbols <, >, =.

1. $\dfrac{4}{6}$ ——?—— $\dfrac{5}{6}$ 2. $\dfrac{4}{8}$ ——?—— $\dfrac{3}{8}$ 3. $\dfrac{4}{10}$ ——?—— $\dfrac{5}{10}$

4. $\dfrac{8}{12}$ ——?—— $\dfrac{5}{12}$ 5. $\dfrac{2}{10}$ ——?—— $\dfrac{2}{10}$ 6. $\dfrac{4}{12}$ ——?—— $\dfrac{3}{12}$

In 7–15, compare each pair of fractions by using the number line. Use the symbols <, >, =.

7. $\dfrac{9}{10}$ __?__ $\dfrac{8}{10}$ 8. $\dfrac{16}{4}$ __?__ $\dfrac{16}{4}$ 9. $\dfrac{8}{12}$ __?__ $\dfrac{9}{12}$

10. $\dfrac{2}{5}$ __?__ $\dfrac{2}{5}$ 11. $\dfrac{3}{32}$ __?__ $\dfrac{1}{32}$ 12. $\dfrac{7}{56}$ __?__ $\dfrac{8}{56}$

13. $\dfrac{4}{20}$ __?__ $\dfrac{5}{20}$ 14. 7 __?__ $\dfrac{7}{1}$ 15. $\dfrac{5}{2}$ __?__ $\dfrac{2}{2}$

In 16–27, determine the larger fraction in each pair. To decide which is larger, use the cross-product method.

16. $\dfrac{1}{2}, \dfrac{2}{3}$ 17. $\dfrac{3}{4}, \dfrac{7}{8}$ 18. $\dfrac{9}{12}, \dfrac{7}{10}$

19. $\dfrac{4}{5}, \dfrac{3}{8}$ 20. $\dfrac{3}{5}, \dfrac{2}{3}$ 21. $\dfrac{3}{4}, \dfrac{4}{5}$

22. $\dfrac{2}{10}, \dfrac{2}{11}$ 23. $\dfrac{1}{9}, \dfrac{2}{8}$ 24. $\dfrac{2}{16}, \dfrac{3}{32}$

25. $\dfrac{9}{8}, \dfrac{8}{7}$ 26. $\dfrac{4}{10}, \dfrac{5}{9}$ 27. $\dfrac{4}{3}, \dfrac{3}{2}$

Given the six points A, B, C, D, E and F on the number line:

28. List the points in order from smallest to largest.

29. List the points in order from largest to smallest.

30. Explain how you know that $2\dfrac{3}{6}$ is greater than $2\dfrac{2}{6}$.

MIXED NUMBERS

The numerator of a fraction may be less than, equal to, or greater than the denominator of the fraction. When the numerator is greater than the denominator, the fraction may be written as a <u>mixed number</u>. A *mixed number* is a number having a whole number part and a fractional part. Different methods may be used to express a fraction as a mixed number.

EXAMPLE 1

Express $\dfrac{23}{4}$ as a mixed number.

Method 1

Step 1	Step 2	Step 3
$\dfrac{20}{4}$ is the largest fraction in $\dfrac{23}{4}$ that equals a whole number.	Rename $\dfrac{20}{4}$ as 5.	Perform the addition.
$\dfrac{23}{4} = \dfrac{20}{4} + \dfrac{3}{4}$	$\dfrac{20}{4} + \dfrac{3}{4} = 5 + \dfrac{3}{4}$	$5 + \dfrac{3}{4} = 5\dfrac{3}{4}$

Method 2

Step 1	Step 2	Step 3
Divide the numerator by the denominator.	Use the quotient for the whole number part.	Write the remainder 3 over the divisor 4.
$\dfrac{23}{4} = \begin{array}{r} 5 \\ 4\overline{)23} \\ \underline{20} \\ 3 \end{array}$	$\begin{array}{r} 5 \\ 4\overline{)23} \\ \underline{20} \\ 3 \end{array}\;5$	$\begin{array}{r} 5 \\ 4\overline{)23} \\ \underline{20} \\ 3 \end{array} = 5\dfrac{3}{4}$

At times, a mixed number is expressed as an improper fraction. This is the opposite of expressing a fraction as a mixed number.

EXAMPLE 2

Express $5\dfrac{3}{4}$ as an improper fraction.

Method 1

Step 1	Step 2	Step 3
5 equals $\dfrac{20}{4}$.	Add the numerators.	Write the sum 23 over the denominator 4.
$5\dfrac{3}{4} = \dfrac{20}{4} + \dfrac{3}{4}$	$20 + 3 = 23$	$\dfrac{23}{4}$

Method 2

Step 1	Step 2	Step 3
Multiply the whole number 5 by the denominator 4.	Add the numerator 3.	Write the sum 23 over the denominator 4.
$5\dfrac{3}{4} = \dfrac{20}{4} + \dfrac{3}{4}$	$20 + 3 = 23$	$\dfrac{23}{4}$

2.6 Exercises

In 1–15, express each fraction as a mixed number. Express the fractional part of the mixed number in lowest terms.

1. $\dfrac{5}{3}$ 2. $\dfrac{7}{3}$ 3. $\dfrac{9}{4}$

4. $\dfrac{11}{2}$ 5. $\dfrac{6}{3}$ 6. $\dfrac{11}{7}$

7. $\dfrac{19}{10}$ 8. $\dfrac{16}{5}$ 9. $\dfrac{13}{3}$

10. $\dfrac{28}{10}$ 11. $\dfrac{42}{32}$ 12. $\dfrac{48}{8}$

13. $\dfrac{121}{100}$ 14. $\dfrac{90}{60}$ 15. $\dfrac{3,250}{1,000}$

In 16–30, express each mixed number as an improper fraction.

16. $1\dfrac{1}{3}$ 17. $2\dfrac{3}{4}$ 18. $5\dfrac{1}{10}$

19. $8\dfrac{1}{2}$ 20. $3\dfrac{2}{3}$ 21. $6\dfrac{5}{6}$

22. $11\dfrac{1}{8}$ 23. $9\dfrac{1}{11}$ 24. $30\dfrac{1}{4}$

25. $33\dfrac{1}{3}$ 26. $90\dfrac{4}{5}$ 27. $21\dfrac{9}{10}$

28. $10\dfrac{3}{32}$ 29. $50\dfrac{1}{4}$ 30. $25\dfrac{9}{16}$

VIAL

FLASKS

APPLICATIONS

Medications may be stored in screw cap or plastic-stoppered glass containers. Glass bottles are used because glass does not react chemically with most drugs. Some drugs must be stored in dark bottles to prevent deterioration when exposed to light. Absorbent materials are sometimes placed in the bottles to remove moisture that may cause deterioration.

When a medication is to be divided into equal portions, it may be placed in vials, flasks, medicine glasses, or minim glasses.

MEDICINE GLASS

DEFINITIONS

- A *vial* is a small bottle.
- A *flask* is a laboratory vessel usually made of glass and having a constricted neck.
- A *medicine glass* is a graduated container. It may be graduated in milliliters, ounces or drams.
- A *minim glass* is another graduated container. It is used to measure smaller portions and is graduated in minims and drams.

DRAM — | — MINIM

MINIM GLASS

2.7 Exercises

1. A screw cap bottle contains a granulated compound. The compound is divided into 9 equal portions and placed in vials. What fractional part of the whole does each portion represent?

2. The content of a plastic-stoppered glass bottle is divided into 325 equal parts. A flask contains 27 of these equal parts. What fractional part of the whole does the 27 parts represent?

3. One nurse works 1/6 of every day in the operating room. Another nurse works 1/3 of every day in the operating room. Which nurse works more time in the operating room?

4. The increase in the length of two babies is compared. Baby A grows 3/8 inch while Baby B grows 5/16 inch. Which baby increases the most?

5. A 250-milligram tablet is divided into 4 equal parts. Each part represents what fractional part of the whole?

6. Orderly A spends 1/4 of his 8-hour shift working in the emergency room. Orderly B spends 3/8 of her 8-hour shift working in the emergency room. Which orderly spends less time per day working in the emergency room?

7. In two different doses, a patient receives 3/20 grain and 3/32 grain of morphine sulfate. Which is the larger dose?

8. A patient was told that 1/3 of the medication he was taking was not effective. If the patient was taking 9 different pills each day, how many pills could be eliminated daily?

9. It was found that 1/5 of the patients entering a clinic could not pay for their treatment.

   a. If 295 patients entered the clinic today, how many could be expected to not pay for their treatment?

   b. If the average cost of treatment for the 295 patients was $47.50 for that day, what total amount went unpaid?

   c. If the clinic decided to pass this cost along to the paying patients, how much more would each paying patient have added to their treatment cost?

10. A partial study was conducted to determine the type of conditions treated at a clinic over a 30-day period. The following was noted:

4/100	measles
127/1000	respiratory conditions
6/250	flu
25/500	broken bones

    a. If 5,000 patients entered the clinic during the 30 days, give the actual number of patients treated in each of the following categories:

measles	_____
respiratory conditions	_____
flu	_____
broken bones	_____

    b. Arrange the conditions from the lowest to highest based on comparison of the fractions.

    c. Why is it difficult to compare the fractions when they are presented as they are in this problem?

    d. Can you suggest a better way to present the data in the form of fractions for easier comparison?

Unit 3

Addition and Subtraction of Common Fractions and Mixed Numbers

Objectives

After studying this unit the student should be able to:
- Find least common denominators.
- Express fractions as equivalent fractions with least common denominators.
- Add fractions and mixed numbers.
- Subtract fractions and mixed numbers.

ADDITION AND SUBTRACTION OF FRACTIONS WITH LIKE DENOMINATORS

To add or subtract fractions, the denominators must be common denominators. *Common denominator* means that the denominator of each fraction is the same, such as $\frac{3}{4}$ and $\frac{2}{4}$. When the denominators are the same, the numerators can be added or subtracted.

EXAMPLE 1

Add $\frac{3}{4} + \frac{2}{4}$.

Step 1	Step 2	Step 3
Since the denominators are the same the fractions can be added.	Add the numerators. $3 + 2 = 5$.	Express the answer in lowest terms. $\frac{5}{4} = \frac{4}{4} + \frac{1}{4}$ *or* $1\frac{1}{4}$
$\frac{3}{4}$ ←Denominators $+\frac{2}{4}$	$\frac{3}{4}$ ←Numerators $+\frac{2}{4}$ $\frac{5}{4}$	$\frac{3}{4}$ $+\frac{2}{4}$ $\frac{5}{4} = 1\frac{1}{4}$

EXAMPLE 2

Subtract $\dfrac{3}{4} - \dfrac{2}{4}$.

Step 1	Step 2
Since the denominators are the same the fractions can be subtracted.	Subtract the numerators. $3 - 2 = 1$.
$\begin{array}{r} \frac{3}{4} \\ -\frac{2}{4} \\ \hline \end{array}$ ←Denominators	$\begin{array}{r} \frac{3}{4} \\ -\frac{2}{4} \\ \hline \frac{1}{4} \end{array}$ ←Numerators

3.1 Exercises

In 1–24, find each sum or difference. Express all answers in lowest terms.

Sum

1. $\dfrac{1}{7}$
   $+\dfrac{3}{7}$

2. $\dfrac{1}{3}$
   $+\dfrac{2}{3}$

3. $\dfrac{7}{8}$
   $+\dfrac{4}{8}$

4. $\dfrac{2}{5}$
   $+\dfrac{5}{5}$

5. $\dfrac{1}{8}$
   $+\dfrac{1}{8}$

6. $\dfrac{4}{3}$
   $+\dfrac{4}{3}$

7. $\dfrac{4}{10} + \dfrac{3}{10}$

8. $\dfrac{2}{4} + \dfrac{2}{4}$

9. $\dfrac{11}{15} + \dfrac{5}{15}$

Perform the operation in parenthesis first.

10. $\left(\dfrac{5}{10} + \dfrac{3}{10}\right) + \dfrac{2}{10}$

11. $\dfrac{8}{16} + \left(\dfrac{3}{16} + \dfrac{2}{16}\right)$

12. $\left(\dfrac{2}{3} + \dfrac{2}{3}\right) + \dfrac{2}{3}$

Difference

13. $\dfrac{9}{8}$
    $-\dfrac{8}{8}$

14. $\dfrac{5}{2}$
    $-\dfrac{3}{2}$

15. $\dfrac{6}{3}$
    $-\dfrac{6}{3}$

16. $\dfrac{12}{8}$

    $-\dfrac{6}{8}$

17. $\dfrac{18}{10}$

    $-\dfrac{3}{10}$

18. $\dfrac{5}{20}$

    $-\dfrac{1}{20}$

19. $\dfrac{4}{10} - \dfrac{3}{10}$

20. $\dfrac{9}{8} - \dfrac{1}{8}$

21. $\dfrac{4}{3} - \dfrac{1}{3}$

Perform the operation in parenthesis first.

22. $\left(\dfrac{4}{6} - \dfrac{2}{6}\right) - \dfrac{1}{6}$

23. $\left(\dfrac{7}{9} - \dfrac{4}{9}\right) - \dfrac{1}{9}$

24. $\left(\dfrac{11}{16} - \dfrac{3}{16}\right) - \dfrac{2}{16}$

25. What is the purpose of the parenthesis in problems 22–24?

26. Is the following statement true or false? Explain. *Hint:* Perform the operation in parenthesis first.

$$\left(\dfrac{4}{3} - \dfrac{2}{3}\right) - \dfrac{1}{3} = \dfrac{4}{3} - \left(\dfrac{2}{3} - \dfrac{1}{3}\right)$$

COMMON DENOMINATORS

When fractions have unlike denominators the common denominator must be found. Least common multiples and prime factorization can be used to find the common denominator.

DEFINITIONS

- The *least common multiple* (LCM) of two or more numbers is the smallest number that has each of the two numbers as a factor. If two numbers are prime numbers, the product is the least common multiple.
- *Prime factorization* is the process of finding the factors of a number using only prime numbers.

EXAMPLE 1

Find the least common multiple of 6 and 8.

The multiples of 6 are 0, 6, 12, 18, 24, 30, 42, 48 . . .
The multiples of 8 are 0, 8, 16, 24, 32, 40, 48, . . .
The common multiples of 6 and 8 are 24 and 48.

The number 24 is the smallest number that is common to both 6 and 8. It is the <u>least common multiple</u> of 6 and 8.

EXAMPLE 2

Use prime factorization to find the least common multiple of 24 and 20.

1. Prime factor each number.
   $24 = 2 \cdot 2 \cdot 2 \cdot 3$
   $20 = 2 \cdot 2 \cdot 5$

2. Circle each factor that appears the greatest number of times in each number.
   $24 = \boxed{2 \cdot 2 \cdot 2} \cdot \boxed{3}$
   $20 = 2 \cdot 2 \cdot \boxed{5}$

3. Multiply the circled factors.
   $2 \cdot 2 \cdot 2 \cdot 3 \cdot 5 = 120$
   The least common multiple is the product of the prime factors, or 120.

EXAMPLE 3

Find the least common denominator (LCD) of $\dfrac{2}{24}$ and $\dfrac{3}{20}$.

The product of the prime factors common to 24 and 20 is 120.

The <u>least common denominator</u> of $\dfrac{2}{24}$ and $\dfrac{3}{20}$ is 120.

EXAMPLE 4

Find the least common denominator of $\dfrac{5}{18}$ and $\dfrac{7}{30}$.

1. Prime factor each denominator
   $18 = 2 \cdot 3 \cdot 3$
   $30 = 2 \cdot 3 \cdot 5$

2. Circle each factor that appears the greatest number of times in each number.
   $18 = 2 \cdot \boxed{3 \cdot 3}$
   $30 = \boxed{2} \cdot 3 \cdot \boxed{5}$

3. Multiply the circled factors.
   $3 \cdot 3 \cdot 2 \cdot 5 = 90$
   The product of the circled factors is 90. Thus, the LCD of 18 and 30 is 90.

3.2 Exercises

In 1–15, find the least common denominator (LCD) for each pair or group of fractions.

1. $\dfrac{1}{6}, \dfrac{1}{8}$
2. $\dfrac{1}{3}, \dfrac{1}{2}$
3. $\dfrac{1}{3}, \dfrac{3}{5}$

4. $\dfrac{2}{10}, \dfrac{4}{5}$
5. $\dfrac{2}{9}, \dfrac{7}{12}$
6. $\dfrac{3}{4}, \dfrac{1}{16}$

7. $\dfrac{3}{8}, \dfrac{4}{32}$
8. $\dfrac{1}{10}, \dfrac{3}{100}$
9. $\dfrac{2}{6}, \dfrac{4}{16}$

10. $\dfrac{1}{2}, \dfrac{5}{8}, \dfrac{3}{4}$
11. $\dfrac{2}{3}, \dfrac{5}{6}, \dfrac{1}{2}$
12. $\dfrac{3}{4}, \dfrac{1}{50}, \dfrac{7}{100}$

13. $\dfrac{1}{4}, \dfrac{5}{2}, \dfrac{1}{16}$ 14. $\dfrac{1}{3}, \dfrac{7}{2}, \dfrac{4}{9}$ 15. $\dfrac{4}{3}, \dfrac{2}{5}, \dfrac{1}{6}$

ADDING AND SUBTRACTING FRACTIONS WITH UNLIKE DENOMINATORS

In adding fractions with unlike denominators the fractions are expressed as equivalent fractions using the least common denominator. Then, the numerators are added.

EXAMPLE 1

Add $\dfrac{1}{4} + \dfrac{5}{6}$.

Step 1	Step 2	Step 3
Since the denominators are different, common denominators must be found.	The least common denominator is 12. It is the smallest number that is divisible by both 4 and 6.	Express the fractions as equivalent fractions using the least common denominator. $4\overline{)12}\overset{3}{} \longrightarrow 3 \times 1 = 3$ $6\overline{)12}\overset{2}{} \longrightarrow 2 \times 5 = 10$
$\dfrac{1}{4}$ $+\dfrac{5}{6}$	$\dfrac{1}{4} = \dfrac{?}{12}$ $+\dfrac{5}{6} = \dfrac{?}{12}$	$\dfrac{1}{4} = \dfrac{3}{12}$ $+\dfrac{5}{6} = \dfrac{10}{12}$

Step 4	Step 5
Add the numerators. $3 + 10 = 13$ Write the numerator 13 over the common denominator 12.	Express the answer in lowest terms. $\dfrac{13}{12} = \dfrac{12}{12} + \dfrac{1}{12}$ $\dfrac{13}{12} = 1 + \dfrac{1}{12} = 1\dfrac{1}{12}$
$\dfrac{1}{4} = \dfrac{3}{12}$ $+\dfrac{5}{6} = \dfrac{10}{12}$ $\dfrac{13}{12}$	$\dfrac{3}{12}$ $+\dfrac{10}{12}$ $\dfrac{13}{12} = 1\dfrac{1}{12}$

Subtracting fractions with unlike denominators is similar to adding fractions with unlike denominators. In subtraction the numerators are subtracted rather than added.

3.3 Exercises

In 1–36, find each sum or difference. Express all answers in lowest terms.

1. $\dfrac{1}{2}$
    $+\dfrac{1}{8}$

2. $\dfrac{1}{10}$
    $+\dfrac{3}{20}$

3. $\dfrac{2}{8}$
    $+\dfrac{3}{16}$

4. $\dfrac{3}{4}$
    $+\dfrac{3}{8}$

5. $\dfrac{4}{3}$
    $+\dfrac{5}{6}$

6. $\dfrac{4}{4}$
    $+\dfrac{10}{8}$

7. $\dfrac{2}{7}+\dfrac{3}{14}$

8. $\dfrac{2}{3}+\dfrac{1}{8}$

9. $\dfrac{4}{8}+\dfrac{3}{20}$

Perform operations in parentheses first.

10. $\dfrac{5}{6}+\left(\dfrac{7}{10}+\dfrac{4}{5}\right)$

11. $\left(\dfrac{7}{8}+\dfrac{1}{6}\right)+\dfrac{5}{12}$

12. $\left(\dfrac{1}{3}+\dfrac{1}{4}\right)+\dfrac{2}{5}$

13. $\dfrac{3}{8}$
     $-\dfrac{1}{4}$

14. $\dfrac{1}{2}$
     $-\dfrac{3}{7}$

15. $\dfrac{2}{5}$
     $-\dfrac{3}{10}$

16. $\dfrac{3}{3}$
     $-\dfrac{1}{6}$

17. $\dfrac{8}{5}$
     $-\dfrac{1}{10}$

18. $\dfrac{5}{3}$
     $-\dfrac{10}{6}$

19. $\dfrac{4}{5}-\dfrac{1}{20}$

20. $\dfrac{11}{6}-\dfrac{2}{3}$

21. $\dfrac{18}{8}-\dfrac{6}{4}$

Perform operations in parentheses first.

22. $\left(\dfrac{3}{4}-\dfrac{1}{7}\right)-\dfrac{1}{2}$

23. $\dfrac{7}{8}-\left(\dfrac{5}{6}-\dfrac{1}{3}\right)$

24. $\left(\dfrac{2}{3}-\dfrac{1}{4}\right)-\dfrac{1}{6}$

Fractions can be added or subtracted using a calculator. Two different methods are modeled below.

EXAMPLE 1

$$\frac{1}{2} + \frac{3}{4} = 1.25$$

1 $\boxed{\div}$ 2 $\boxed{+}$ 3 $\boxed{\div}$ 4 $\boxed{=}$

EXAMPLE 2

$$\frac{3}{4} - \frac{1}{8} = 0.625$$

3 $\boxed{\div}$ 4 $\boxed{-}$ 1 $\boxed{\div}$ 8 $\boxed{=}$

In 25–36, use a calculator to find each sum or difference. Leave the answer in decimal form.

25. $\dfrac{5}{8} + \dfrac{3}{4}$

26. $\dfrac{1}{2} + \dfrac{3}{10}$

27. $\dfrac{3}{5} + \dfrac{7}{2}$

28. $\dfrac{13}{25} + \dfrac{1}{8}$

29. $\dfrac{11}{4} + \dfrac{9}{20}$

30. $\dfrac{9}{10} + \dfrac{5}{8}$

31. $\dfrac{7}{2} - \dfrac{1}{2}$

32. $\dfrac{5}{2} - \dfrac{2}{5}$

33. $\dfrac{9}{10} - \dfrac{3}{4}$

34. $\dfrac{17}{20} - \dfrac{3}{10}$

35. $\dfrac{9}{4} - \dfrac{7}{25}$

36. $\dfrac{4}{5} - \dfrac{8}{10}$

ADDITION AND SUBTRACTION OF MIXED NUMBERS

Addition of mixed numbers is exactly the same as addition of fractions except that the whole numbers must also be added.

EXAMPLE 1

Add $4\frac{1}{4} + 2\frac{5}{6}$.

Step 1	Step 2	Step 3
Find a common denominator for the fractions.	The least common denominator is 12. It is the smallest number that is divisible by both 4 and 6.	Express the fractions as equivalent fractions using the least common denominator. $4\overline{)12}\,^{3} \longrightarrow 3 \times 1 = 3$ $6\overline{)12}\,^{2} \longrightarrow 2 \times 5 = 10$
$\begin{array}{r} 4\frac{1}{4} \\ + 2\frac{5}{6} \\ \hline \end{array}$	$\begin{array}{r} 4\frac{1}{4} = 4\frac{?}{12} \\ + 2\frac{5}{6} = 2\frac{?}{12} \\ \hline \end{array}$	$\begin{array}{r} 4\frac{1}{4} = 4\frac{3}{12} \\ + 2\frac{5}{6} = 2\frac{10}{12} \\ \hline \end{array}$

Step 4	Step 5	Step 6
Add the fractions. $\frac{3}{12} + \frac{10}{12} = \frac{13}{12}$	Add the whole numbers. $4 + 2 = 6$	Express the answer in lowest terms. $6\frac{13}{12} = 6 + \frac{12}{12} + \frac{1}{12}$ $6 + 1\frac{1}{12} = 7\frac{1}{12}$
$\begin{array}{r} 4\frac{1}{4} = 4\frac{3}{12} \\ + 2\frac{5}{6} = 2\frac{10}{12} \\ \hline \frac{13}{12} \end{array}$	$\begin{array}{r} 4\frac{1}{4} = 4\frac{3}{12} \\ + 2\frac{5}{6} = 2\frac{10}{12} \\ \hline 6\frac{13}{12} \end{array}$	$4\frac{1}{4} + 2\frac{5}{6} = 7\frac{1}{12}$

Subtraction of mixed numbers involves the same process as addition. In order to subtract the fractional part, it is sometimes necessary to express the whole number part as a mixed number.

EXAMPLE 2

Subtract $4\dfrac{1}{4} - 2\dfrac{5}{6}$.

Step 1	Step 2	Step 3
Find a common denominator for the fractions.	The least common denominator for 4 and 6 is 12.	Express as equivalent fractions. $4\,\overline{)12}\!\!\overset{3}{}\longrightarrow 3 \times 1 = 3$ $6\,\overline{)12}\!\!\overset{2}{}\longrightarrow 2 \times 5 = 10$
$4\dfrac{1}{4}$ $-\,2\dfrac{5}{6}$ _____	$4\dfrac{1}{4} = 4\dfrac{?}{12}$ $-\,2\dfrac{5}{6} = 2\dfrac{?}{12}$ _____	$4\dfrac{1}{4} = 4\dfrac{3}{12}$ $-\,2\dfrac{5}{6} = 2\dfrac{10}{12}$ _____

Step 4	Step 5	Step 6
Express the whole number 4 as the mixed number $3\dfrac{12}{12}$. $4\dfrac{3}{12} = 3 + \dfrac{12}{12} + \dfrac{3}{12} = 3\dfrac{15}{12}$	Subtract the numerators. $15 - 10 = 5$	Subtract the whole numbers. $3 - 2 = 1$
$4\dfrac{3}{12} = 3\dfrac{15}{12}$ $-\,2\dfrac{10}{12} = 2\dfrac{10}{12}$ _____	$3\dfrac{15}{12}$ $-\,2\dfrac{10}{12}$ _____ $\dfrac{5}{12}$	$3\dfrac{15}{12}$ $-\,2\dfrac{10}{12}$ _____ $1\dfrac{5}{12}$

3.4 Exercises

In 1–30, find each sum or difference. Express all answers in lowest terms.

1. $8\frac{1}{5}$

  $+\ 6\frac{2}{5}$

2. $3\frac{1}{3}$

  $+\ 5\frac{3}{4}$

3. $2\frac{3}{4}$

  $+\ 4\frac{8}{16}$

4. $9\frac{3}{6}$

  $+\ 11\frac{4}{12}$

5. $30\frac{7}{8}$

  $+\ 1\frac{3}{4}$

6. $4\frac{3}{5}$

  $+\ 24\frac{9}{10}$

7. $8\frac{1}{12}$

  $+\ 3\frac{5}{36}$

8. $2\frac{1}{16}$

  $+\ 5\frac{3}{32}$

9. $5\frac{1}{2}$

  $+\ 9\frac{7}{10}$

10. $4\frac{4}{5}+13\frac{1}{3}$

11. $10\frac{1}{7}+8\frac{2}{3}$

12. $4\frac{4}{5}+21\frac{2}{10}$

Perform operations in parenthesis first.

13. $\left(3\frac{1}{2}+2\frac{1}{4}\right)+1\frac{1}{3}$

14. $5\frac{1}{8}+\left(4\frac{1}{4}+5\frac{5}{8}\right)$

15. $2\frac{1}{6}+\left(3\frac{2}{3}+4\frac{3}{6}\right)$

16. $3\frac{11}{12}$

  $-\ 2\frac{2}{10}$

17. $10\frac{3}{4}$

  $-\ 4\frac{1}{8}$

18. $16\frac{4}{5}$

  $-\ 9\frac{3}{10}$

19. $18\frac{37}{100}$

  $-\ 3\frac{1}{10}$

20. 5

  $-\ 2\frac{1}{4}$

21. $37\frac{2}{3}$

  $-\ 14\frac{3}{4}$

22. $11\frac{4}{32}$

  $-\ 2\frac{3}{8}$

23. $53\frac{13}{100}$

  $-\ 12\frac{3}{25}$

24. $16\frac{2}{3}$

  $-\ 5$

25. $18\frac{1}{2}-9\frac{7}{10}$

26. $25-4\frac{4}{10}$

27. $18\frac{3}{4}-7$

Perform operations in parenthesis first.

28. $\left(20\frac{3}{4} - 2\frac{1}{8}\right) - 9\frac{1}{2}$ 29. $42\frac{7}{10} - \left(5\frac{5}{8} - 2\frac{1}{4}\right)$ 30. $\left(35\frac{1}{2} - 14\frac{1}{4}\right) - 11$

 Mixed numbers can be added or subtracted using a calculator. Two different examples are modeled below.

EXAMPLE 1

$5\frac{1}{4} + 3\frac{1}{2} = 8.75$

$5 \boxed{+} 1 \boxed{\div} 4 \boxed{+} 3 \boxed{+} 1 \boxed{\div} 2 \boxed{=}$

EXAMPLE 2

$5\frac{1}{4} - 3\frac{1}{2} = 1.75$

$5 \boxed{+} 1 \boxed{\div} 4 \boxed{-} \boxed{(} \; 3 \boxed{+} 1 \boxed{\div} 2 \boxed{)} \boxed{=}$

In 31–42, use a calculator to find each sum or difference. Leave the answer in decimal form.

31. $3\frac{1}{2} + 5\frac{3}{4}$ 32. $8\frac{4}{5} + 6\frac{12}{25}$ 33. $9\frac{3}{10} + 4\frac{7}{8}$

34. $3\frac{1}{8} + 2\frac{7}{10}$ 35. $14\frac{11}{20} + 2\frac{9}{20}$ 36. $1\frac{12}{16} + 3\frac{3}{4}$

37. $5\frac{1}{2} - 3\frac{3}{10}$ 38. $4\frac{5}{8} - 1\frac{10}{16}$ 39. $2\frac{3}{4} - 1\frac{9}{10}$

40. $6\frac{3}{5} - 3\frac{5}{8}$ 41. $9\frac{7}{20} - 2\frac{9}{10}$ 42. $8\frac{10}{25} - 7\frac{8}{20}$

COMPARING FRACTIONS USING COMMON DENOMINATORS

To compare two fractions that have a common denominator, compare the numerators. The larger fraction has the greater numerator.

EXAMPLE 1

Compare $\frac{1}{2}$ and $\frac{3}{5}$.

The common denominator is 10.

Express each fraction as an equivalent fraction.

$\frac{1}{2} = \frac{?}{10}$ $2\overline{)10}^{\;5} \longrightarrow 5 \times 1 = 5$ $\frac{1}{2} = \frac{5}{10}$

$\frac{3}{5} = \frac{?}{10}$ $5\overline{)10}^{\;2} \longrightarrow 2 \times 3 = 6$ $\frac{3}{5} = \frac{6}{10}$

3.5 Exercises

In 1–6, express each fraction as an equivalent fraction using the given denominator. Compare the two fractions. Use the symbols, <, >, =, to show the relationship.

1. $\dfrac{1}{2} = \dfrac{?}{4} ; \dfrac{3}{4} = \dfrac{?}{4}$

   $\dfrac{1}{2} \underline{\quad ? \quad} \dfrac{3}{4}$

2. $\dfrac{2}{3} = \dfrac{?}{6} ; \dfrac{5}{6} = \dfrac{?}{6}$

   $\dfrac{2}{3} \underline{\quad ? \quad} \dfrac{5}{6}$

3. $\dfrac{1}{3} = \dfrac{?}{12} ; \dfrac{1}{4} = \dfrac{?}{12}$

   $\dfrac{1}{3} \underline{\quad ? \quad} \dfrac{1}{4}$

4. $\dfrac{2}{5} = \dfrac{?}{30} ; \dfrac{3}{6} = \dfrac{?}{30}$

   $\dfrac{2}{5} \underline{\quad ? \quad} \dfrac{3}{6}$

5. $\dfrac{3}{4} = \dfrac{?}{20} ; \dfrac{2}{5} = \dfrac{?}{20}$

   $\dfrac{3}{4} \underline{\quad ? \quad} \dfrac{2}{5}$

6. $\dfrac{1}{10} = \dfrac{?}{100} ; \dfrac{10}{100} = \dfrac{?}{100}$

   $\dfrac{1}{10} \underline{\quad ? \quad} \dfrac{10}{100}$

In 7–15, classify each fraction as less than (<), equal to (=) or greater than (>) $\dfrac{1}{2}$.

7. $\dfrac{7}{16} \underline{\quad\quad} \dfrac{1}{2}$

8. $\dfrac{4}{5} \underline{\quad\quad} \dfrac{1}{2}$

9. $\dfrac{12}{24} \underline{\quad\quad} \dfrac{1}{2}$

10. $\dfrac{1}{2} \underline{\quad\quad} \dfrac{3}{2}$

11. $\dfrac{1}{2} \underline{\quad\quad} \dfrac{3}{8}$

12. $\dfrac{1}{2} \underline{\quad\quad} \dfrac{2}{6}$

13. $\dfrac{1}{3} \underline{\quad\quad} \dfrac{1}{2}$

14. $\dfrac{5}{1} \underline{\quad\quad} \dfrac{1}{2}$

15. $\dfrac{1}{2} \underline{\quad\quad} \dfrac{5}{10}$

In 16–30, rename each pair of fractions using a common denominator. Compare the numerators of each pair. The larger fraction has the greater numerator. Name the larger fraction for each pair of fractions.

16. $\dfrac{4}{5}, \dfrac{2}{3}$

17. $\dfrac{5}{12}, \dfrac{4}{6}$

18. $\dfrac{2}{3}, \dfrac{3}{2}$

19. $\dfrac{9}{3}, \dfrac{9}{4}$

20. $\dfrac{1}{2}, \dfrac{3}{4}$

21. $\dfrac{13}{20}, \dfrac{4}{5}$

22. $3\dfrac{1}{2}, 3\dfrac{1}{4}$

23. $\dfrac{10}{5}, \dfrac{3}{2}$

24. $\dfrac{1}{6}, \dfrac{2}{5}$

25. $\dfrac{2}{3}, \dfrac{2}{4}$

26. $\dfrac{5}{3}, 2\dfrac{1}{2}$

27. $5\dfrac{1}{5}, \dfrac{26}{4}$

28. $\dfrac{3}{8}, \dfrac{3}{4}$

29. $\dfrac{11}{2}, 5$

30. $6\dfrac{1}{3}, 5\dfrac{2}{3}$

In 31–36, rename each group of fractions using a common denominator. Compare the numerators. The smallest fraction has the smallest numerator. Name the smallest fraction for each group of fractions.

31. $\dfrac{1}{2}, \dfrac{1}{6}, \dfrac{1}{3}$

32. $\dfrac{2}{50}, \dfrac{1}{20}, \dfrac{3}{100}$

33. $\dfrac{9}{32}, \dfrac{3}{8}, \dfrac{2}{16}$

34. $\dfrac{5}{8}, \dfrac{3}{4}, \dfrac{1}{2}$

35. $\dfrac{4}{5}, \dfrac{11}{50}, \dfrac{13}{25}$

36. $\dfrac{7}{2}, \dfrac{7}{4}, \dfrac{7}{3}$

APPLICATIONS

Medication may be in the form of a liquid or a solid (crystals, powders, or tablets). Liquid medication may be measured in ounces, milliliters, or pints. Crystals, powders, and tablets are weighed in grains, grams, or milligrams.

3.6 Exercises

1. David drinks 1/3 ounce of medication and 2 hours later he drinks 4/5 ounce. How much medication does David consume during this period?

2. A patient is given 2 1/2 grains of medication followed by 1 2/3 grains. What is the total dosage the patient receives?

3. A medical assistant is responsible for maintaining a record of laboratory supplies. When purchased, a bottle contains 7 1/2 ounces of liquid. When the medical assistant conducts an inventory, only 3 3/4 ounces of liquid remain. How much liquid has been removed?

4. An operating room technician assist surgeons before, during, and after an operation. One of the technician's responsibilities is to set up the operating room with instruments, equipment, and fluids that may be needed. The surgeon requests that the operating technician supply 10 liters of dextrose solution. After the operation, the operating room technician finds that 6 2/3 liters of dextrose solution remain. How many liters of solution were used for the operation?

5. In two doses, a patient receives 1/20 grain and 2/32 grain of morphine sulfate. What is the total dosage that the patient receives?

6. A new baby grew 3/8 inch in January. In February the baby grew 7/16 inch. How many total inches did the baby grow during January and February?

7. A prescription calls for a first dose of 2 1/4 grains followed three hours later by a dose of 1 1/2 grain. The second dose is how many grains less than the first dose?

8. The flow rate of oxygen, through a mask, was measured at 5 L/min (liters per minute). As the patient's condition changed the flow rate was adjusted.

    Give the rate of flow of oxygen.

Change	Flow
a. decrease the 5 L/min flow by 1/3 L/min	_____ L/min
b. decrease the answer found in **a** by 1/2 L/min	_____ L/min
c. increased the answer found in **b** by 1/3 L/min	_____ L/min
d. increased the answer found in **c** by 1/2 L/min	_____ L/min
e. increased the answer found in **d** by 2/3 L/min	_____ L/min
f. decrease the answer found in **e** by 1 L/min	_____ L/min

Unit 4

Multiplication and Division of Common Fractions and Mixed Numbers

Objectives

After studying this unit the student should be able to:
- **Multiply fractions and mixed numbers.**
- **Divide fractions and mixed numbers.**

MULTIPLICATION OF FRACTIONS

Multiplication of fractions can be illustrated by using the concept that fractions are part of a whole.

EXAMPLE 1

What part of a whole is $\frac{1}{2}$ of $\frac{1}{3}$?

$\frac{1}{3}$ of the total region is dotted.

Darken $\frac{1}{2}$ of the $\frac{1}{3}$ region.

$\frac{1}{6}$ of the total region is darkened.

The mathematical notation is: $\frac{1}{2}$ of $\frac{1}{3} = \frac{1}{2} \cdot \frac{1}{3} = \frac{1}{6}$.

EXAMPLE 2

What part of the total region is $\frac{1}{2}$ of $\frac{2}{3}$?

   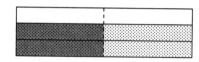

$\frac{2}{3}$ of the total region is dotted.

Darken $\frac{1}{2}$ of the $\frac{2}{3}$ region.

$\frac{2}{6}$ *or* $\frac{1}{3}$ of the total region is darkened.

The mathematical notation is: $\frac{1}{2}$ of $\frac{2}{3} = \frac{1}{2} \cdot \frac{2}{3}$ *or* $\frac{1}{3}$.

EXAMPLE 3

What part of the whole is $\frac{1}{2}$ of $\frac{3}{3}$?

$\frac{3}{3}$ of the total region is dotted.

Darken $\frac{1}{2}$ of the $\frac{3}{3}$ region.

$\frac{3}{6}$ *or* $\frac{1}{2}$ of the total region is darkened.

The mathematical notation is: $\frac{1}{2}$ of $\frac{3}{3} = \frac{1}{2} \cdot \frac{3}{3} = \frac{3}{6}$ *or* $\frac{1}{2}$.

- To multiply two fractions find the product of the numerators, then find the product of the denominators. This may be written:

$$\frac{a}{b} \cdot \frac{c}{d} = \frac{a \cdot c}{b \cdot d} = \frac{ac}{bd}$$

Note: Letters written side-by-side imply multiplication.
    Thus, no multiplication symbol (dot) is needed.

EXAMPLE 4

Multiply $\dfrac{2}{5} \cdot \dfrac{1}{3}$.

Multiplication of fractions involves multiplying numerators then multiplying denominators.

Step 1	**Step 2**	**Step 3**
Multiply the numerators. $2 \cdot 1 = 2$	Multiply the denominators. $5 \cdot 3 = 15$	Express the answer, or product, as a fraction. $\dfrac{2}{15}$
$\dfrac{2}{5} \cdot \dfrac{1}{3} = \dfrac{2 \cdot 1}{?} = \dfrac{2}{?}$	$\dfrac{2}{5} \cdot \dfrac{1}{3} = \dfrac{2 \cdot 1}{5 \cdot 3} = \dfrac{2}{15}$	$\dfrac{2}{5} \cdot \dfrac{1}{3} = \dfrac{2}{15}$

The product of fractions is usually expressed in lowest terms. The product may be expressed in lowest terms after the multiplication is performed.

EXAMPLE 5

Multiply $\dfrac{2}{5} \cdot \dfrac{3}{4}$.

Step 1	**Step 2**
Multiply the numerators. $2 \cdot 3 = 6$ Multiply the denominators. $5 \cdot 4 = 20$	Express the answer in lowest terms. $\dfrac{6}{20} = \dfrac{(6 \div 2)}{(20 \div 2)} = \dfrac{3}{10}$ $\dfrac{6}{20} = \dfrac{3}{10}$
$\dfrac{2}{5} \cdot \dfrac{3}{4} = \dfrac{2 \cdot 3}{5 \cdot 4} = \dfrac{6}{20}$	$\dfrac{2}{5} \cdot \dfrac{3}{4} = \dfrac{6}{20}$ or $\dfrac{3}{10}$

Fractions may also be expressed in lowest terms before multiplying. This process involves dividing a numerator and denominator by the same number. Dividing a numerator and a denominator by the same number is commonly called *cancellation*.

EXAMPLE 6

Find $\dfrac{2}{3}$ of $\dfrac{6}{8}$.

$\dfrac{2}{3}$ of $\dfrac{6}{8}$ means $\dfrac{2}{3} \cdot \dfrac{6}{8} = \dfrac{2 \cdot 6}{3 \cdot 8}$.

Step 1	Step 2	Step 3
Prime factor the numerator. $2 \cdot 6 = 2 \cdot 2 \cdot 3$ Prime factor the denominator. $3 \cdot 8 = 3 \cdot 2 \cdot 2 \cdot 2$	Reduce or cancel common prime factors.	Multiply factors in the numerator. $1 \cdot 1 \cdot 1 = 1$ Multiply factors in the denominator. $1 \cdot 1 \cdot 1 \cdot 2 = 2$
$\dfrac{2 \cdot 6}{3 \cdot 8} = \dfrac{2 \cdot 2 \cdot 3}{3 \cdot 2 \cdot 2 \cdot 2}$	$\dfrac{\overset{1}{\cancel{2}} \cdot \overset{1}{\cancel{2}} \cdot \overset{1}{\cancel{3}}}{\underset{1}{\cancel{3}} \cdot \underset{1}{\cancel{2}} \cdot \underset{1}{\cancel{2}} \cdot 2}$	$\dfrac{1 \cdot 1 \cdot 1}{1 \cdot 1 \cdot 1 \cdot 2} = \dfrac{1}{2}$

4.1 Exercises

In 1–24, find the product of each pair of fractions. When possible, express the fractions in lowest terms before multiplying. Express all answers in lowest terms. (Remember that *of* usually indicates multiplication.)

1. $\dfrac{1}{2}$ of 50

2. $\dfrac{1}{3}$ of 45

3. $\dfrac{1}{10}$ of 100

4. $\dfrac{1}{2}$ of $\dfrac{1}{3}$

5. $\dfrac{1}{3}$ of $\dfrac{1}{2}$

6. $\dfrac{1}{2}$ of $\dfrac{3}{5}$

7. $\dfrac{1}{4}$ of $\dfrac{3}{7}$

8. $\dfrac{1}{5}$ of $\dfrac{2}{3}$

9. $\dfrac{2}{3}$ of $\dfrac{3}{3}$

10. $\dfrac{1}{2}$ of $\dfrac{2}{4}$

11. $\dfrac{3}{4} \cdot \dfrac{2}{5}$

12. $\dfrac{1}{8} \cdot \dfrac{2}{3}$

13. $\dfrac{1}{16} \cdot \dfrac{2}{5}$

14. $\dfrac{4}{3} \cdot \dfrac{3}{4}$

15. $\dfrac{1}{2} \cdot \dfrac{8}{5}$

16. $\dfrac{7}{12} \cdot \dfrac{6}{7}$

17. $\dfrac{3}{8} \cdot \dfrac{1}{2}$

18. $\dfrac{14}{5} \cdot \dfrac{10}{7}$

19. $\dfrac{1}{3} \cdot \dfrac{2}{2}$

20. $\dfrac{2}{9} \cdot \dfrac{4}{7}$

21. $\dfrac{5}{6} \cdot \dfrac{2}{3}$

22. $\dfrac{24}{5} \cdot \dfrac{5}{10}$

23. $\dfrac{5}{10} \cdot \dfrac{10}{5}$

24. $\dfrac{2}{3} \cdot \dfrac{3}{5}$

In 25–30, make each expression true by writing <, =, or >.

25. $\dfrac{2}{5} \cdot \dfrac{1}{4} \;\Box\; \dfrac{1}{2}$ 26. $\dfrac{1}{4} \cdot 0 \;\Box\; \dfrac{1}{4}$ 27. $\dfrac{1}{8} \cdot \dfrac{3}{4} \;\Box\; \dfrac{1}{8}$

28. $\dfrac{3}{4} \cdot \dfrac{2}{9} \;\Box\; \dfrac{1}{6}$ 29. $\dfrac{7}{10} \cdot \dfrac{2}{3} \;\Box\; \dfrac{8}{15}$ 30. $\dfrac{7}{8} \cdot \dfrac{4}{5} \;\Box\; \dfrac{9}{10}$

 Fractions can be multiplied with a calculator using two different methods.

EXAMPLE 1

$\dfrac{1}{2} \cdot \dfrac{3}{4} = 0.375$

1 ☒ 3 ÷ ⑃ 2 ☒ 4 ⑄ ⊟ 0.375

EXAMPLE 2

$\dfrac{1}{2} \cdot \dfrac{3}{4} = 0.375$

1 ÷ 2 ☒ 3 ÷ 4 ⊟ 0.375

In 31–39, find each product with a calculator using either Example 1 or Example 2.

31. $\dfrac{3}{4} \cdot \dfrac{2}{3}$ 32. $\dfrac{2}{5}$ of $\dfrac{3}{2}$ 33. $\dfrac{5}{9} \cdot \dfrac{9}{10}$

34. $\dfrac{7}{8}$ of $\dfrac{8}{20}$ 35. $\dfrac{6}{5} \cdot \dfrac{5}{6}$ 36. $\dfrac{7}{8}$ of 0

37. $\dfrac{1}{4}$ of 15   38. $\dfrac{3}{8} \cdot \dfrac{3}{2}$ 39. $\dfrac{1}{3}$ of 12

40. Give an example where the numerator of the product (answer) of two fractions is prime but the product is not in lowest terms.

MULTIPLICATION OF MIXED NUMBERS

When mixed numbers are multiplied, the mixed numbers are first expressed as fractions. Whole numbers are also expressed as fractions. The same process as in multiplication of fractions can then be performed.

EXAMPLE 1

Multiply $2 \cdot 3\frac{1}{4}$.

Step 1	Step 2	Step 3
Express the first number as a fraction. $2 = \frac{2}{1}$ Express the second number as a fraction. $3\frac{1}{4} = \frac{(3 \cdot 4) + 1}{4} = \frac{13}{4}$	Multiply the numerators. $2 \cdot 13 = 26$ Multiply the denominators. $1 \cdot 4 = 4$	Express the answer as a mixed number. $\frac{26}{4} = \frac{(26 \div 2)}{(4 \div 2)} = \frac{13}{2}$ $13 \div 2 = 2\overline{)13}$ $= 6\frac{1}{2}$
$2 \cdot 3\frac{1}{4} = \frac{2}{1} \cdot \frac{13}{4}$	$\frac{2}{1} \cdot \frac{13}{4} = \frac{2 \cdot 13}{1 \cdot 4} = \frac{26}{4}$	$2 \cdot 3\frac{1}{4} = 6\frac{1}{2}$

Notice that in Example 2 the fractions are simplified before being multiplied.

EXAMPLE 2

Multiply $3\frac{3}{4} \cdot 6\frac{1}{3}$.

Step 1	Step 2	Step 3	Step 4
Express the first number as a fraction. $3\frac{3}{4} = \frac{(3 \cdot 4) + 3}{4} = \frac{15}{4}$ Express the second number as a fraction. $6\frac{1}{3} = \frac{(6 \cdot 3) + 1}{3} = \frac{19}{3}$	Simplify. Divide 3 and 15 by 3.	Multiply the numerators. $5 \cdot 19 = 95$ Multiply the denominators. $4 \cdot 1 = 4$	Express the answer as a mixed number. $95 \div 4 = 23\frac{3}{4}$
$3\frac{3}{4} \cdot 6\frac{1}{3} = \frac{15}{4} \cdot \frac{19}{3}$	$\frac{15}{4} \cdot \frac{19}{3} = \frac{\overset{5}{\cancel{15}} \cdot 19}{4 \cdot \underset{1}{\cancel{3}}}$	$\frac{5 \cdot 19}{4 \cdot 1} = \frac{95}{4}$	$3\frac{3}{4} \cdot 6\frac{1}{3} = \frac{95}{4} = 23\frac{3}{4}$

Mixed numbers may also be expressed as fractions in lowest terms before multiplying.

EXAMPLE 3

Multiply $2\frac{2}{5} \cdot 6\frac{1}{4}$.

Step 1	Step 2	Step 3	Step 4
Express the numbers as fractions. $$2\frac{2}{5} = \frac{(2 \cdot 5) + 2}{5} = \frac{12}{5}$$ $$6\frac{1}{4} = \frac{(6 \cdot 4) + 1}{4} = \frac{25}{4}$$	Prime factor the numerator. $12 \cdot 25 = 2 \cdot 2 \cdot 3 \cdot 5 \cdot 5$ Prime factor the denominator. $5 \cdot 2 \cdot 2$	Reduce or cancel common prime factors.	Multiply factors in numerator. $1 \cdot 1 \cdot 3 \cdot 5 \cdot 1 = 15$ Multiply factors in denominator. $1 \cdot 1 \cdot 1 = 1$
$$2\frac{2}{5} \cdot 6\frac{1}{4} = \frac{12}{5} \cdot \frac{25}{4}$$	$$\frac{12 \cdot 25}{5 \cdot 4} = \frac{2 \cdot 2 \cdot 3 \cdot 5 \cdot 5}{5 \cdot 2 \cdot 2}$$	$$\frac{\overset{1}{\cancel{2}} \cdot \overset{1}{\cancel{2}} \cdot 3 \cdot 5 \cdot 5}{\underset{1}{\cancel{5}} \cdot \underset{1}{\cancel{2}} \cdot \underset{1}{\cancel{2}}}$$	$$\frac{1 \cdot 1 \cdot 3 \cdot 5 \cdot 1}{1 \cdot 1 \cdot 1}$$ $$= \frac{15}{1} \text{ or } 15$$

4.2 Exercises

In 1–24, find each product. When possible, express the fractions in lowest terms before multiplying.

1. $1\frac{1}{3} \cdot \frac{2}{5}$

2. $3\frac{1}{4} \cdot \frac{1}{6}$

3. $4\frac{2}{8} \cdot \frac{2}{3}$

4. $4\frac{1}{2} \cdot \frac{1}{4}$

5. $1\frac{1}{3} \cdot 2\frac{3}{4}$

6. $3\frac{1}{2} \cdot 4\frac{2}{3}$

7. $4\frac{1}{3} \cdot 5\frac{1}{2}$

8. $3\frac{1}{2} \cdot 4\frac{3}{8}$

9. $3\frac{1}{5} \cdot 5\frac{1}{3}$

10. $3\frac{1}{5} \cdot 2\frac{3}{4}$

11. $1\frac{5}{6} \cdot 2\frac{1}{4}$

12. $2\frac{7}{8} \cdot 2\frac{1}{4}$

13. $3\frac{1}{3} \cdot 10$

14. $2\frac{5}{6} \cdot 4\frac{1}{3}$

15. $2\frac{2}{5} \cdot 0$

16. $1\frac{1}{10} \cdot 3\frac{2}{5}$

17. $4\frac{1}{4} \cdot \frac{4}{17}$

18. $2\frac{4}{10} \cdot \frac{20}{6}$

19. $20\frac{5}{8} \cdot 2\frac{4}{5}$

20. $0 \cdot 8\frac{1}{3}$

21. $11\frac{1}{5} \cdot 2\frac{1}{4}$

22. $8\frac{5}{6} \cdot 3\frac{1}{5}$

23. $4\frac{1}{4} \cdot 2\frac{2}{9}$

24. $6\frac{5}{8} \cdot 4\frac{3}{4}$

Mixed numbers can be multiplied with a calculator using two different methods.

EXAMPLE 1

$$4\frac{2}{5} \cdot 2\frac{1}{4} = 9.9$$

$\boxed{(}\ 4\ \boxed{+}\ 2\ \boxed{\div}\ 5\ \boxed{)}\ \boxed{\times}\ \boxed{(}\ 2\ \boxed{+}\ 1\ \boxed{\div}\ 4\ \boxed{)}\ \boxed{=}\ 9.9$

EXAMPLE 2

$$4\frac{2}{5} \cdot 2\frac{1}{4} = \frac{22}{5} \cdot \frac{9}{4} = 9.9$$

$22\ \boxed{\div}\ 5\ \boxed{\times}\ 9\ \boxed{\div}\ 4\ \boxed{=}\ 9.9$

In 25–33, find each product with a calculator using either Example 1 or Example 2.

25. $15 \cdot 2\frac{3}{4}$　　　　26. $3\frac{1}{2} \cdot 1\frac{1}{5}$　　　　27. $7\frac{2}{10} \cdot 2\frac{1}{2}$

28. $5\frac{1}{5} \cdot 4\frac{3}{6}$　　　　29. $2\frac{1}{6} \cdot 18$　　　　30. $3\frac{3}{4} \cdot 10\frac{1}{3}$

31. $1\frac{1}{4} \cdot 2\frac{2}{5} \cdot 2\frac{1}{2}$　　　　32. $3\frac{4}{10} \cdot 8 \cdot 4\frac{1}{2}$　　　　33. $6\frac{3}{4} \cdot 5\frac{1}{5} \cdot 0$

34. Explain why leaving out parenthesis, when using Example 1, to multiply mixed numbers with a calculator gives an incorrect product.

35. When multiplying a whole number and a fraction, is the product greater than, equal to, or less than the whole number? Explain.

DIVISION OF FRACTIONS

　　Division of fractions can also be illustrated by using the concept that fractions are part of a whole.

EXAMPLE 1

How many halves are there in 4?

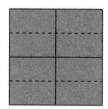

　　4 regions　　　　　　　　　　　Each of the 4 regions is divided in half.
　　　　　　　　　　　　　　　　　8 regions are formed.

The mathematical notation is: 4 divided by $\frac{1}{2} = 4 \div \frac{1}{2} = 8$.

EXAMPLE 2

How many fourths are there in 3?

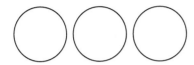   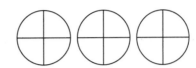

3 regions   Each of the 3 regions is divided into fourths.
                              12 regions are formed.

The mathematical notation is: 3 divided by $\frac{1}{4}$ = 3 ÷ $\frac{1}{4}$ = 12.

4.3 Exercises

In 1–6, find each answer. Use diagrams if necessary.

1. How many thirds are there in 2?

    $2 \div \frac{1}{3} = ?$

2. How many fifths are there in 4?

    $4 \div \frac{1}{5} = ?$

3. How many fourths are there in 4?

    $4 \div \frac{1}{4} = ?$

4. How many tenths are there in 2?

    $2 \div \frac{1}{10} = ?$

5. How many eighths are there in 3?

    $3 \div \frac{1}{8} = ?$

6. How many sixths are there in 1?

    $1 \div \frac{1}{6} = ?$

7.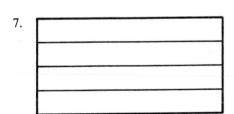

    a. How many $\frac{1}{4}$s are there in 1?

    b. How many $\frac{2}{4}$s are there in 1?

8.

    a. How many $\frac{1}{8}$s are there in 1?

    b. How many $\frac{2}{8}$s are there in 1?

    c. How many $\frac{4}{8}$s are there in 1?

9.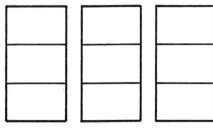

a. How many $\frac{1}{3}$s are there in 3?

b. How many $\frac{2}{3}$s are there in 3?

c. How many $\frac{3}{3}$s are there in 3?

The division process can be generalized by using reciprocals or multiplicative inverses.

• Two numbers whose products equal 1 are reciprocals or multiplicative inverses of each other.

EXAMPLE 3

The numbers $\frac{9}{1}$ and $\frac{1}{9}$ are reciprocals since $\frac{9}{1} \cdot \frac{1}{9}$.

The expressions $\frac{a}{b}$ and $\frac{b}{a}$ are reciprocals since $\frac{a}{b} \cdot \frac{b}{a} = 1$.

4.4 Exercises

In 1–6, find each reciprocal.

1. $\frac{1}{3}$

2. 7

3. $\frac{9}{5}$

4. 22

5. 0

6. $\frac{1}{0}$

In 7–18, complete.

7. $\frac{3}{5} \cdot \dfrac{\square}{\square} = 1$

8. $\dfrac{\square}{\square} \cdot \dfrac{8}{3} = 1$

9. $4 \cdot \dfrac{\square}{\square} = 1$

10. $2\frac{1}{2} \cdot \dfrac{\square}{\square} = 1$

11. $\dfrac{\square}{\square} \cdot 5\frac{1}{4} = 1$

12. $\dfrac{\square}{\square} \cdot 2 = 1$

	Mixed Numbers	Improper Fraction (a)	Decimal Value (b)	Reciprocal of Improper Fraction (c)	Decimal Value of Reciprocal (d)
13.	$7\frac{1}{2}$	_____	_____	_____	_____
14.	$6\frac{3}{10}$	_____	_____	_____	_____
15.	$1\frac{7}{8}$	_____	_____	_____	_____
16.	$2\frac{4}{5}$	_____	_____	_____	_____
17.	$10\frac{1}{4}$	_____	_____	_____	_____
18.	$3\frac{2}{5}$	_____	_____	_____	_____

19. What symbol on your calculator finds the reciprocal of a number?

20. Explain two ways that your calculator can be used to find the reciprocal of a number.

Using reciprocals, the division process can be generalized this way.

• Dividing by a fraction is the same as multiplying by its reciprocal. This may be written:

$$\text{If } \frac{a}{b} \text{ and } \frac{c}{d} \text{ are fractions, and } \frac{c}{d} \neq 0, \text{ then}$$

$$\frac{a}{b} \div \frac{c}{d} = \frac{a}{b} \cdot \frac{d}{c} = \frac{ad}{bc}.$$

EXAMPLE 4

Divide $\frac{1}{3} \div \frac{5}{8}$.

Division of fractions involves finding the reciprocal of the divisor and then multiplying.

	Fraction	Reciprocal
	$\frac{5}{8}$	$\frac{8}{5}$

Step 1	**Step 2**	**Step 3**
The reciprocal of $\frac{5}{8}$ is $\frac{8}{5}$	Multiply the numerators. $1 \cdot 8 = 8$ Multiply the denominators. $3 \cdot 5 = 15$	Express the answer in lowest terms. $\frac{8}{15}$ is in lowest terms.
$\frac{1}{3} \div \frac{5}{8} = \frac{1}{3} \cdot \frac{8}{5}$	$\frac{1}{3} \cdot \frac{8}{5} = \frac{1 \cdot 8}{3 \cdot 5} = \frac{8}{15}$	$\frac{1}{3} \div \frac{5}{8} = \frac{8}{15}$

As in multiplication, the fractions can be expressed in lowest terms before performing the operation. Expressing the fraction in lowest terms is done <u>after</u> the reciprocal of the divisor is found.

EXAMPLE 5

Divide $\frac{8}{12} \div \frac{4}{3}$.

Step 1	**Step 2**	**Step 3**
The reciprocal of $\frac{4}{3}$ is $\frac{3}{4}$. Change the division problem into a multiplication problem.	Prime factor the numerator. $8 \cdot 3 = 2 \cdot 2 \cdot 2 \cdot 3$ Prime factor the denominator. $12 \cdot 3 = 2 \cdot 2 \cdot 3 \cdot 3$	Cancel common prime factors. Multiply factors in numerator. $1 \cdot 1 \cdot 2 \cdot 1 = 2$ Multiply factors in denominator. $1 \cdot 1 \cdot 3 \cdot 1 = 3$
$\frac{8}{12} \div \frac{4}{3} = \frac{8}{12} \cdot \frac{3}{4} = \frac{8 \cdot 3}{12 \cdot 4}$	$\frac{8 \cdot 3}{12 \cdot 4} = \frac{2 \cdot 2 \cdot 2 \cdot 3}{2 \cdot 2 \cdot 3 \cdot 3}$	$\frac{\overset{1}{\cancel{2}} \cdot \overset{1}{\cancel{2}} \cdot 2 \cdot \overset{1}{\cancel{3}}}{\underset{1}{\cancel{2}} \cdot \underset{1}{\cancel{2}} \cdot 3 \cdot \underset{1}{\cancel{3}}} = \frac{2}{3}$

Note: Since the fractions have already been expressed in lowest terms, no simplification is necessary. If the product (answer) is an improper fraction it should be changed to a mixed number.

4.5 Exercises

In 1–20, find each quotient. Express answers as mixed numbers in lowest terms.

1. $\dfrac{1}{6} \div \dfrac{2}{5}$ 2. $\dfrac{4}{9} \div \dfrac{1}{3}$ 3. $\dfrac{1}{2} \div \dfrac{3}{4}$ 4. $\dfrac{10}{1} \div \dfrac{2}{5}$

5. $\dfrac{3}{4} \div \dfrac{10}{7}$ 6. $\dfrac{6}{7} \div \dfrac{1}{8}$ 7. $\dfrac{3}{5} \div \dfrac{3}{4}$ 8. $\dfrac{2}{5} \div \dfrac{4}{10}$

9. $\dfrac{1}{2} \div \dfrac{1}{4}$ 10. $22 \div \dfrac{11}{2}$ 11. $\dfrac{6}{7} \div \dfrac{2}{14}$ 12. $\dfrac{3}{6} \div \dfrac{2}{3}$

13. $2 \div \dfrac{4}{3}$ 14. $\dfrac{18}{3} \div 5$ 15. $0 \div \dfrac{2}{3}$ 16. $\dfrac{4}{20} \div \dfrac{1}{5}$

17. $\dfrac{3}{10} \div \dfrac{2}{5}$ 18. $\dfrac{1}{2} \div 6$ 19. $\dfrac{3}{8} \div \dfrac{1}{2}$ 20. $\dfrac{5}{8} \div 0$

 Fractions can be divided with a calculator using two different methods.

EXAMPLE 1

$\dfrac{5}{6} \div \dfrac{2}{3} = 1.25$

5 $\boxed{\div}$ 6 $\boxed{\div}$ $\boxed{(}$ 2 $\boxed{\div}$ 3 $\boxed{)}$ $\boxed{=}$ 1.25

EXAMPLE 2

$\dfrac{5}{6} \div \dfrac{2}{3} = \dfrac{5}{6} \cdot \dfrac{3}{2} = 1.25$

5 $\boxed{\div}$ 6 $\boxed{\times}$ 3 $\boxed{\div}$ 2 $\boxed{=}$ 1.25

In 21–28, find each product with a calculator using either Example 1 or Example 2.

21. $\dfrac{3}{5} \div \dfrac{1}{5}$ 22. $14 \div \dfrac{7}{5}$ 23. $\dfrac{4}{10} \div \dfrac{4}{5}$ 24. $\dfrac{5}{8} \div \dfrac{5}{8}$

25. $\dfrac{3}{5} \div \dfrac{4}{7}$ 26. $\dfrac{3}{4} \div 5$ 27. $\dfrac{6}{5} \div \dfrac{2}{3}$ 28. $\dfrac{1}{2} \div \dfrac{1}{10}$

In 29–36, without actually working the problem, mentally predict if each quotient will be less than one, equal to one, or greater than one.

29. $\dfrac{1}{2} \div \dfrac{3}{4}$ 30. $\dfrac{3}{4} \div \dfrac{1}{2}$ 31. $4 \div \dfrac{1}{4}$ 32. $\dfrac{1}{4} \div 4$

33. $\dfrac{5}{6} \div \dfrac{5}{6}$ 34. $\dfrac{5}{6} \div \dfrac{6}{5}$ 35. $10 \div \dfrac{1}{20}$ 36. $\dfrac{1}{20} \div 10$

37. What can be said about two quotients whose product is one?

38. Explain why leaving out parenthesis, when using Example 1, to divide fractions with a calculator gives an incorrect quotient.

DIVISION OF MIXED NUMBERS

When mixed numbers are used in division, the mixed numbers are first expressed as fractions. Expressing whole numbers as fractions also simplifies the process. The same procedure as in division of fractions can then be performed.

EXAMPLE 1

Divide $2\frac{1}{6} \div 3\frac{1}{3}$.

Step 1	Step 2
Express the first number as a fraction. $$2\frac{1}{6} = \frac{(2 \cdot 6) + 1}{6} = \frac{13}{6}$$ Express the second number as a fraction. $$3\frac{1}{3} = \frac{(3 \cdot 3) + 1}{3} = \frac{10}{3}$$	The reciprocal of $\frac{10}{3}$ is $\frac{3}{10}$. Change the division problem into a multiplication problem.
$$2\frac{1}{6} \div 3\frac{1}{3} = \frac{13}{6} \div \frac{10}{3}$$	$$\frac{13}{6} \div \frac{10}{3} = \frac{13}{6} \cdot \frac{3}{10}$$

Step 3	Step 4
Simplify. Divide 3 and 6 by 3.	Multiply the numerators and the denominators. $$\frac{13 \cdot 1}{2 \cdot 10} = \frac{13}{20}$$
$$\frac{13}{6} \cdot \frac{3}{10} = \frac{13 \cdot \overset{1}{\cancel{3}}}{\underset{2}{\cancel{6}} \cdot 10}$$	$$2\frac{1}{6} \div 3\frac{1}{3} = \frac{13}{20}$$

After finding the reciprocal of the divisor, the fractions may be expressed in lowest terms by cancelling common factors before multiplying.

EXAMPLE 2

Divide $4\frac{4}{5} \div 1\frac{3}{5}$.

Step 1	Step 2
Express the numbers as fractions. $$4\frac{4}{5} = \frac{(4 \cdot 5) + 4}{5} = \frac{24}{5}$$ $$1\frac{3}{5} = \frac{(1 \cdot 5) + 3}{5} = \frac{8}{5}$$	The reciprocal of $\frac{8}{5}$ is $\frac{5}{8}$. Change the division problem into a multiplication problem.
$$4\frac{4}{5} \div 1\frac{3}{5} = \frac{24}{5} \div \frac{8}{5}$$	$$\frac{24}{5} \div \frac{8}{5} = \frac{24}{5} \cdot \frac{5}{8} = \frac{24 \cdot 5}{5 \cdot 8}$$

Step 3	Step 4
Prime factor the numerator. $24 \cdot 5 = 2 \cdot 2 \cdot 2 \cdot 3 \cdot 5$ Prime factor the denominator. $5 \cdot 8 = 5 \cdot 2 \cdot 2 \cdot 2$	Cancel common prime factors. Multiply factors in numerator. $1 \cdot 1 \cdot 1 \cdot 3 \cdot 1 = 3$ Multiply factors in denominator. $1 \cdot 1 \cdot 1 \cdot 1 = 1$
$$\frac{24 \cdot 5}{5 \cdot 8} = \frac{2 \cdot 2 \cdot 2 \cdot 3 \cdot 5}{5 \cdot 2 \cdot 2 \cdot 2}$$	$$\frac{\overset{1}{\cancel{2}} \cdot \overset{1}{\cancel{2}} \cdot \overset{1}{\cancel{2}} \cdot 3 \cdot \overset{1}{\cancel{5}}}{\underset{1}{\cancel{5}} \cdot \underset{1}{\cancel{2}} \cdot \underset{1}{\cancel{2}} \cdot \underset{1}{\cancel{2}}} = \frac{3}{1} = 3$$

4.6 Exercises

In 1–21, find each quotient. Express answers as mixed numbers in lowest terms.

1. $1\frac{1}{2} \div 2\frac{1}{4}$

2. $3\frac{1}{3} \div 4\frac{2}{9}$

3. $6\frac{3}{5} \div 2\frac{1}{2}$

4. $4\frac{7}{8} \div 2\frac{1}{6}$

5. $2\frac{2}{3} \div 1\frac{1}{2}$

6. $2\frac{3}{4} \div 1\frac{2}{3}$

7. $16\frac{2}{3} \div 3\frac{1}{3}$

8. $100 \div 3\frac{1}{3}$

9. $4\frac{3}{8} \div 2\frac{5}{6}$

10. $\frac{1}{28} \div 7\frac{1}{2}$

11. $18\frac{1}{2} \div 1$

12. $15 \div 2\frac{1}{2}$

13. $4\frac{1}{3} \div 3$

14. $1 \div 8\frac{1}{3}$

15. $0 \div 3\frac{1}{4}$

16. $3\dfrac{1}{3} \div 1\dfrac{2}{3}$ 17. $\dfrac{7}{8} \div 3\dfrac{1}{2}$ 18. $2\dfrac{5}{8} \div \dfrac{2}{7}$

19. $18 \div 3\dfrac{3}{5}$ 20. $6\dfrac{2}{3} \div 10$ 21. $8\dfrac{1}{5} \div 0$

  Mixed numbers can be divided with a calculator using two different methods.

EXAMPLE 1

$$6\dfrac{1}{2} \div 2\dfrac{1}{2} = 2.6$$

$\boxed{(}\;6\;\boxed{+}\;1\;\boxed{\div}\;2\;\boxed{)}\;\boxed{\div}\;\boxed{(}\;2\;\boxed{+}\;1\;\boxed{\div}\;2\;\boxed{)}\;\boxed{=}\;2.6$

EXAMPLE 2

$$6\dfrac{1}{2} \div 2\dfrac{1}{2} = \dfrac{13}{2} \div \dfrac{5}{2} = \dfrac{13}{2} \cdot \dfrac{2}{5} = 2.6$$

$13\;\boxed{\div}\;2\;\boxed{\times}\;2\;\boxed{\div}\;5\;\boxed{=}\;2.6$ *or* $13\;\boxed{\times}\;2\;\boxed{\div}\;\boxed{(}\;2\;\boxed{\times}\;5\;\boxed{)}\;\boxed{=}\;2.6$

In 22–27, find each quotient with a calculator using either Example 1 or Example 2.

22. $8 \div 2\dfrac{2}{3}$ 23. $1\dfrac{3}{5} \div 1\dfrac{1}{4}$ 24. $\dfrac{12}{8} \div 1\dfrac{1}{2}$

25. $7\dfrac{1}{4} \div 7\dfrac{2}{8}$ 26. $3\dfrac{1}{5} \div 4$ 27. $11\dfrac{2}{3} \div 0$

In 28–33, insert parentheses as needed to make each statement true.

28. $2\dfrac{1}{4} \div \dfrac{3}{4} \div 3 \cdot 1\dfrac{7}{9} = 16$ 29. $2\dfrac{1}{4} \div \dfrac{3}{4} \div 3 \cdot 1\dfrac{7}{9} = 1\dfrac{7}{9}$

30. $2\dfrac{1}{4} \div \dfrac{3}{4} \div 3 \cdot 1\dfrac{7}{9} = \dfrac{9}{16}$ 31. $25 \div \dfrac{3}{4} \cdot 1\dfrac{1}{3} \cdot 6\dfrac{1}{4} = 4$

32. $6 \div 2 \div 9 \div \dfrac{3}{4} = \dfrac{1}{4}$ 33. $6 \div 2 \div 9 \div \dfrac{3}{4} = 36$

34. Write three division problems involving fractions where the quotient of the two fractions is less than one, equal to one, or greater than one.

35. Explain why leaving out parenthesis, when using Example 1, to divide mixed numbers with a calculator gives an incorrect quotient.

APPLICATIONS

Health care assistants are a vital part of the medical field. *Health care assistants* maintain the control of medications, drugs, and medical records by keeping an accurate inventory. The assistant occupations enable the health care field to run smoothly.

Medical records clerks are one type of assistant. The medical records clerk is responsible for assembling, in sequence, information for medical records. They may also gather statistical data for reports.

Medical assistants help physicians examine and treat patients. The medical assistant may be responsible for arranging instruments and equipment and for checking office and laboratory supplies.

A *medical laboratory assistant* aids the medical technologist in the laboratory. The medical laboratory assistant is responsible for cleaning and sterilizing laboratory equipment, glassware, and instruments. The medical laboratory assistant may prepare solutions following standard laboratory formulas.

4.7 Exercises

1. A medical records clerk gathers data about patients with broken bones. It is found that 1/3 of the medical records are for individuals with broken bones. A total of 14,391 records are in the files. How many records are for individuals with broken bones?

2. How many 1/2-gram doses can be obtained from a 5-gram vial of medication?

3. A medical laboratory assistant prepares a solution using a standard laboratory formula. The solution in the flask weighs 6 1/3 ounces. What is the weight of 3/5 of the solution?

4. How many fourths can be produced from one scored medication tablet?

5. A medical assistant fills a storage cabinet with 14 bottles of dextrose. Each bottle is filled with 16 2/3 ounces of the solution. How many ounces of this solution are in the cabinet?

6. An inventory is conducted for six departments to determine the amount of ethyl alcohol presently available. Using the following data calculate the total amount of ethyl alcohol.

Department A	$24\frac{5}{6}$ ounces
Department B	$17\frac{2}{3}$ ounces
Department C	$19\frac{1}{2}$ ounces
Department D	$42\frac{3}{5}$ ounces
Department E	$9\frac{1}{7}$ ounces
Department F	$14\frac{7}{8}$ ounces

7. Several containers were found to contain the following amounts of saline solution. Complete the chart to find the total amount of saline solution.

Container	Number of Containers	Number of Ounces
$16\frac{1}{3}$ ounces	3	
$23\frac{1}{5}$ ounces	6	
$9\frac{6}{7}$ ounces	4	
TOTAL		

Unit 5

Section One
Applications to Health Work

Objectives

After studying this unit the student should be able to:
- **Use the basic operations of fractions to solve health work problems.**

A *drug* is a substance or a mixture of substances which have been found to have a definite value in the detection, prevention, or treatment of diseases. Drugs may be used to cure diseases, assist the body in overcoming diseases, or help prevent diseases. Drugs may also be used to help diagnose a disease.

In administering medication, accuracy is of the utmost importance. Errors in the amount of medication may have fatal results. Medications that are a mixture of two or more substances must be accurately prepared; liquid medication must be accurately measured.

The ways in which drugs are administered are: oral, injection, intramuscular, and topical.

DEFININTIONS

- Administering a drug *orally* means that the drug is taken via the mouth. If the drug is to be given orally, it is usually in the form of a liquid, tablet, or powder.
- Administering a drug by *injection* means that the drug enters the body through the blood system. For injection purposes, the medication must be in a liquid form. The medication is administered by using a syringe.
- An *intramuscular* administration of a drug means that the drug is placed directly into the muscle. A syringe is used for this purpose and the preparation must be in a liquid form.
- A *topical* application of a drug means that the medication is placed on the external body. Topical drugs cannot be administered internally. Topical drugs are usually in a liquid, powder, or semisolid form.

The portion of a drug to be administered at one time is the *dose*. The total quantity that is to be administered is the *dosage*. Drug dosage depends on:
- The weight, sex, and age of the patient.
- The disease being treated.
- How the drug is to be administered.
- The patient's tolerance of the drug.

In calculating doses and dosages, these terms are used.

DEFININTIONS

- An *average dose* is the amount of medication which has been proven most effective with minimal toxic effects.
- An *initial dose* is the first dose.
- A *maximum dose* is the largest amount of medication which can be safely administered at one time.
- A *lethal dose* is the amount that could cause death to a patient.

In administering doses of medication, more than one tablet may be used or a scored tablet may have to be divided. Liquid preparations may have to be divided into equal portions or additional portions may have to be made. Care must always be taken when preparing the medications and administering the drugs.

5.1 Exercises

1. A physician prescribes a 90-milligram dosage of thyroid tablets. Each tablet contains 30 milligrams, or 1/2 grain, of medication.

   a. What fractional part of the prescribed dosage does one tablet represent?

   b. How many 30-milligram tablets would be needed to administer the prescribed dosage to the patient?

   c. How many total grains are present in the total dosage?

2. A patient is allergic to natural thyroid preparations. The physician prescribes a synthetic thyroid preparation which is available in 25-microgram tablets. The prescribed dose is 37 1/2 micrograms. How many tablets are needed?

3. An unlocked medical cabinet contains 67 different compounds. A total of 23 are used for skin conditions and are applied topically. A total of 15 are used for intestinal disorders and are administered orally. A registered nurse removes the chemical compounds used for intestinal disorders. What fractional part of the remaining chemicals are used for skin disorders?

4. In a community hospital, a medical records technician is responsible for preparing statistical reports. The medical records clerk assists the technician in the gathering of data that is needed. The clerk finds that the hospital is composed of 60 people. There are 3 maintenance personnel, 2 secretaries, and 1 billing clerk. Find the fractional part represented by each group.

   a. maintenance personnel
   b. medical secretaries
   c. billing clerks
   d. others

5. A health service administrator determines that a medical staff of 127 people must be decreased by approximately 1/12. How many staff positions must be eliminated?

6. Each day, a doctor administers 2 1/2 milligrams of medication to a patient. There are 25 milligrams of medication left in the bottle. How many more 2 1/2 milligram doses can the doctor administer from the bottle?

7. One poison is known to cause death in amounts as small as 1/250 ounce. A solution of 4/5-ounce poison in 1 gallon of distilled water will produce the lethal results. If this solution is divided into equal amounts, how many lethal doses can be prepared? *Note:* 1 gallon = 128 ounces.

8. Death may be caused by 1/20 ounce of arsenic. How many lethal doses are represented in 8/9 ounce?

9. A pharmacist prepares a mixture containing six grams of compound *A* and three grams of compound *B*.

    a. What part of the mixture is compound *A?*

    b. What part of the mixture is compound *B?*

10. A powder to be used topically is prepared by mixing 5 grams of compound *C* and 7 grams of compound *D*.

    a. How many grams does the mixture contain?
    b. What fractional part of the mixture is compound *C?*
    c. What fractional part of the mixture is compound *D?*

11. A medical laboratory assistant prepares containers of a saline solution. Each container has 4 3/8 ounces of solution in it. How many ounces of solution are present in

    a. 3 1/2 containers?
    b. 1 3/5 containers?
    c. 2/3 container?
    d. 17 1/3 containers?

12. A certain medical assistant is allowed to administer injections. The assistant has a vial of medication containing 2 ounces of medication. The average dose is 3/5 ounce. How many full 3/5-ounce doses can be obtained from the vial?

13. An average dose of medicine is 3/4 ounce. One patient is given an initial dose of 2/5 of the 3/4-ounce dose. A second patient is given an initial dose of 2 1/2 times the 3/4-ounce dose.

    a. How many ounces of medicine does the first patient receive?

    b. How many ounces of medicine does the second patient receive?

14. A *health services administrator* makes management decisions about personnel, space requirements, and budgets. The administrator determines that for the coming year, the proposed budget must be decreased by $237,000. The decrease is 1/15 of the proposed budget. How much is the proposed budget?

15. Two file cabinets for medical records are the same size. One cabinet is 4/5 full and the other is 1/4 full. A medical records clerk wants to combine the cabinets. Can the contents of these two cabinets be combined? Explain.

16. Ann receives a gradually decreasing dose of medication. She drinks 3/4 ounce of medication at 1 p.m.; 1/2 ounce at 3 p.m.; and 1/4 ounce at 6 p.m. What is the total amount of medication that Ann consumes during this time period?

17. A registered nurse removes two scored medication tablets from the same container. One patient is given 1/2 tablet and another is given 1/4 tablet. Which patient is given the larger dose?

18. A dietitian knows that a box containing 566 grams of cereal is composed of 100 grams protein, 400 grams carbohydrates, and 20 grams fat. Find the amounts present in 1/9 of the box.

    a. protein
    b. carbohydrates
    c. fat
    d. other substances

19. A pharmacist in a hospital weighs a capsule of medication. The capsule weighs 4/90 ounce.

    a. What is the weight of three capsules?

    b. What is the weight of a capsule containing 1/2 the amount of this same medication?

20. Margaret has a condition requiring the administration of colchicine. The maximum dose of colchicine is 1/50 grain; each tablet contains 1/100 grain of medication.

    a. How many tablets are in the maximum dose?

    b. How many maximum doses could be obtained from a bottle containing 100 tablets of colchicine?

21. Two vials of liquid, one 2/3 full and a second 3/8 full, are to be combined in a third vial of the same size. Will the third vial hold the combined contents?

22. A nurse has a vial containing 8 ounces of medication. The average dose is 1/2 ounce. After 5 doses how many ounces of medication are left in the vial?

23. A powder is to be prepared by mixing 7 grams of substance *A* and 3 grams of substance *B*.

    a. Substance *A* is what fractional part of the mixture?

    b. Substance *B* is what fractional part of the mixture?

24. Flasks, each containing 6 2/5 ounces of an iodine solution, are prepared in a lab. How many ounces of solution are present in:

    a. 1/4 flask?

    b. 2 1/2 flasks?

25. The $125,000 supply budget for maintenance in a nursing home must be reduced by 1/10. How much money will remain in the budget for supplies?

SECTION 1: Summary, Review and Study Guide

VOCABULARY

common denominator
common factor
common multiple
cross-product
denominator
difference
divisible
equivalent fractions
factors
factor tree
fraction
greatest common factor (GCF)

improper fraction
least common denominator (LCD)
least common multiple (LCM)
lowest terms
multiple
multiplicative inverse
numerator
product
prime number
quotient
reciprocal
sum

CONCEPTS, SKILLS AND APPLICATIONS

Objectives With Study Guide

Upon completion of Section 1, you should be able to:

- **Use divisibility rules for 2, 3, 4, 5, 6, 8, 9, 10.**

  Is 210 divisible by 2, 5 and 10?

  2:   the units digit, 0, is even. Thus, 210 is divisible by 2.

  5:   the units digit is 0. Thus, 210 is divisible by 5.

  10: the units digit is 0. Thus, 210 is divisible by 10.

- **Compare whole numbers.**

  Is $4 \cdot 2 \cdot 5$ <u> <, =, > </u> $5 \cdot 9$?

  Since 40 is less than 45

  $4 \cdot 2 \cdot 5 < 5 \cdot 9$.

- **Prime factor a number. Prime factor 48.**

$$
\begin{array}{r}
3 \\
2\overline{)6} \\
2\overline{)12} \\
2\overline{)24} \\
2\overline{)48}
\end{array}
$$

  The prime factorization of 48 is $2 \cdot 2 \cdot 2 \cdot 2 \cdot 3$.

- **Find the common multiple.**

  Find three common multiples of 6 and 9.

  Using counting:

  6 = 6, 12, 18, 24, 30, 36, 42, 48, 54 . . .
  9 =   18, 27, 36, 45, 54 . . .

  Using prime factoring:
  6 = 2 · 3
  9 = 3 · 3
  $2 \cdot 3 \cdot 3 = 18$ the least common multiple is the product of the prime factors.

- **Express fractions in lowest terms.**

  Reduce $\dfrac{12}{28}$

$$\frac{12}{28} = \frac{\overset{1}{\cancel{2}} \cdot \overset{1}{\cancel{2}} \cdot 3}{\underset{1}{\cancel{2}} \cdot \underset{1}{\cancel{2}} \cdot 7} = \frac{3}{7}$$

- **Express fractions as equivalent fractions.**

  Find 4 fractions equivalent to $\dfrac{2}{5}$.

$$\left(\frac{2}{3}\right)\left(\frac{2}{2}\right) = \frac{4}{6}; \left(\frac{2}{3}\right)\left(\frac{3}{3}\right) = \frac{6}{9};$$

$$\left(\frac{2}{3}\right)\left(\frac{4}{4}\right) = \frac{8}{12}; \left(\frac{2}{3}\right)\left(\frac{5}{5}\right) = \frac{10}{15}$$

$$\frac{2}{3} = \frac{4}{6} = \frac{6}{9} = \frac{8}{12} = \frac{10}{15}$$

- **Compare fraction values.**

  Are $\dfrac{2}{3}$ and $\dfrac{9}{12}$ equivalent fractions?

$$\frac{2}{3} < \frac{9}{12}$$

- **Find least common denominators.**

  Use prime factoring to help find the LCD of $\dfrac{3}{8}$ and $\dfrac{5}{6}$.

  $8 = 2 \cdot 2 \cdot 2$

  $6 = 3 \cdot 2$

  $\text{LCM} = \text{LCD} = 2 \cdot 2 \cdot 2 \cdot 3 = 24$

- **Express fractions as equivalent fractions with least common denominators (LCD).**

  Find fractions equivalent to $\dfrac{3}{8}$ and $\dfrac{5}{6}$ containing the LCD.

  $\dfrac{3}{8} = \dfrac{a}{24}$ $\qquad\qquad$ $\dfrac{5}{6} = \dfrac{10}{b}$

  $8a = 3 \cdot 24$ $\qquad\qquad$ $5b = 6 \cdot 10$

  $8a = 72$ $\qquad\qquad$ $5b = 60$

  $a = 9$ $\qquad\qquad$ $b = 12$

  $\dfrac{3}{8} = \dfrac{9}{24}$ $\qquad\qquad$ $\dfrac{5}{6} = \dfrac{10}{12}$

- **Add fractions and mixed numbers.**

  Find the sum. $\dfrac{3}{8} + \dfrac{5}{6}$

  $$\dfrac{3}{8} = \dfrac{9}{24}$$

  $$+ \dfrac{5}{6} = \dfrac{20}{24}$$

  $$\dfrac{29}{24} \ \ or \ \ 1\dfrac{5}{24}$$

  Find the sum. $3\dfrac{1}{2} + 2\dfrac{4}{5}$

  $$3\dfrac{1}{2} = 3\dfrac{5}{10}$$

  $$+ 2\dfrac{4}{5} = 2\dfrac{8}{10}$$

  $$5\dfrac{13}{10} \ = \ 6\dfrac{3}{10}$$

- **Subtract fractions and mixed numbers.**

  Find the difference. $\dfrac{5}{9} - \dfrac{3}{6}$

  $$\dfrac{5}{9} = \dfrac{10}{18}$$

  $$- \dfrac{3}{6} = \dfrac{9}{18}$$

  $$\dfrac{1}{18}$$

  Find the difference. $7\dfrac{1}{3} - 3\dfrac{1}{2}$

  $$7\dfrac{1}{3} = 7\dfrac{2}{6} = 6\dfrac{8}{6}$$

  $$- 3\dfrac{1}{2} = 3\dfrac{3}{6} = 3\dfrac{3}{6}$$

  $$3\dfrac{5}{6}$$

- **Multiply fractions and mixed numbers.**

  Find the product. $\dfrac{2}{6} \times \dfrac{3}{4}$

  $$\dfrac{2}{6} \times \dfrac{3}{4} = \dfrac{\overset{1}{\cancel{2}} \cdot \overset{1}{\cancel{3}}}{\underset{1}{\cancel{2}} \cdot \underset{1}{\cancel{3}} \cdot 2 \cdot 2} = \dfrac{1}{4}$$

Find the product. $3\frac{1}{3} \times 2\frac{2}{5}$

$$3\frac{1}{3} \times 2\frac{2}{5} = \frac{10}{3} \times \frac{12}{5} = \frac{\cancel{5} \cdot 2 \cdot 2 \cdot 2 \cdot \cancel{3}}{\cancel{3} \cdot \cancel{5}} = 8$$

- **Divide fractions and mixed numbers.**

Find the quotient. $\frac{3}{8} \div \frac{5}{4}$

$$\frac{3}{8} \div \frac{5}{4} = \frac{3}{8} \times \frac{4}{5} = \frac{3 \cdot \cancel{2} \cdot \cancel{2}}{\cancel{2} \cdot \cancel{2} \cdot 2 \cdot 5} = \frac{3}{10}$$

Find the quotient. $3\frac{2}{6} \div 1\frac{1}{5}$

$$3\frac{2}{6} \div 1\frac{1}{5} = \frac{20}{6} \div \frac{6}{5} = \frac{20}{6} \times \frac{5}{6} = \frac{\cancel{2} \cdot \cancel{2} \cdot 5 \cdot 5}{3 \cdot \cancel{2} \cdot 3 \cdot \cancel{2}} = \frac{25}{9} = 2\frac{7}{9}$$

Review

Major concepts and skills to be mastered for the Section 1 Test.

In 1–10, complete each sentence. Choose terms from the vocabulary list.

1. _____?_____ means that there is no remainder in the quotient.

2. If the greatest common factor (GCF) of both the numerator and denominator of a fraction is one, the fraction is in _____?_____ .

3. Numbers being multiplied to find a product are called _____?_____ .

4. _____?_____ fractions name the same point on a number line.

5. Two numbers are _____?_____ if their product is one.

6. When subtracting the answer is called the _____?_____ .

7. The _____?_____ is the solution to a multiplication problem.

8. The _____?_____ is the top number in a fraction.

9. The _____?_____ is the least common multiple of the denominator of two or more fractions.

10. The bottom number in a fraction is the _____?_____ .

In 11–14, decide if each number is divisible by 2, 5 or 10.

11. 39   12. 900   13. 75   14. 333

15. If a number is divisible by 5, is it also divisible by 10? Explain.

In 16–18, compare: Use <, =, or >.

16. $82 \div 2$ _____ $7 \cdot 6$

17. $200 + 50 \div 5$ _____ $(200 + 50) \div 5$

18. $5(10 \div 2)$ _____ $(5 \cdot 10) \div 2$

In 19–22, prime factor.

19. 333   20. 23   21. 1015   22. 88

In 23 and 24, find two common multiples.

23. 8 and 10   24. 12 and 8

In 25 and 26, find the least common multiple (LCM).

25. 12 and 15   26. 10 and 25

In 27–30, express each fraction in lowest terms.

27. $\dfrac{6}{24}$ 28. $\dfrac{18}{12}$ 29. $\dfrac{70}{50}$ 30. $\dfrac{8}{9}$

In 31–34, find three equivalent fractions for each given fraction.

31. $\dfrac{5}{9}$ 32. $\dfrac{3}{8}$ 33. $\dfrac{4}{3}$ 34. $\dfrac{8}{5}$

In 35 and 36, compare.

35. Is $\dfrac{3}{4} = \dfrac{6}{12}$?   36. Is $\dfrac{7}{2} = \dfrac{12}{4}$?

In 37 and 38, find the least common denominator (LCD).

37. $\dfrac{4}{6}$ and $\dfrac{5}{9}$ 38. $\dfrac{6}{5}$ and $\dfrac{8}{12}$

In 39 and 40, find the appropriate equivalent fractions.

39. $\dfrac{2}{3} = \dfrac{?}{18}$ 40. $\dfrac{3}{5} = \dfrac{12}{?}$

In 41–44, add. Reduce if necessary.

41. $\dfrac{2}{3} + \dfrac{3}{4}$ 42. $\dfrac{1}{3} + \dfrac{1}{5}$ 43. $5\dfrac{1}{4} + 4\dfrac{5}{12}$ 44. $2\dfrac{3}{8} + 7\dfrac{4}{5}$

In 45–48, subtract. Reduce if necessary.

45. $\dfrac{7}{8} - \dfrac{3}{4}$ 46. $\dfrac{11}{16} - \dfrac{10}{32}$ 47. $11\dfrac{1}{2} - 8\dfrac{3}{4}$ 48. $2\dfrac{4}{5} - 1\dfrac{5}{10}$

In 49–52, multiply. Reduce if necessary.

49. $\dfrac{7}{8} \times \dfrac{4}{5}$ 50. $\dfrac{5}{12} \times \dfrac{3}{5}$ 51. $1\dfrac{2}{3} \times 1\dfrac{8}{10}$ 52. $1\dfrac{2}{7} \times 3\dfrac{2}{3}$

In 53–56, divide. Reduce if necessary.

53. $\dfrac{5}{9} \div \dfrac{4}{3}$ 54. $\dfrac{7}{16} \div \dfrac{3}{4}$ 55. $2\dfrac{4}{8} \div 1\dfrac{1}{4}$ 56. $3\dfrac{2}{3} \div 2\dfrac{1}{3}$

SECTION 1 TEST

1. If a number is divisible by 10, it is also divisible by 5. Explain.

In 2–5, compare. Use <, =, >.

2. $400 - 3\,(60)$ _____ $200 + (40 \div 2)$

3. $3 \cdot 15$ _____ $45 \cdot 0$

4. $2\,(35 \div 7)$ _____ $5\,(90) - 40\,(11)$

5. $0 \div 25$ _____ $0 \cdot 15$

In 6–13, circle YES if the answer is prime or NO if the answer is not prime.

Expression	Answer	Prime
$(12 \cdot 13) - (11 \cdot 13)$	6. _____	7. Y or N
$(7 \cdot 25) + 28 + 52$	8. _____	9. Y or N
$4\,(12 + 6) \div 3$	10. _____	11. Y or N
$(1744 \div 4) \div 4$	12. _____	13. Y or N

In 14–16, find the least common multiple (LCM).

14. 10 and 12   15. 4 and 8   16. 9 and 8

In 17–19, express each fraction in lowest terms.

17. $\dfrac{4}{5}$ 18. $\dfrac{9}{21}$ 19. $\dfrac{8}{20}$

In 20–22, find three equivalent fractions for each given fraction.

20. $\dfrac{3}{8}$ 21. $\dfrac{4}{3}$ 22. $\dfrac{2}{5}$

In 23–25, compare.

23. Is $\dfrac{10}{12} = \dfrac{5}{6}$?   24. Is $\dfrac{1}{4} = \dfrac{4}{16}$?   25. Is $\dfrac{2}{3} = \dfrac{3}{4}$?

In 26–28, find the least common denominator (LCD).

26. $\dfrac{6}{10}$ and $\dfrac{3}{4}$ 27. $\dfrac{1}{8}$ and $\dfrac{2}{6}$ 28. $\dfrac{4}{7}$ and $\dfrac{2}{3}$

In 29 and 30, find the appropriate equivalent fractions.

29. $\dfrac{4}{5} = \dfrac{x}{20}$ 30. $\dfrac{7}{8} = \dfrac{14}{c}$

In 31–34, find the sum. Reduce if necessary.

31. $\dfrac{1}{2} + \dfrac{1}{3}$ 32. $\dfrac{5}{6} + \dfrac{2}{3}$ 33. $4\dfrac{1}{8} + 5\dfrac{3}{4}$ 34. $10\dfrac{1}{6} + 5\dfrac{5}{6}$

In 35–38, find the difference. Reduce if necessary.

35. $\dfrac{6}{10} - \dfrac{3}{5}$ 36. $\dfrac{5}{8} - \dfrac{1}{2}$ 37. $10\dfrac{5}{6} - 8\dfrac{1}{3}$ 38. $4\dfrac{1}{2} - 2\dfrac{7}{8}$

In 39–42, find the product. Reduce if necessary.

39. $\dfrac{3}{6} \times \dfrac{4}{4}$ 40. $\dfrac{8}{12} \times \dfrac{3}{4}$ 41. $1\dfrac{2}{3} \times 2\dfrac{2}{5}$ 42. $4\dfrac{2}{10} \times 3\dfrac{1}{3}$

In 43–46 find the quotient. Reduce if necessary.

43. $\dfrac{1}{2} \div \dfrac{3}{2}$ 44. $\dfrac{0}{10} \div \dfrac{2}{5}$ 45. $1\dfrac{3}{5} \div \dfrac{4}{5}$ 46. $3\dfrac{2}{8} \div 2\dfrac{1}{2}$

SOLVE THE FOLLOWING APPLICATION PROBLEMS

The operational budget for the clinic has been set at $125,000 per month.

47. What fractional part of the annual operating cost does this represent?

48. What is the total operational cost for the year?

Expenditure of the budget has been set as follows: 3/5 salaries and benefits, 1/10 supplies, 1/10 equipment lease/purchase, 3/20 liability insurance and 1/20 miscellaneous.

49. Is the total of the projected expenditures <, =, or > than 5/5?

Calculate, to the nearest dollar, the annual expenditure for:

50. Salaries and benefits

51. Supplies

52. Equipment lease/purchase

53. Liability insurance

54. Miscellaneous

55. Total

Records indicate that an average of 2,500 patients are served each month.

56. What fractional part of the operational cost must be paid by each patient, assuming all patients receive comparable services?

57. Calculate the actual dollar amount that each patient must pay to cover the monthly operational cost.

58. If the cost to each patient is higher than other clinics providing comparable services, how can this clinic reduce the cost to each patient without reducing services or losing money?

A patient is administered medication that produces its maximum effect in two hours. Beyond this point the effect of the medication declines at a rate of 1/7 each 30 minutes the first 1 1/2 hours, 1/14 each 30 minutes the next 2 1/2 hours and 1/20 each 30 minutes until completely gone from the patient's system.

59. After reaching the maximum effect on the patient, two hours after taking the medication, how long will it be before all medication is gone from the patient's system? Show your calculations.

The pulse rate of a patient is measured during a period of rest and again after a five-minute exercise. The beginning rate is 72 beats per minute and 110 beats per minute immediately following the exercise. The heart rate is monitored each 5 minutes after completion of the exercise to determine the rate of recovery to the original heart beat level.

60. Indicate the fractional increase as the heart rate moved from 72 beats per minute to 110 beats per minute.

61. After a period of one minute the heart beat decreased from 110 beats per minute to 92 beats per minute. Indicate the fractional decrease that occurred.

62. After a period of two minutes the heart beat had decreased from 110 beats per minute to 83 beats per minute. Indicate the fractional decrease that occurred.

63. After a period of three minutes the heart beat had decreased from 110 beats per minute to 77 beats per minute. Indicate the fractional decrease that occurred.

64. After a period of four minutes the heart beat had decreased from 110 beats per minute to 75 beats per minute. Indicate the fractional decrease that occurred.

65. After a period of five minutes the heart beat had decreased from 110 beats per minute to 70 beats per minute. Indicate the fractional decrease that occurred.

The patient is prescribed 4 ounces of liquid medication. The dosage is to be taken in declining amounts. The directions indicate that the patient is to take 1/2 ounce each hour for the first two hours, 1/3 ounce each hour for the next four hours and 1/4 ounce each hour until the remaining medication is less than 1/4 ounce. Assume there is no waste at each medication.

66. How many hours will be required to use the medication to meet the prescription directions. Show your calculations.

Section Two

Decimal Fractions

Unit 6

Introduction to Decimal Fractions

Objectives

After studying this unit the student should be able to:

- **Express a common fraction with a denominator which is a multiple of ten as a decimal fraction.**
- **Express a decimal fraction as a mixed number with a denominator which is a multiple of ten.**
- **Compare decimal fractions.**

DECIMAL FRACTIONS

A decimal fraction is used to express a common fraction whose denominators are multiples of 10, such as 10; 100; 1,000; 10,000; 100,000; and 1,000,000.

EXAMPLE 1

The fraction $\frac{2}{5}$ is equivalent to $\frac{4}{10}$, $\frac{40}{100}$ and $\frac{400}{1,000}$. These denominators are multiples of 10.

$\frac{4}{10} = 0.4$ This is read "4 tenths."

$\frac{40}{100} = 0.40$ This is read "40 hundredths."

$\frac{400}{1,000} = 0.400$ This is read "400 thousandths."

Similarly, any decimal fraction can be expressed as a common fraction with a denominator as a power of ten.

EXAMPLE 2

five tenths	0.5	$\dfrac{5}{10}$
seven hundredths	0.07	$\dfrac{7}{100}$
eleven thousandths	0.011	$\dfrac{11}{1,000}$
two hundred nineteen ten-thousandths	0.0219	$\dfrac{219}{10,000}$
forty-three hundred-thousandths	0.00043	$\dfrac{43}{100,000}$
eight hundred seventeen millionths	0.000817	$\dfrac{817}{1,000,000}$

DECIMAL FRACTIONS IN THE PLACE-VALUE SYSTEM

The base ten place-value system is a deciaml system. The place-value system is extended to give meaning to decimals. The decimal point separates the whole numbers from the decimal fractions.

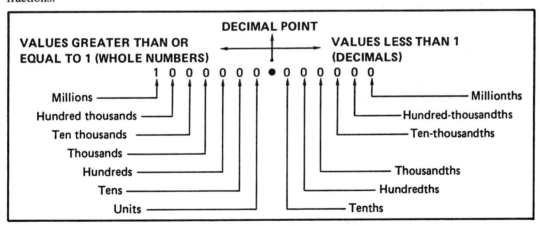

EXAMPLE 1

The sum $86 + \dfrac{3}{10} + \dfrac{7}{100} + \dfrac{4}{1,000}$ can be written as a decimal using the place-value system.

$86 \qquad + \qquad \dfrac{3}{10} \qquad + \qquad \dfrac{7}{100} \qquad + \qquad \dfrac{4}{1,000}$ is written

TENS	UNITS		TENTHS		HUNDREDTHS		THOUSANDTHS
8	6	•	3		7		4

EXAMPLE 2

Express 4.517 as a mixed number. Use the sum of fractions whose denominators are multiples of ten.

$$4.517 = 4 + \frac{5}{10} + \frac{1}{100} + \frac{7}{1,000} = 4 + \frac{500}{1,000} + \frac{10}{1,000} + \frac{7}{1,000} = 4\frac{517}{1,000}$$

6.1 Exercises

In 1–8, write, in words, each decimal.

1. 0.47 2. 8.095 3. 25.0090 4. 0.0509

5. 14.00750 6. 0.0078 7. 121.8719 8. 117.9103

In 9–14, write, using numerals, each expression.

9. Sixteen hundredths

10. Nine and eight thousandths

11. Twenty-five and seventeen ten-thousandths

12. Thirty seven ten-thousandths

13. Seven and ninety-nine thousandths

14. Six hundred fifty-two and one ten-thousandth

In 15–30, express each quantity as a decimal.

15. $\frac{3}{10}$ 16. $\frac{21}{100}$ 17. $\frac{137}{1,000}$ 18. $\frac{1,571}{10,000}$

19. $\frac{9}{1,000}$ 20. $\frac{7}{100}$ 21. $\frac{31}{10,000}$ 22. $\frac{1}{1,000}$

23. $4 + \frac{2}{10}$ 24. $5 + \frac{13}{100}$ 25. $6 + \frac{21}{100}$ 26. $23 + \frac{137}{1,000}$

27. $42 + \frac{8,131}{10,000}$ 28. $11 + \frac{17}{1,000}$ 29. $10 + \frac{107}{10,000}$ 30. $0 + \frac{107}{10,000}$

In 31–34, express each decimal as a fraction or mixed number. Use the sum of fractions whose denominators are powers of ten.

31. 3.462 32. 0.981 33. 17.803 34. 9.012

In 35–46, express each decimal as a mixed number. Do not express in lowest terms.

35. 5.9 36. 7.31 37. 8.90 38. 2.5

39. 42.657 40. 14.057 41. 92.705 42. 1.0057

43. 17.00570 44. 111.00072 45. 9.7005 46. 15.82090

COMPARING DECIMAL FRACTIONS

Decimals can be grouped on a number line.

EXAMPLE 1

Give the decimal name for each point.

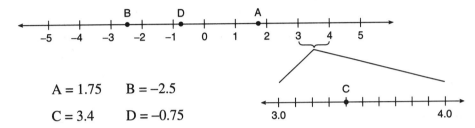

A = 1.75   B = −2.5

C = 3.4   D = −0.75

When two numbers are graphed on a number line, the number on the right is the larger of the two.

EXAMPLE 2

Compare 1.59 and −2.99.

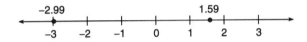

Since 1.59 is to the right of −2.99, 1.59 > −2.99.

Comparing decimals is similar to comparing whole numbers. Starting from the place value to the far left, each number is compared. This process continues until unequal values are found in corresponding place values.

EXAMPLE 3

Which number is smaller: 3.1478 or 3.1468?

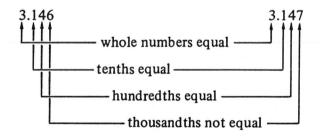

Since, in the thousandths place, 6 < 7, 3.1468 < 3.1478.

EXAMPLE 4

Which number is larger: 0.279 or 0.27?

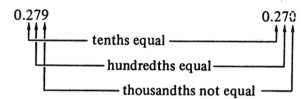

Since 9 > 0, 0.279 > 0.27.

6.2 Exercises

In 1–4 , use words to write each decimal.

1. 4.715   2.   0.43   3.   517.208   4.   0.0057

In 5–10, use numbers to represent each written expression.

5. Four and seventeen thousandths

6. Thirty-four ten-thousandths

7. One hundred seventy-one and five hundredths

8. Twenty-three and thirty ten-thousandths

9. Three and nine hundred-thousandths

10. Name the decimal that corresponds approximately to the point indicated by each letter.

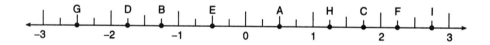

A. _____   B. _____   C. _____

D. _____   E. _____   F. _____

G. _____   H. _____   I. _____

In 11 and 12, order from smallest to largest.

11. 0.04007, 0.04, 0.40009, 4.005, 0.004007

12. 7.9999, 8.0001, 7.99998, 8.00900, 8.10006

In 13–27, compare each pair of decimals. Use the symbols <, >, or = .

13. 0.56 __?__ 0.59   14. 0.68 __?__ 0.0680   15. 0.325 __?__ 0.327

16. 2.398 __?__ 2.389   17. 3.765 __?__ 2.76518   18. 4.001 __?__ 3.999

19. 0.3 __?__ 0.99   20. 0.005 __?__ 0.012   21. 0.19 __?__ 0.989

22. 7.304 __?__ 7.3   23. 0.009 __?__ 0.010   24. 4.7690 __?__ 4.769

25. 0.0099 __?__ 0.100   26. 4.342 __?__ 4.0342   27. 0.32 __?__ 0.32001

28. Respond to the statement that 3.69 is greater than 3.8 because 9 is greater than 8.

29. The numbers 8,751, 6,192, 3,404, 2,967 and 1,835 are ordered from largest to smallest. Without rearranging the numbers, place a decimal point in each number so that the resulting numbers will be ordered from smallest to largest.

30. Explain how comparing decimals is similar to comparing whole numbers.

APPLICATIONS

A patient in a hospital is given 0.4 grams of medication instead of 0.04 gram. The patient dies. The cause of death might read: MISPLACED DECIMAL. This tragic example of carelessness — or perhaps lack of knowledge — with decimals can occur.

Decimals are a vital part of the health care field. Each health care worker should understand and use decimals carefully. Preparing medications, comparing and administering doses, and compiling statistical reports are only a few examples of instances where decimals are used.

6.3 Exercises

1. Two patients receive the same medication. The first patient receives 0.125 gram and the second receives 0.325 gram. Which patient receives the larger dose?

2. *Occupational therapists* design and direct vocational, educational and recreational activities for handicapped people. There are about 9,600 occupational therapists in the United States of which 4/10 work with the emotionally handicapped and 6/10 work with the physically handicapped. Express these fractional parts as decimals.

3. During an operation, a patient receives 2 8/10 pints of blood. Express the amount of blood as a decimal.

4. *Chiropractic* is based on the principle that a person's health is determined mainly by the nervous system. *Chiropractors* treat their patients by manipulating parts of the body to correct interferences in the nervous system. More than 4/10 of the total chiropractors are located in California, Michigan, Missouri, New York, Pennsylvania, and Texas. Express this fraction as a decimal.

5. A medical record technician calculates that 0.323 of all optometrists are women and that 0.332 of all veterinarians are women. Which health career has more women?

6. Megan grew 1.34 centimeters since her last checkup. Jamie grew 1.43 centimeters in the same time period. Which one grew the most?

7. An inventory of the ethyl alcohol supply found the following amounts on hand: 24 5/6 ounces, 17 2/3 ounces, 19 1/2 ounces, 42 3/5 ounces, 9 1/7 ounces, and 14 7/8 ounces. Convert the fractions to decimal values and calculate the total ethyl alcohol on hand. Carry this calculation out to the nearest thousandth.

8. A microscope was used to measure the diameter of several cells. Arrange the following measurements from largest to smallest: 0.0148 mm, 0.0269 mm, 0.0158 mm, 0.0301 mm, 0.0159 mm, 0.0219 mm, 0.0169285 mm.

9. A medication is prepared in a liquid form. If a total of 950 mL has been prepared and is used at the rate of 8.250 mL per hour:

   a. How many hours will the medication allow a patient to take a full dosage?
   b. How much medication will remain?

Unit 7

Rounding Decimal Numbers and Finding Equivalents

Objectives

After studying this unit the student should be able to:
- **Round decimal numbers to any required number of places.**
- **Express common fractions as decimal fractions.**
- **Express decimal fractions as common fractions.**

After solving a problem using decimal numbers the answer may have more decimal places than are needed. In these cases, the numbers may be rounded. *Rounding* is an expression used when referring to an approximation for a number.

ROUNDING DECIMAL NUMBERS

To round a decimal number:

- Determine the number of decimal places required.

- If the digit directly to the right of the last decimal place required is less than 5, the last required digit is not altered.

- If the digit directly to the right of the last decimal place required is 5 or larger, the last required digit is increased by one.

The symbol $\approx$ is used in rounding numbers. It means *is approximately equal to.*

EXAMPLE 1

The place-value names for the number 5,250.849 are:

Thousands	Hundreds	Tens	Units	Tenths	Hundredths	Thousandths
5	2	5	0	8	4	9

This number is rounded to various places.

Round to the nearer hundredth.

5,250.849   Check the digit to the right of the hundredths place. Since 9 > 5, the preceding digit is increased by 1.

      5,250.849 $\approx$ 5,250.85 (hundedths)

Round to the nearer tenth.

5,250.849   Check the digit to the right of the tenths place. Since 4 < 5, the preceding digit is not altered.

      5,250.849 $\approx$ 5,250.8 (tenths)

Round to the nearer whole number.

5,250.849   Check the digit to the right of the units place. Since 8 > 5, the preceding digit is increased by 1.

   5,250.849 ≈ 5,251 (units)

Round to the nearer ten.

5,250.849   Check the digit to the right of the tens place. Since 0 < 5, the preceding digit is not altered.

   5,250.849 ≈ 5,250 (tens)

Round to the nearer hundred.

5,250.849   Check the digit to the right of the hundreds place. Since 5 = 5, the preceding digit is increased by 1.

   5,250.849 ≈ 5,300 (hundreds)

Round to the nearer thousand.

5,250.849   Check the digit to the right of the thousands place. Since 2 < 5, the preceding digit is not altered.

   5,250.849 ≈ 5,000 (thousands)

EXAMPLE 2

Round 8.99 to the nearer tenth.

8.99   Check the digit to the right of the tenths place. Since 9 > 5, the preceding digit is increased by 1. Because the tenths digit becomes 0, the ones digit has to be be increased by 1.

   8.99 ≈ 9.0 (tenths)

7.1 Exercises

In 1–54, round each number to the indicated place.

Thousandth

1. 2.8925	2. 74.51930	3. 0.43276
4. 0.01875	5. 100.0264	6. 36.0552
7. 0.2509	8. 1.2170	9. 73.6050

Hundredth

10. 16.321	11. 0.895	12. 4.7166
13. 0.0875	14. 0.0074	15. 3.1416
16. 74.2785	17. 0.3704	18. 17.00347

Tenth

19. 5.74	20. 289.86	21. 0.68
22. 3.512	23. 706.475	24. 23.07
25. 0.21794	26. 29.059	27. 0.00243

Whole Number (ones)

28. 3.719	29. 0.1	30. 29.61
31. 44.099	32. 317.268	33. 9.09
34. 40.17	35. 410.71	36. 0.692

Ten

37. 273	38. 145.9	39. 308.29
40. 1,872.65	41. 29.31	42. 1,607.48
43. 470.91	44. 502.67	45. 2,106.39

Hundred

46. 6,234.9	47. 1,760.81	48. 3,909
49. 1,200.675	50. 15,650.125	51. 3,100.875
52. 6,849	53. 9,025.6	54. 1,989.1

In 55–60, round each number to the place-value position that is underlined.

55. 432.92$\underline{1}$5	56. 31.7$\underline{5}$09	57. 1,323.$\underline{0}$91
58. 0.$\underline{7}$18	59. $\underline{0}$.50	60. 0.$\underline{0}$097

61. Give a real life example where it would always be best to round up.

62. Scan your local newspaper's front page for headlines that contain numbers. Is each number rounded or exact? Explain how you decided.

EXPRESSING COMMON FRACTIONS AS DECIMAL FRACTIONS

Expressing a decimal as a fraction is not difficult since the denominator is a power of ten, for example: $0.13 = \dfrac{13}{100}$. To express a common fraction as a decimal, an equivalent fraction whose denominator is 10, 100 or another power of ten needs to be found. For simple fractions, this can be done mentally.

EXAMPLE 1

Using mental math, express $\dfrac{1}{2}$ as a decimal.

$$\overset{(\times 5)}{\underset{(\times 5)}{\dfrac{1}{2} = \dfrac{5}{10}}} = 0.5$$

Thus, the decimal of $\dfrac{1}{2}$ is 0.5.

As noted in Section 1, Unit 1, equivalent fractions form an equation. For example, $\dfrac{1}{2} = \dfrac{5}{10}$. By using cross-products, an unknown number (denoted by the variable n) can be found.

EXAMPLE 2

Using cross-products, express $\dfrac{3}{4}$ as a decimal.

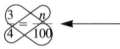

$\dfrac{3}{4} = \dfrac{n}{100}$ ← ——————— the denominator of 100 was chosen because

4 is a factor of 100.

$4(n) = 3(100)$ Cross-multiply.

$4n = 300$ Simplify.

$\dfrac{4n}{4} = \dfrac{300}{4}$ Divide each side by 4.

$n = 75$

Thus, $\dfrac{3}{4} = \dfrac{75}{100} = 0.75$.

 Finding the correct power of ten using the cross-product method to find equivalent fractions can become increaingly difficult. For this reason, division is usually used to express fractions as decimals (numerator ÷ denominator).

EXAMPLE 3

Using division, express $\dfrac{5}{8}$ as a decimal.

$\dfrac{5}{8}$ *means* $5 \div 8$.

Pencil and Paper   Calculator

$$\begin{array}{r} 0.625 \\ 8\,)\overline{5.000} \\ \underline{48} \\ 20 \\ \underline{16} \\ 40 \\ \underline{40} \\ 0 \end{array}$$

$5 \boxed{\div}\, 8 \boxed{=}\, 0.625$

Thus, $\dfrac{5}{8} = 0.625$.

7.2 Exercises

In 1–42, use the method indicated to express each fraction as a decimal.

Use mental math to find a set of equivalent fractions.

1. $\dfrac{4}{5}$ ⎰˟ ⎱ ——— =

2. $\dfrac{13}{50}$ ⎰˟ ⎱ ——— =

2. $\dfrac{1}{4}$ ⎰˟ ⎱ ——— =

4. $\dfrac{17}{25}$ ⎰˟ ⎱ ——— =

5. $\dfrac{1}{20}$ ⎰˟ ⎱ ——— =

6. $\dfrac{7}{500}$ ⎰˟ ⎱ ——— =

Use Cross-products.

7. $\dfrac{2}{5} = \dfrac{n}{10}$ 8. $\dfrac{14}{20} = \dfrac{n}{100}$ 9. $\dfrac{2}{50} = \dfrac{n}{100}$

10. $\dfrac{3}{8} = \dfrac{n}{1,000}$ 11. $\dfrac{19}{25} = \dfrac{n}{100}$ 12. $\dfrac{9}{16} = \dfrac{n}{10,000}$

Use Division.

13. $\dfrac{3}{5} \rightarrow 5\overline{)3}$ 14. $\dfrac{7}{8} \rightarrow 8\overline{)7}$ 15. $\dfrac{3}{200} \rightarrow 200\overline{)3}$

13. $\dfrac{5}{2} \rightarrow 2\overline{)5}$ 14. $\dfrac{171}{500} \rightarrow 500\overline{)171}$ 15. $\dfrac{9}{20} \rightarrow 20\overline{)9}$

13. $\dfrac{32}{25} \rightarrow 25\overline{)32}$ 14. $\dfrac{7}{16} \rightarrow 16\overline{)7}$ 15. $\dfrac{25}{2} \rightarrow 2\overline{)25}$

Use Either Method.

22. $\dfrac{1}{2}$ 23. $\dfrac{5}{8}$ 24. $\dfrac{9}{10}$

25. $\dfrac{13}{25}$ 26. $2\dfrac{3}{4}$ 27. $\dfrac{11}{25}$

28. $\dfrac{7}{2}$ 29. $\dfrac{21}{80}$ 30. $\dfrac{11}{5}$

31. $\dfrac{23}{50}$ 32. $\dfrac{11}{16}$ 33. $\dfrac{9}{40}$

34. $\dfrac{13}{8}$ 35. $\dfrac{121}{100}$ 36. $\dfrac{17}{20}$

37. $\dfrac{3}{75}$ 38. $\dfrac{115}{250}$ 39. $\dfrac{3}{32}$

40. $\dfrac{0}{100}$ 41. $\dfrac{1}{64}$ 42. $\dfrac{13}{4}$

REPEATING DECIMAL FRACTIONS

When expressing a fraction, whose numerator and denominator are whole numbers, as a decimal, only one of two things can happen.

- The decimal *terminates*. This means that the last remainder is 0.
- The decimal *repeats*. This means that the remainders repeat again and again.

EXAMPLE 1

Express $\frac{1}{4}$ as a decimal.

Pencil and Paper

$$\frac{1}{4} = 4 \overline{)\begin{array}{r} 0.25 \\ 1.00 \\ \underline{8} \\ 20 \\ \underline{20} \\ 0 \end{array}}$$

Calculator

$$\frac{1}{4} = 1 \boxed{\div} 4 \boxed{=} \; 0.25$$

Since the last remainder is 0, the decimal equivalent 0.25 is a *terminating decimal*.

EXAMPLE 2

Express $\frac{2}{3}$ as a decimal.

Pencil and Paper

$$\frac{2}{3} = 3 \overline{)\begin{array}{r} 0.666\ldots \\ 2.000 \\ \underline{18} \\ 20 \\ \underline{18} \\ 20 \\ \underline{18} \\ 2 \end{array}}$$

 Calculator

Some calculators truncate (cut off at a certain place value position).

$$\frac{2}{3} = 2 \boxed{\div} 3 \boxed{=} \; \begin{array}{l} 0.6666666 \; or \\ 0.6666667 \end{array}$$

Some calculators round

Since the remainder, 2, repeats again and again, the decimal equivalent 0.666 . . . is a *repeating decimal*.

In a repeating decimal, the three dots or the calculator display filled with a repeating pattern show that the indicated pattern continues indefinitely. Another way to represent a repeating decimal is to place a bar over the digit or group of digits that repeat. For example, 0.666 . . . or 0.6666666 can be written as $0.\overline{6}$. This shows that the digit 6 repeats indefinitely and is read *"point 6 repetend."*

The period of a repeating decimal is the number of digits in the repeating block.

EXAMPLE 3

Express $\dfrac{2}{11}$ as a decimal. Find its period, and write it using a repetend bar.

Pencil and Paper

$$\dfrac{2}{11} = 11 \overline{)\begin{array}{l} 0.1818 \\ 2.0000 \end{array}}$$

$$\begin{array}{r} \underline{11} \\ 90 \\ \underline{88} \\ 20 \\ \underline{11} \\ 90 \\ \underline{88} \\ 2 \end{array}$$

Calculator

$$\dfrac{2}{11} = 2 \boxed{\div} 11 \boxed{=} \ 0.1818182$$

$0.181818\ldots$ has a period of 2 and is written as $0.\overline{18}$.

7.3 Exercises

In 1–8, find the repeating decimal for each fraction. Use proper notation to indicate that the decimal repeats.

1. $\dfrac{7}{3}$ 2. $\dfrac{7}{12}$ 3. $\dfrac{1}{6}$ 4. $\dfrac{6}{11}$

5. $\dfrac{4}{15}$ 6. $\dfrac{1}{7}$ 7. $\dfrac{5}{9}$ 8. $\dfrac{11}{13}$

In 9–26, use the bar notation to represent those repeating decimals which have been expressed with three dots. Use the three dot notation to represent those repeating decimals which have been expressed with the bar. Find the period for each repeating decimal.

9. $0.\overline{5}$ 10. $5.7\overline{4}$ 11. $0.\overline{02}$

12. $0.\overline{004}$ 13. $0.0\overline{37}$ 14. $9.72317231\ldots$

15. $32.608608\ldots$ 16. $1.243243\ldots$ 17. $0.55\ldots$

18. $0.5833\ldots$ 19. $0.\overline{18}$ 20. $41.28\overline{9}$

21. $4.\overline{27}$ 22. $1.\overline{142857}$ 23. $15.\overline{0}$

24. $0.166\ldots$ 25. $0.4545\ldots$ 26. $0.428571428571\ldots$

At times, when using a calculator, it is difficult to tell if a decimal is terminating or repeating. Consider the following example.

 Using a calculator, express $\dfrac{3}{13}$ as a decimal.

$$\dfrac{3}{13} = 3 \boxed{\div} 13 \boxed{=} \ 0.2307692$$

The following steps can be used to tell if the decimal terminates or repeats.

Step 1	Step 2		
		Step 3	**Step 4**
Write down the quotient from the display.	Clear the calculator.	Enter the number written down in Step 1.	Multiply the number in Step 3 by the divisor in the original fraction. ⊠
0.2307692			=

If the decimal is terminating, the product in Step 4 will equal exactly the numerator of the original fraction. If the decimal is repeating, the product in Step 4 will approximate the numerator of the original fraction.

27. Using your calculator, decide which of the following fractions are terminating and which are repeating.

$$\frac{1}{2}, \frac{1}{3}, \frac{1}{4}, \frac{1}{5}, \frac{1}{6}, \frac{1}{7}, \frac{1}{8}, \frac{1}{9}, \frac{1}{10}, \frac{1}{11}, \frac{1}{12}, \frac{1}{13}, \frac{1}{14}, \frac{1}{15}, \frac{1}{16}$$

28. Explain how you know if your calculator rounds or truncates answers.

APPLICATIONS

When calculating the amount of medication to be administered, values that are too small may be meaningless for the kinds of measurements that are to be used. A calculation which results in 3.003 tablets to be administered is meaningless since it would be almost impossible to find out what 0.003 tablet is. A tablet can not be divided into such a small portion. Conversely, a calculation which results in 0.003 grams of medication to be placed into a solution would not be meaningless.

When confronted with a "rounding situation," use common sense. Always follow the directions of the doctor or the supervisor in the laboratory or pharmacy. If in doubt, consult a pharmacist or a technician in that particular field.

7.4 Exercises

1. Through calculations it is determined that 4.6893200 grams of glucose are required in the preparation of a solution. The available balance can weigh to thousandths of a gram. Find the number of grams of glucose used in preparing the solution.

2. The lengths of four babies are compared. These birth lengths are recorded: 43.5 centimeters; 44.125 centimeters; 42.025 centimeters; 46.65 centimeters. Round each measurement to the number of decimal places in the least precise measurement.

3. A standard adult dose for a specific medication is 0.025 liter. The central supply has 5.692 liters on hand. Using this information, the number of human doses is found to be 227.68. Can 227.68 doses be rounded to 228 doses?

4. The instructions are to prepare 1 700 milliliters of an alcohol solution containing one-third ethyl alcohol and two-thirds distilled water.

    a. Find the number of milliliters of ethyl alcohol required to prepare the solution.

    b. Find the number of milliliters of distilled water in the preparation.

    c. The graduated measuring device measures to 0.01 milliliter. Find the number of milliliters of ethyl alcohol used.

5. Five individuals were weighed with the following results: 14.65 kilograms, 21.15 kilograms, 13.09 kilograms, 15.16 kilograms, and 17.26 kilograms.

   a. What is the total weight of all five individuals?

   b. What is the average weight of these five individuals?

      (1) To the nearest one hundredth of a kilogram?

      (2) To the nearest tenth of a kilogram?

6. The death rate in a community is found to be one person per 1,500 annually. If the current population is 152,375, how many people can be anticipated to die in the next 12 months?

7. It has been determined that an average of 35 cotton swabs are used per patient in one week. These swabs are purchased in packages of 100 each. If the patient load is determined to average 62 patients per week, how many packages of swabs must be ordered to provide a twenty-six week supply?

8. Give the decimal equivalents for each of the following cell measurements:

   A. 1/8 mm _____ mm

   B. 1/27 mm _____ mm

   C. 4/6 mm _____ mm

   D. 3/48 mm _____ mm

9. Section enrollment in basic health care classes was compared between two different schools. The following data was compiled:

School A

Section 1	23 students	14 students
Section 2	19 students	25 students
Section 3	17 students	19 students
Section 4	21 students	None

School B

   a. What is the average number of students per section in School A? In School B?

   b. What is the average number of students per section when the classes of the two schools are combined?

   c. Review the answers in a and b above and round the answer to the nearer whole number.

Unit **8**

Basic Operations
With Decimal Fractions

Objectives

After studying this unit the student should be able to:
- **Add decimal fractions.**
- **Subtract decimal fractions.**
- **Multiply decimal fractions.**
- **Divide decimal fractions.**

Remember that different symbols express the same decimal. For example 0.5 can be written 0.50, 0.500, 0.5000, etc. The number 1 can be written as 1, 1.0, 1.00, etc. If no decimal point is shown, the point is understood to be to the right of the last digit. Keep this fact in mind when adding and subtracting decimals.

ADDITION OF DECIMAL FRACTIONS

In adding decimals, using zeros as placeholders reduces the possibility of errors. Zeros are used so that all the values have the same number of places to the right of the decimal point. This does not affect the value of the numbers.

EXAMPLE 1

Add 36.014 + 7.28 + 14.9.

Step 1	Step 2	Step 3
Align the decimal points.	Use zeros as placeholders so that each value has the same number of decimal places.	Add as in whole numbers. The decimal point in the answer is aligned with the decimal points in the values that are added.
36.014 7.28 + 14.9	36.014 7.280 + 14.900	36.014 7.280 + 14.900 58.194

SUBTRACTION OF DECIMAL FRACTIONS

In subtraction as in addition, decimal points are aligned and zeros are used as placeholders.

EXAMPLE 1

Find the difference in the equation. $62.7 - 29.38 = n$.

Step 1	Step 2	Step 3
Align the decimal points.	Use zeros as placeholders so that each value has the same number of decimal places.	Subtract as in whole numbers. The decimal point in the answer is aligned with the decimal points in the values that are subtracted.
62.7 $-$ 29.38	62.70 $-$ 29.38	62.70 $-$ 29.38 33.32 $n = 33.32$

8.1 Exercises

In 1–24, find each sum or difference. Where needed, use zeros as placeholders.

1.
$$\begin{array}{r} 0.12 \\ 0.45 \\ +\ 0.14 \end{array}$$

2.
$$\begin{array}{r} 16.7 \\ 0.23 \\ +\ 0.08 \end{array}$$

3.
$$\begin{array}{r} 25.37 \\ 6.05 \\ +\ 7.2 \end{array}$$

4.
$$\begin{array}{r} 3.478 \\ 2.32 \\ +\ 17.9 \end{array}$$

5.
$$\begin{array}{r} 4.216 \\ 0.35 \\ +\ 1.2 \end{array}$$

6.
$$\begin{array}{r} 45 \\ 0.78 \\ +\ 13.2 \end{array}$$

7.
$$\begin{array}{r} 32.6 \\ 16.73 \\ +\ 0.003 \end{array}$$

8.
$$\begin{array}{r} 1.003 \\ 0.0379 \\ +\ 16.4 \end{array}$$

9.
$$\begin{array}{r} 3.8 \\ -\ 2.9 \end{array}$$

10.
$$\begin{array}{r} 7.0 \\ -\ 5.28 \end{array}$$

11.
$$\begin{array}{r} 92.37 \\ -\ 57.56 \end{array}$$

12.
$$\begin{array}{r} 18.97 \\ -\ 9.248 \end{array}$$

13.
$$\begin{array}{r} 13.403 \\ -\ 9.7 \end{array}$$

14.
$$\begin{array}{r} 4.702 \\ -\ 3.278 \end{array}$$

15.
$$\begin{array}{r} 15.2 \\ -\ 0.003 \end{array}$$

16.
$$\begin{array}{r} 1.0 \\ -\ 0.789 \end{array}$$

17. $2.6 + 1.5 + 0.48$

18. $0.45 + 2.6 + 0.7$

19. $28.2 - 8.28$

20. $0.8 - 0.77$

21. $4.56 + 21.7 + 0.08$

22. $6 + 4.25 + 0.008$

23. $123.6 - 15.74$

24. $5.87 - 3.242$

In 25–32, solve each equation.

25. $18.7 + 0.124 + 7.0 = t$

26. $x = 6.654 + 0.398 + 14.822$

27. $2.01 - 0.998 = n$

28. $R = 0.01 - 0.001$

29. $x = 132.1 + 17.813 + 0.999$

30. $37.41 + 13.3 + 7.345 = t$

31. $b = 189.01 - 0.987$

32. $10 - 0.0012 = a$

33. Explain how adding and subtracting decimals compares with adding and subtracting whole numbers.

MULTIPLICATION OF DECIMAL FRACTIONS

Decimal fractions are an expression for fractions whose denominators are powers of ten. Multiplication of decimals can be modeled by using fractions whose denominators are powers of ten.

EXAMPLE 1

Find $\frac{1}{2}$ of 1.

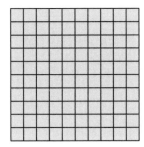

Shade the total region.

Darken $\frac{1}{2}$ or 0.5 of the total region.

50 regions in 100 are darkened.

The mathematical notation is:
$$\frac{1}{2} \cdot 1 = \frac{1}{2} \text{ or } \frac{50}{100}$$

$$\frac{5}{10} \cdot 1 = \frac{5}{10} \text{ or } \frac{50}{100}$$

$$0.5 \cdot 1 = 0.5 \text{ or } \frac{50}{100}$$

EXAMPLE 2

Find $\frac{1}{2}$ of $\frac{1}{2}$.

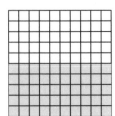

Shade $\frac{1}{2}$ or 0.5 of the total region.

Darken $\frac{1}{2}$ or 0.5 of the $\frac{1}{2}$ (0.5) region.

25 regions in 100 are darkened.

The mathematical notation is: $\dfrac{1}{2} \bullet \dfrac{1}{2} = \dfrac{1}{4}$ or $\dfrac{25}{100}$

$$\dfrac{5}{10} \bullet \dfrac{5}{10} = \dfrac{25}{100}$$

$$0.5 \bullet 0.5 = 0.25 \; or \; \dfrac{25}{100}$$

EXAMPLE 3

Find $\dfrac{1}{10}$ of $\dfrac{1}{10}$.

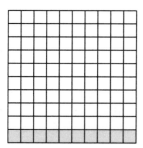   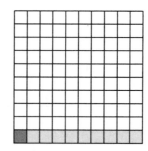

Shade $\dfrac{1}{10}$ or 0.1 of the total region.   Darken $\dfrac{1}{10}$ or 0.1 of the $\dfrac{1}{10}$ region.

1 region in 100 are darkened.

The mathematical notation is: $\dfrac{1}{10} \bullet \dfrac{1}{10} = \dfrac{1}{100}$

$$0.1 \bullet 0.1 = 0.01 \; or \; \dfrac{1}{100}$$

The notation for multiplication of decimals can be generalized this way:

Common Fractions   **Decimal Fractions**

$\dfrac{1}{10} \bullet \dfrac{1}{10} = \dfrac{1}{100}$

0.1 • 0.1 = 0.01
(tenths × tenths = hundredths)

$\dfrac{1}{10} \bullet \dfrac{1}{100} = \dfrac{1}{1,000}$

0.1 • 0.01 = 0.001
(tenths × hundredths = thousandths)

$\dfrac{1}{10} \bullet \dfrac{1}{1,000} = \dfrac{1}{10,000}$

0.1 • 0.001 = 0.0001
(tenths × thousandths = ten-thousandths)

Note: The number of decimal places in the product depends on the number of decimal places in each factor. The number of decimal places in the product is the sum of the decimal places in the factors.

EXAMPLE 4

Multiply 2.3 • 45.

Step 1	Step 2	Step 3
Multiply as in whole numbers.	Count the number of decimal places in each factor.	The number of decimal places in the product is the sum of the decimal places in the factors. Locate the decimal point to indicate that number.
$$\begin{array}{r} 2.3 \\ \times\ 45 \\ \hline 115 \\ 92\ \ \\ \hline 1035 \end{array}$$	$$\begin{array}{r} 2.3 \\ \times\ 45 \\ \hline 1035 \end{array}$$ $\boxed{\begin{array}{r}1\\+\ 0\\\hline 1\end{array}}$	$$\begin{array}{r} 2.3 \\ \times\ 45 \\ \hline 103.5 \end{array}$$ ↑——1 decimal place

EXAMPLE 5

Find the product n in the equation 2.3 • 45 = n.

Step 1	Step 2	Step 3
Multiply as in whole numbers.	Count the number of decimal places in each factor.	The number of decimal places in the product is the sum of the decimal places in the factors. Locate the decimal point to indicate that number.
$$\begin{array}{r} 2.3 \\ \times\ 4.5 \\ \hline 115 \\ 92\ \ \\ \hline 1035 \end{array}$$	$$\begin{array}{r} 2.3 \\ \times\ 4.5 \\ \hline 1035 \end{array}$$ $\boxed{\begin{array}{r}1\\+\ 1\\\hline 2\end{array}}$	$$\begin{array}{r} 2.3 \\ \times\ 4.5 \\ \hline 10.35 \end{array}$$ ↑——2 decimal places $n = 10.35$

8.2 Exercises

In 1–14, find each product.

1. $$\begin{array}{r} 4.8 \\ \times\ 3 \\ \hline \end{array}$$
2. $$\begin{array}{r} 0.48 \\ \times\ 3 \\ \hline \end{array}$$
3. $$\begin{array}{r} 0.048 \\ \times\ 3 \\ \hline \end{array}$$
4. $$\begin{array}{r} 0.0048 \\ \times\ 3 \\ \hline \end{array}$$

5. $$\begin{array}{r} 5.2 \\ \times\ 31 \\ \hline \end{array}$$
6. $$\begin{array}{r} 5.2 \\ \times\ 3.1 \\ \hline \end{array}$$
7. $$\begin{array}{r} 0.52 \\ \times\ 3.1 \\ \hline \end{array}$$
8. $$\begin{array}{r} 5.2 \\ \times\ 0.031 \\ \hline \end{array}$$

9. 72 • 0.13
10. 34.95 • 0.96
11. 0.007 • 0.008

12. 251 • 0.15
13. 15.34 • 31.3
14. 0.371 • 4.24

In 15–20, solve each equation.

15. $0.75 \cdot 83.60 = b$ 16. $a = 6.3 \cdot 0.0006$ 17. $c = 421.25 \cdot 416.40$

18. $11.009 \cdot 31.7 = x$ 19. $0.001 \cdot 10.001 = t$ 20. $q = 8.75 \cdot 3.81$

When performing computations with more than one operation, mathematicians have agreed upon the order in which calculations should be performed.

Order of Operations
1. Compute inside parenthesis.
2. Calculate powers.
3. Multiply and divide from left to right.
4. Add and subtract from left to right.

The following two examples illustrate order of operations using pencil and paper as well as a scientific calculator. Scientific calculators are programmed to follow the order of operations.

EXAMPLE 6

  Compute $2 (3 + 5) - 4 \cdot 3$. When using a calculator be sure to enter each symbol in order from left to right.

Pencil and Paper

$2(3 + 5) - 4 \cdot 3$	*Add 3 + 5*
$2(8) - 4 \cdot 3$	*Multiply by 2*
$16 - 4 \cdot 3$	*Multiply 4 · 3*
$16 - 12$	*Subtract 12 from 16*
$= 4$	

Calculator

$2 (3 + 5) - 4 \cdot 3 = 4$

$2 \boxed{\times} \boxed{(} 3 \boxed{+} 5 \boxed{)} \boxed{-} 4 \boxed{\times} 3 \boxed{=} 4$

The calculator doesn't understand that 2 ( ) means multiply. You have to insert the $\boxed{\times}$ symbol.

EXAMPLE 7

Compute $(6.5) (4) + (8.7 - 3.7)^3$.

Pencil and Paper

$(6.5) (4) + (8.7 - 3.7)^3$	*Subtract 3.7 from 8.7*
$(6.5) (4) + (5)^3$	*Evaluate 5^3 (5 · 5 · 5)*
$(6.5) (4) + 125$	*Multiply 6.5 · 4*
$26 + 125$	*Add 26 + 125*
$= 151$	

Calculator

$(6.5) (4) + (8.7 - 3.7)^3 = 151$

$6.5 \boxed{\times} 4 \boxed{+} \boxed{(} 8.7 \boxed{-} 3.7 \boxed{)} \boxed{Y^X} 3 \boxed{=} 151$

$\boxed{Y^X}$ or $\boxed{X^Y}$ is the universal power key. The calculator multiplies 5 · 5 · 5.

In 21–32, evaluate each expression.

21. $3 + 4.6(5.3) - 19$ 22. $59 - (2) (3.8) + 16$ 23. $0.05(27.3) + 5.8(11.1)$

24. $(18.2) (0) + 15.2$ 25. $81.75 - 16.2(3.7) + 9$ 26. $72 - (12 + 0.05) (5)$

27. $3(4^5)$ 28. $52 - 3^3$ 29. $5^1 - 1^5$

30. $2(3^4) + 4(2^3)$ 31. $3^5 + 2.7(1^4)$ 32. $2^5 - 4(8) + 3(15^2)$

33. Place the decimal point in one of the factors so that the product is 717.4.

$$17935 \cdot 400$$

34. Give an example where the product of two decimal factors is less than both factors.

35. Explain why the decimal point moves to the right in the product when multiplying by a power of ten.

DIVISION OF DECIMAL FRACTIONS

Division of decimals can also be illustrated using fractions whose denominators are powers of ten.

EXAMPLE 1

Find the quotient of $4 \div \dfrac{1}{2}$.

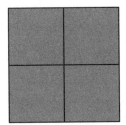

4 regions

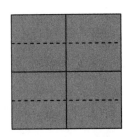

Each of the 4 regions is divided in half
or 0.5.
8 regions are formed.

The mathematical notation is: $4 \div \dfrac{1}{2} = 4 \times \dfrac{2}{1}$ *or* 8.

$$4 \div 0.5 = 8$$

Division of decimals can be performed by first finding equivalent fractions which have whole numbers as numerators and denominators. Then, the division of whole numbers is performed.

EXAMPLE 2

Divide 75.50 by 2.5.

In symbols, 75.50 divided by 2.5 means $75.50 \div 2.5$, which can be written as $\dfrac{75.50}{2.5}$.

$$\frac{75.50}{2.5} \times \frac{10}{10} = \frac{755}{25}$$

Recall that the parts in a division problem are:

$$\overset{quotient}{divisor)\overline{dividend}}$$

When checking a division problem:

$$\begin{array}{r} quotient \\ \times \ divisor \\ \hline dividend \end{array}$$

$$\begin{array}{r} 30.2 \\ 25\overline{)755.0} \\ \underline{75} \\ 05 \\ \underline{0} \\ 50 \\ \underline{50} \\ 0 \end{array}$$

EXAMPLE 3

Divide 75.50 ÷ 0.25.

$75.50 \div 0.25$ means $\dfrac{75.50}{0.25}$.

$\dfrac{75.50}{0.25} \times \dfrac{100}{100} = \dfrac{7{,}550}{25}$

$$
\begin{array}{r}
302 \\
25\,\overline{)7{,}550} \\
75 \\
\overline{05} \\
0 \\
\overline{50} \\
50 \\
\overline{0}
\end{array}
$$

EXAMPLE 4

Divide 75.50 ÷ 0.025.

$75.50 \div 0.025$ means $\dfrac{75.50}{0.025}$.

$\dfrac{75.50}{0.025} \times \dfrac{1{,}000}{1{,}000} = \dfrac{75{,}500}{25}$

$$
\begin{array}{r}
3{,}020 \\
25\,\overline{)75{,}500} \\
75 \\
\overline{0\,5} \\
0\,0 \\
\overline{50} \\
50 \\
\overline{00} \\
00 \\
\overline{0}
\end{array}
$$

Note: All solutions should be checked. The solution may be checked by multiplying the quotient by the divisor. The result of this calculation should be the dividend.

Notice that in finding the equivalent fractions, the form of 1 is a power of ten. Each time a number is multiplied by a power of ten, the place value of the number increases or decreases by the same power. Zeros serve as placeholders in the multiplication of powers of ten and the division process.

EXAMPLE 5

Divide 1.178 ÷ 0.69. Round the quotient to the nearer hundredth.

Step 1	Step 2	Step 3
Multiply both numbers by 100 and locate the decimal point in the answer.	Divide to the thousandths place.	Since 7 > 5, the preceding digit is increased by 1.
$0.69.\,\overline{)1.17.80}$ $69\,\overline{)117.8}$	$\begin{array}{r} 1.707 \\ 69\,\overline{)117.800} \\ 69 \\ \overline{48\,8} \\ 48\,3 \\ \overline{500} \\ 483 \\ \overline{17} \end{array}$	$\begin{array}{r} 1.707 \\ 69\,\overline{)117.800} \end{array}$ $1.178 \div 0.69 = 1.71$ (rounded)

EXAMPLE 6

Find the quotient n in the equation $14 \div 1.5 = n$. Round the quotient n to the nearer tenth.

Step 1	Step 2	Step 3
Multiply both numbers by 10 and locate the decimal point in the answer.	Divide to the hundredths place.	Since $3 < 5$, the preceding digit is not altered.

8.3 Exercises

In 1–9, find each quotient.

1. $3\overline{)21.9}$

2. $5\overline{)4.25}$

3. $7\overline{)0.714}$

4. $12\overline{)24.36}$

5. $15\overline{)4.575}$

6. $28\overline{)0.5684}$

7. $0.2\overline{)31}$

8. $0.004\overline{)0.82040}$

9. $2.1\overline{)4.221}$

In 10–18, solve each equation.

10. $c = 246.123 \div 123$

11. $132.3 \div 0.3 = r$

12. $306 \div 0.51 = a$

13. $6.8 \div 0.34 = x$

14. $d = 7.44 \div 248$

15. $t = 3618 \div 0.006$

16. $7{,}919.924 \div 1.48 = y$

17. $42{,}840 \div 0.612 = p$

18. $a = 1{,}696.424 \div 21.2$

In 19–21, divide. Round to the nearer hundredth.

19. $0.08\overline{)1.573}$

20. $6.1\overline{)0.178}$

21. $0.135\overline{)0.2946732}$

In 22–24, divide. Round to the nearer thousandth.

22. $172 \div 0.003$

23. $57.74 \div 2.02$

24. $14.78234 \div 0.09$

In 25–30, use the relationship between multiplication and division to complete each of the following:

	Dividend	**Divisor**	**Quotient**
25.	1.2060	0.040	
26.	19,603.92		5.001
27.		813	0.04
28.	649.6		20,300
29.	756.0	0.0035	
30.	15.42		0.03

Recall the Order of Operations mathematicians have agreed upon.

1. Compute inside parenthesis.
2. Calculate powers.
3. Multiply and divide from left to right.
4. Add and subtract from left to right.

The following two examples illustrate order of operations using pencil and paper as well as a scientific calculator.

EXAMPLE 7

 Compute $10(8 - 5) \div 2.5 + 17.8$.

Pencil and Paper

$10(8 - 5) \div 2.5 + 17.8$	*Subtract 5 from 8*
$10(3) \div 2.5 + 17.8$	*Multiply 10(3)*
$30 \div 2.5 + 17.8$	*Divide 30 by 2.5*
$12 + 17.8$	*Add 12 + 17.8*
$= 29.8$	

Calculator

$10(8 - 5) \div 2.5 + 17.8$ $10 \boxed{\times} \boxed{(} 8 \boxed{-} 5 \boxed{)} \boxed{\div} 2.5 \boxed{+} 17.8 \boxed{=} 29.8$

EXAMPLE 8

Compute $(9 + 7) \div 2(6) + 4(3^4 + 21)$.

Pencil and Paper

$(9 + 7) \div 2(6) + 4(3^4 + 21)$	*Add 9 + 7; Evaluate 3^4 + 21*
$16 \div 2(6) + 4(102)$	*Divide 16 by 2*
$8(6) + 4(102)$	*Multiply 8(6); Multiply 4(102)*
$48 + 408$	*Add*
$= 456$	

Calculator

$(9 + 7) \div 2(6) + 4(3^4 + 21)$

$\boxed{(}\ 9\ \boxed{+}\ 7\ \boxed{)}\ \boxed{\div}\ 2\ \boxed{\times}\ 6\ \boxed{+}\ 4\ \boxed{\times}\ \boxed{(}\ 3\ \boxed{Y^X}\ 4\ \boxed{+}\ 21\ \boxed{)}\ \boxed{=}\ 456$

  In 31–39, evaluate each expression using pencil and paper. Then check using a scientific calculator.

31. $3(7^2 + 1) \div 10$

32. $7(3.5) + 11 \div 2$

33. $32 \div 4 - 16 \div 2$

34. $15^2 - 10(20) - (50 \div 2)15$

35. $4(9 - 2)^3 \div 2 + 5(14 - 2.2)$

36. $(10.25 - 2.5^2)^3 \div 2^5$

37. $[(10 + 5)3] \div 3^2$

38. $5(9 - 3^2) + 3^2(5)$

39. $10.5^3 \div (10.5)(2) - 3(55 - 15)$

40. Explain why the decimal point moves to the left in the quotient when dividing by a power of ten.

APPLICATIONS

When administering medications certain basic procedures are followed. These procedures ensure that the correct amount of medication is administered to the correct patient at the correct time.

When administering any type of medication, there are certain general considerations.

- Avoid handling the medications with the fingers. Use the cap of the container and a medicine glass when administering pills or capsules.
- Never try to divide a tablet that is not scored.
- When administering a part of a scored tablet, always throw away the unused part.
- While pouring a medication, take care not to contaminate the bottle.
- Never pour unused liquids back into the stock bottle or container.
- Always replace bottle caps after pouring the medicine.
- Check to see if the correct dose is being given. If the dose is not the same as what is ordered, calculate the correct dose.

8.4 Exercises

1. A patient receives this series of doses of medication: 1.25 milliliters; 0.5 milliliter; 2.125 milliliters; and 1 milliliter. What is the total dosage?

2. A total of 4.5 milliliters of medication is in a vial. An injection of 1.125 milliliters is withdrawn into a syringe. How much remains in the vial?

3. A container holds 12.5 milliliters of streptomycin sulfate. How many 1.25-milliliter doses can be administered from the container?

4. Fifteen people are each to receive 1.25-milliliter doses of streptomycin sulfate. How many milliliters of streptomycin sulfate are needed?

5. Small vials are purchased at a cost of $0.1062 each. Assuming no discount is given, how much does the clinic pay for 15 dozen vials of this size?

6. The hospital places a request for individual servings of meat. The supplier provides this meat at a cost of $0.4289 per serving. The cost of preparing and serving this meat to the patient is an additional $0.1364 per serving. What is the cost to the hospital for 250 servings of meat?

7. The mass of a patient is recorded over a thirty-day period at ten-day intervals.

Start	88.904 kilograms
10th day	87.772 kilograms
20th day	86.592 kilograms
30th day	84.036 kilograms

   a. Find the mass loss over the 30-day period.

   b. If the loss occurs uniformly for each day, find the loss in mass for one 24-hour period. Round the answer to the nearer thousandth kilogram.

   c. Find the average mass for the four recorded masses. Average = total mass divided by number of recorded masses.

8. If 1.5 milliliters of an experimental drug is administered daily to each of 12 patients, how many milliliters of the drug will be used in 30 days?

9. You are to prepare 1,500 milliliters of an alcohol solution containing 0.3 ethyl alcohol and 0.7 distilled water.

   a. How many milliliters of distilled water will be used?

   b. How many milliliters of ethyl alcohol are used?

10. For a $5.00 fee, a person can receive a flu shot at a local clinic. It costs the clinic $0.50 per shot for the vaccine. Yesterday, the clinic gave 178 people flu shots. What was the clinic's gross profit for the day?

11. A hemacytometer is used to determine the bacterial population in one milliliter of a liquid sample. The population is found to be 1,000,000 cells per milliliter. Complete the following chart indicating the population of the bacteria per milliliter when diluted with sterile, distilled water.

   Original sample = 1,000,000 cells per mL

   Sterile, distilled water = 0 cells per mL

	Size of Sample mL	Sterile, Distilled Water in mL	Bacterial Population per mL
	1 mL	0 mL	1,000,000 cells
A.	0.1 mL	9.9 mL	_____ cells
B.	0.01 mL	9.99 mL	_____ cells
C.	1 mL	9 mL	_____ cells
D.	5 mL	95 mL	_____ cells
E.	25 mL	75 mL	_____ cells
F.	25 mL	975 mL	_____ cells

12. Measure the distance from the center of the bridge of the nose to the center of the eye for six different individuals.

   a. Record each measurement to the nearest tenth of a centimeter.
   Note: The marks between each centimeter represent one-tenth (0.1) of a centimeter.

   b. Calculate the average distance of the measurements to the nearest tenth of a centimeter.

Unit 9

Exponents and Scientific Notation

Objectives

After studying this unit the student should be able to:
- Multiply and divide powers of ten.
- Express numbers using scientific notation.
- Perform multiplication using scientific notation.

EXPONENTS

Numbers that are multiplied to find a product are factors.

EXAMPLE 1

$4 \times 5 = 20$
$4 \cdot 5 = 20$
$(4)(5) = 20$

Notice that three different symbols can be used to denote multiplication. These symbols will be used interchangeably to denote multiplication.

EXAMPLE 2

$(10)(10) = 100$
$10 \times 10 \times 10 = 1,000$
$10 \cdot 10 \cdot 10 \cdot 10 = 10,000$

Notice that in each case the factors are the same. The number 10 is used as a factor 2, 3, and 4 times respectively. Each product is a power of ten.

DEFINITIONS

- A *base* is the number used as a factor.
- An *exponent* is the number of times the base is used as a factor.
- A *power* refers to an expression with exponents.
- The mathematical notation for expressing an exponential number is:

$$b^n = P$$

where b means the base; n means the exponent; and P means the product.

EXAMPLE 3

10^4

BASE
(number used as a factor)

EXPONENT
(number of times the base is used as a factor)

10^4 means $10 \cdot 10 \cdot 10 \cdot 10$ *or* 10,000.

It is read "ten to the fourth power" or "the fourth power of ten."

EXAMPLE 4

6^3

BASE
(number used as a factor)

EXPONENT
(number of times the base is used as a factor)

6^3 means $6 \cdot 6 \cdot 6$ *or* 216.
It is read "six to the third power" or "the third power of six," or "six cubed."
In the base ten place-value system, each place can be represented as a power of 10.

EXAMPLE 5

The thousands place can be represented by 10^3. This means $10 \cdot 10 \cdot 10$ *or* 1,000.

EXAMPLE 6

The hundredths place can be represented by $\frac{1}{10^2}$. This means $\frac{1}{10} \cdot \frac{1}{10}$ *or* $\frac{1}{100}$.

The number $\frac{1}{10^2}$ can also be represented by 10^{-2}. This is read "ten to the negative second

power." The negative number indicates that the product is a fraction.

- Any *negative exponent* may be expressed as:
  $b^{-n} = \frac{1}{b^n}$, where *n* is any whole number and b ≠ 0

$10^{-15} \qquad \frac{1}{10^{15}}$

EXAMPLE 7

The units place can be represented by 10^0 *or* 1. It is read "ten to the zero power."

- Any number raised to the *zero power* is <u>one</u>.

  This is expressed as: $b^0 = 1$, where $b \neq 0$. The symbol 0^0 is undefined.

  The following chart illustrates the usage of powers of ten in the place-value system.

Place-Value Name	TEN THOUSANDS	THOUSANDS	HUNDREDS	TENS	UNITS	TENTHS	HUNDREDTHS	THOUSANDTHS	TEN-THOUSANDTHS
Powers Of Ten	10^4	10^3	10^2	10^1	10^0	10^{-1}	10^{-2}	10^{-3}	10^{-4}
Number The Power Represents	$10 \cdot 10 \cdot 10 \cdot 10$ 10,000	$10 \cdot 10 \cdot 10$ 1,000	$10 \cdot 10$ 100	10 10	1 1	$\dfrac{1}{10}$ $\dfrac{1}{10}$	$\dfrac{1}{10 \cdot 10}$ $\dfrac{1}{100}$	$\dfrac{1}{10 \cdot 10 \cdot 10}$ $\dfrac{1}{1,000}$	$\dfrac{1}{10 \cdot 10 \cdot 10 \cdot 10}$ $\dfrac{1}{10,000}$

9.1 Exercises

In 1–6, express each set of fractions using exponents.

1. $4 \cdot 4$

2. $2 \cdot 2 \cdot 2 \cdot 2 \cdot 2$

3. $10 \cdot 10 \cdot 10 \cdot 10$

4. $9 \cdot 9 \cdot 9 \cdot 9$

5. $7 \cdot 7 \cdot 7$

6. $6 \cdot 6 \cdot 6 \cdot 6$

In 7–15, name the base and the exponent.

7. 10^5

8. 5^1

9. 10^{100}

10. 2^3

11. 10^9

12. 9^0

13. 4^{-3}

14. Y^7

15. 2^{-5}

In 16–24, compare. Insert <, =, or > to make each statement true.

16. 2^3 _____ 3^2

17. 5^2 _____ 2^5

18. 8^0 _____ 1

19. 1^5 _____ 5^1

20. 3^3 _____ 27

21. 2^4 _____ 4^2

22. 1 _____ 100^0

23. 9^1 _____ 3^2

24. 8^2 _____ 2^6

In 25–40, use positive exponents to rewrite those expressions having negative exponents. Use negative exponents to rewrite those expressions having positive exponents.

25. 10^{-3}

26. 7^{-1}

27. 4^{-5}

28. 10^{-4}

29. $\dfrac{1}{3^5}$

30. $\dfrac{1}{10^1}$

31. $\dfrac{1}{24}$

32. $\dfrac{1}{3^2}$

33. 8^{-2}

34. $\dfrac{1}{10^5}$

35. 5^{-3}

36. $\dfrac{1}{9^3}$

37. $\dfrac{1}{1000^4}$

38. 9^{-6}

39. $\dfrac{1}{3^1}$

40. 10^{-10}

 In 41–49, use a calculator, and trial and error, to find the value of the letter in each equation.

41. $16 = 2^n$

42. $p = 8^2$

43. $5^n = 125$

44. $b^3 = 27$

45. $3^n = 243$

46. $p = 3^4$

47. $32 = b^5$

48. $5^n = 625$

49. $100^n = 1$

Expressions involving positive or negative exponents can be evaluated using a calculator and the universal power key.

$\dfrac{1}{2^3} = 0.125$ $2^{-3} = 0.125$

1 $\boxed{\div}$ $\boxed{(}$ 2 $\boxed{Y^X}$ 3 $\boxed{)}$ $\boxed{=}$ 0.125   2 $\boxed{Y^X}$ 3 $\boxed{+/-}$ $\boxed{=}$ 0.125

In 50–53, complete.

	Expression	Evaluate	Write the expression using a negative exponent	Evaluate
50.	$\dfrac{1}{4^2}$			
51.	$\dfrac{1}{2^5}$			
52.	$\dfrac{1}{10^2}$			
53.	$\dfrac{1}{5^2}$			

54.   Explain how the $\boxed{1/x}$ key could be used to evaluate the expressions in questions 50 through 53.

USING EXPONENTS IN MULTIPLICATION AND DIVISION

Numbers expressed as powers that have the same bases can be multiplied. Using your knowledge of exponents compare the expressions $10^2 \cdot 10^4$ and $10^5 \cdot 10^1$.

$10^2 \cdot 10^4 = (10 \cdot 10)(10 \cdot 10 \cdot 10 \cdot 10) = 10^6$

$10^5 \cdot 10^1 = (10 \cdot 10 \cdot 10 \cdot 10 \cdot 10)(10) = 10^6$

Since each expression equals 10^6, $10^2 \cdot 10^4 = 10^5 \cdot 10^1$.

Note: $10^2 \cdot 10^4 = 10^{2+4}$ and $10^5 \cdot 10^1 = 10^{5+1} = 10^6$. Thus, to simplify work, the following rule has been devised for multiplying powers with the same bases.

- Multiplication of powers with the same bases can be expressed as:

$$b^X \cdot b^Y = b^{X+Y}$$

This means that in multiplication of powers, when the bases are the same, the exponents are added. The exponent of the product is the sum of the exponents of the factors.

EXAMPLE 1

Find the product of $5^3 \cdot 5^4$ using the meaning of exponents and the rule for multiplication of powers.

Meaning of Exponents   **Rule for Multiplication of Powers**

$5^3 \cdot 5^4 = \underline{(5 \cdot 5 \cdot 5)(5 \cdot 5 \cdot 5 \cdot 5)}$ $5^3 \cdot 5^4 = 5^{3+4} = 5^7$

$= 5^7$ *7 factors of* 5 $= 78,125$

$= 78,125$

Numbers expressed as powers that have the same bases can also be divided. Using your knowledge of exponents compare the expressions $10^6 \div 10^2$ *and* $10^5 \div 10^1$.

$$10^6 \div 10^1 = \frac{10^6}{10^2} = \frac{10 \cdot 10 \cdot 10 \cdot 10 \cdot 10 \cdot 10}{10 \cdot 10} = \frac{\overset{1}{\cancel{10}} \cdot \overset{1}{\cancel{10}} \cdot 10 \cdot 10 \cdot 10 \cdot 10}{\underset{1}{\cancel{10}} \cdot \underset{1}{\cancel{10}}} = 10^4$$

$$10^5 \div 10^1 = \frac{10^5}{10^1} = \frac{10 \cdot 10 \cdot 10 \cdot 10 \cdot 10}{10} = \frac{\overset{1}{\cancel{10}} \cdot 10 \cdot 10 \cdot 10 \cdot 10}{\underset{1}{\cancel{10}}} = 10^4$$

Since each expression equals 10^4, $10^6 \div 10^2 = 10^5 \div 10^1$.

Note: $10^6 \div 10^2 = 10^{6-2}$ *and* $10^5 \div 10^1 = 10^4$. Thus, to simplify work, the following rule has been devised for dividing powers with the same bases.

- Division of powers with the same bases can be expressed as:

$$b^X \div b^Y = b^{X-Y} \, if \, X > Y$$

$$b^X \div b^Y = \frac{1}{b^{Y-X}} \, if \, Y > X$$

$$b^X \div b^Y = 1 \, if \, X = Y$$

This means that in division of powers, when the bases are the same, the exponents are subtracted. The exponent of the quotient is the difference of the exponents.

EXAMPLE 2

Find the quotient of $3^7 \div 3^5$ using the meaning of exponents and the rule for division of powers.

Meaning of Exponents

$$3^7 \div 3^5 = \frac{3^7}{3^5} = \frac{\overset{1}{\cancel{3}} \cdot \overset{1}{\cancel{3}} \cdot \overset{1}{\cancel{3}} \cdot \overset{1}{\cancel{3}} \cdot \overset{1}{\cancel{3}} \cdot 3 \cdot 3}{\underset{1}{\cancel{3}} \cdot \underset{1}{\cancel{3}} \cdot \underset{1}{\cancel{3}} \cdot \underset{1}{\cancel{3}} \cdot \underset{1}{\cancel{3}}} = 3^2$$

$$= 9 \quad \underbrace{\qquad\qquad}_{\text{2 factors of 3}}$$

Rule for Division of Powers

$$3^7 \div 3^5 = 3^{7-5} = 3^2$$

$$= 9$$

The preceding two rules work for variables as well as numbers.

EXAMPLE 3

Simplify $a^3 \cdot a^5$ *and* $t^8 \div t^2$ using exponents.

$$a^3 \cdot a^5 = a^{3+5} = a^8 \qquad\qquad t^8 \div t^2 = t^{8-2} = t^6$$

At times, a number or variable expressed as a power is raised to another power, such as $(10^2)^3$. This expression is read "ten to the second power raised to the third power."

This means that the number 10^2 is raised to the third power. The value of this expression can be found using the meaning of exponents and the rule for multiplication of powers.

$$(10^2)^3 = (10^2)(10^2)(10^2) \quad 3 \, factors \, of \, 10^2$$

$$(10^2)(10^2)(10^2) = 10^{2+2+2} = 10^6$$

Note: $(10^2)^3 = 10^6$, the product of the exponents, $2 \times 3 = 6$, is the exponent in the answer.

In general, raising a power to a power can be expressed as:

$$(b^X)^Y = b^{X \cdot Y}$$

This means that in raising a power to a power, multiplication is used. The exponent of the result is the product of the exponents.

EXAMPLE 4

Simplify $(4^2)^5$ *and* $(a^3)^4$ using exponents.

$(4^2)^5 = 4^{2 \cdot 5} = 4^{10}$ $\qquad\qquad\qquad$ $(a^3)^4 = a^{3 \cdot 4} = a^{12}$

$\qquad\qquad\quad = 1,048,576$

It is possible to raise a product to a power.

EXAMPLE 5

Simplify $(a^2b)^3$ using exponents.

$$(a^2b)^3 \quad = (a^2b)\,(a^2b)\,(a^2b)$$
$$= (a^2 \cdot a^2 \cdot a^2)\,(b^1 \cdot b^1 \cdot b^1)$$
$$= a^{2+2+2}b^{1+1+1}$$
$$= a^6b^3$$

Note: Each factor is raised to the third power; $(a^2)^3$ *and* $(b)^3$.

• Raising a product to a power can be expressed as:

$$(a^mb)^n = a^{mn}b^n$$

This means that each factor is raised to the power.

9.2 Exercises

In 1–6, find each product. Write each answer using exponents.

1. $10^1 \cdot 10^2$ $\qquad\qquad$ 2. $4^3 \cdot 4^3$ $\qquad\qquad$ 3. $5^0 \cdot 5^4$

4. $(0.5^2)\,(0.5)^5$ $\qquad\quad$ 5. $x^3 \cdot x^4$ $\qquad\qquad$ 6. $a^2 \cdot a^6$

In 7–12, find each quotient. Write each answer using exponents.

7. $3^4 \div 3^2$ $\qquad\qquad$ 8. $10^5 \div 10^5$ $\qquad\qquad$ 9. $5^0 \div 5^2$

10. $8^0 \div 8^0$ $\qquad\qquad$ 11. $c^6 \div c^3$ $\qquad\qquad$ 12. $t^5 \div t^1$

In 13–18, find each power. Write each answer using exponents.

13. $(5^2)^3$ $\qquad\qquad$ 14. $(7^3)^4$ $\qquad\qquad$ 15. $(2^3)^5$

16. $(3^4)^3$ $\qquad\qquad$ 17. $(d^5)^2$ $\qquad\qquad$ 18. $(y^4)^2$

In 19–24, find the power of each product. Write each answer using exponents.

19. $(2a)^3$ $\qquad\qquad$ 20. $(x^2y^3)^2$ $\qquad\qquad$ 21. $(a^3b^4)^4$

22. $(4^2y^3)^2$ $\qquad\qquad$ 23. $(5r^2t)^3$ $\qquad\qquad$ 24. $(3c^2d^3)^0$

In 25–33, compare each expression. Insert <, =, or > to make each statement true.

25. (2^3) _____ 2^5

26. $5^3 \cdot 5^3$ _____ 25^6

27. $10^6 \div 10^2$ _____ $10^5 \cdot 10^0$

28. $2^5 \cdot 2^{-5}$ _____ 2^{10}

29. 9^2 _____ $(3^2)^2$

30. $(2^{-2})^2$ _____ 2^0

31. $(0.7^4)(0.7^{-1})$ _____ $(0.7^1)(0.7^3)$

32. 4^3 _____ $(2^2)^3$

33. $(10^5)^5$ _____ $(10^{12})^2$

 A calculator can be very useful when evaluating expressions. Note the following when evaluating expressions using a calculator.

- Substitute numbers in for their respective variables.
- Enter each symbol in your calculator from left to right.

EXAMPLE 6

Evaluate each expression when $a = 3$, $b = 6$ and $c = 2$.

$5a + 4b$

$5(3) + 4(6)$

$5 \boxed{\times} 3 \boxed{+} 4 \boxed{\times} 6 \boxed{=} 39$

$7b^3 - 4c$

$7(6^3) - 4(2)$

$7 \boxed{\times} 6 \boxed{y^x} 3 \boxed{-} 4 \boxed{\times} 2 \boxed{=} 1{,}504$

$(c^4 + b^2)^3$

$(2^4 + 6^2)^3$

$\boxed{(} 2 \boxed{y^x} 4 \boxed{+} 6 \boxed{y^x} 2 \boxed{)} \boxed{y^x} 3 \boxed{=} 140{,}608$

Write the original problem. Show the substitution step. Evaluate each expression when $a = 5$, $b = 4$ and $c = 3$.

34. $(ab)^2 + 9(7)$

35. $(ab \div 2)^3 - 4c$

36. $17.5c - 2a^2$

37. $abc - (2b)^2$

38. $40b \div b^2$

39. $4a^3 + 2(a + b) - bc$

40. $b(ac)^3 \div 15 - 100$

41. $143.2 - 6a + 3b^2$

42. $(abc)^2 \div abc$

Note how parenthesis alter the usual order of operations in problems 43 through 45.

43. $(600 \div b)c + 3a^4$

44. $600 \div (bc) + 3a^4$

45. $(600 \div bc + 3)a^4$

SCIENTIFIC NOTATION

Numbers may be expressed in different ways without changing their value. For example, 90,000 could be written as 300×300 or $9{,}000{,}000 \div 100$ or 9.0×10^4. 90,000 written as 9.0×10^4 is expressed using scientific notation.

Scientific notation is a special way to express very large or very small numbers using powers of ten. A number is said to be expressed in scientific notation when it is written as a product of a number from 1 to 10 (not including 10) and a power of 10. The number is written with the decimal point after the first non-zero digit.

EXAMPLE 1

Using scientific notation, the number 8.5 acquires different values depending on the power of ten.

Number		Scientific Notation
8,500	$8.5 \cdot 1{,}000$	8.5×10^3
850.0	$8.5 \cdot 100$	8.5×10^2
85.00	$8.5 \cdot 10$	8.5×10^1
8.5	$8.5 \cdot 1$	8.5×10^0
0.85	$8.5 \cdot 0.1$	8.5×10^{-1}
0.085	$8.5 \cdot 0.01$	8.5×10^{-2}
0.0085	$8.5 \cdot 0.001$	8.5×10^{-3}

To express a number in decimal form using scientific notation:

- Place the decimal point after the first non-zero digit.
- Determine the power of ten that is needed. (How many places was the decimal point moved?)
- Write the power of ten using an exponent.
  a. Positive exponent if the decimal point was moved left.
  b. Negative exponent if the decimal point was moved right.

EXAMPLE 2

Express both 238,400 *and* 0.000597 using scientific notation.

$$238{,}400 = 2.38400$$
$$= 2.384 \cdot 100{,}000$$
$$= 2.384 \times 10^5$$

$$0.000597 = 0.0005.97$$
$$= 5.97 \cdot \frac{1}{10{,}000}$$
$$= 5.97 \times 10^{-4}$$

To express a number in scientific notation as a number in standard decimal form:

- Determine the value that the power of ten represents.
- Multiply the value that the power of ten represents by the number from 1 to 10. Move the decimal point right if the exponent is positive and left if the exponent is negative.

EXAMPLE 3

Express 6.203×10^4 *and* 7.13×10^{-5} using standard decimal form.

$$6.203 \times 10^4 = 6.203 \cdot 10{,}000$$
$$= 62{,}030$$

$$7.13 \times 10^{-5} = 7.13 \cdot \frac{1}{100{,}000}$$
$$= 0.0000713$$

Note: The decimal point is moved four places to the right.

Note: The decimal point is moved five places to the left.

9.3 Exercises

In 1–6, complete.

1. $5,280 = 5.28 \times 10^{\square}$ 2. $5.280 = 5.28 \times 10^{\square}$ 3. $0.05280 = 5.28 \times 10^{\square}$

4. $52.80 = \boxed{} \times 10^1$ 5. $5.28 \times 10^{-3} = \boxed{}$ 6. $5.28 \times 10^4 = \boxed{}$

In 7–18, express using scientific notation.

7. 342   8. 0.072   9. 20,000,000

10. 41.8   11. 1,000   12. 0.173000

13. 628,000   14. 0.000342   15. 2.73

16. 0.01   17. 1,000,000   18. 0.0000001

In 19–27, express using standard decimal form.

19. 2.7×10^0 20. 3.8×10^{-1} 21. 6.3×10^{-4}

22. 1.1×10^5 23. 5.314×10^{-6} 24. 3.85×10^2

25. 8.92×10^{-4} 26. 1.000×10^5 27. 7.35×10^0

 Scientific calculators automatically change from standard decimal notation (called floating point notation) into scientific notation when displaying very large or very small numbers.

28. For your particular calculator try to predict where numbers will be displayed using scientific notation.

29. Enter the following powers of 10 using Universal Power Key $\boxed{Y^X}$ to actually see where your calculator begins displaying numbers in scientific notation.

Number	Display	Number	Display
10^1	*100*	10^{-1}	
10^2		10^{-2}	
10^3		10^{-3}	
10^4		10^{-4}	
10^5		10^{-5}	
10^6		10^{-6}	
10^7		10^{-7}	
10^8		10^{-8}	
10^9		10^{-9}	
10^{10}		10^{-10}	

30. Explain the value in writing very large or very small numbers using scientific notation.

Using scientific notation facilitates multiplication. The place value, and the placement of decimal point is regulated by the powers of ten. Thus, the multiplication of the numbers and powers of ten can be done separately. All results are expressed using scientific notation.

Examples 1–4 model how this is done using pencil and paper as well as a calculator.

EXAMPLE 1

 Express the product of $500 \cdot 5000$ using scientific notation.

$500 \cdot 5000 = (5 \times 10^2) \cdot (5 \times 10^3)$.

Pencil and Paper

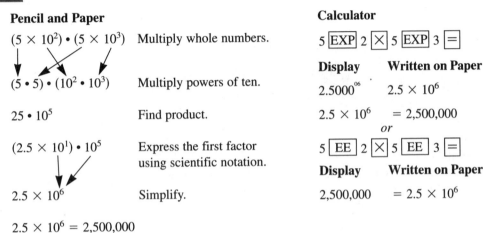

$(5 \times 10^2) \cdot (5 \times 10^3)$ Multiply whole numbers.

$(5 \cdot 5) \cdot (10^2 \cdot 10^3)$ Multiply powers of ten.

$25 \cdot 10^5$ Find product.

$(2.5 \times 10^1) \cdot 10^5$ Express the first factor using scientific notation.

2.5×10^6 Simplify.

$2.5 \times 10^6 = 2,500,000$

Calculator

$5\ \boxed{\text{EXP}}\ 2\ \boxed{\times}\ 5\ \boxed{\text{EXP}}\ 3\ \boxed{=}$

Display	**Written on Paper**
2.5000^{06}	2.5×10^6
2.5×10^6	$= 2,500,000$

or

$5\ \boxed{\text{EE}}\ 2\ \boxed{\times}\ 5\ \boxed{\text{EE}}\ 3\ \boxed{=}$

Display	**Written on Paper**
$2,500,000$	$= 2.5 \times 10^6$

Note: Most brands of calculators either use $\boxed{\text{EXP}}$ or $\boxed{\text{EE}}$ to represent scientific notation. The process of multiplication using exponents is modeled in Example 1. The multiplication is $10^2 \cdot 10^3 = 10^{2+3}$ *or* 10^5. The product 25 is expressed using scientific notation. This leads to the final answer of 2.5×10^6 *or* $2,500,000$.

EXAMPLE 2

 Express the product of $0.004 \cdot 600,000$ using scientific notation.

$0.004 \cdot 600,000 = (4 \times 10^{-3}) \cdot (6 \times 10^5)$

Pencil and Paper

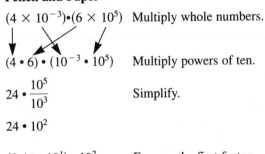

$(4 \times 10^{-3}) \cdot (6 \times 10^5)$ Multiply whole numbers.

$(4 \cdot 6) \cdot (10^{-3} \cdot 10^5)$ Multiply powers of ten.

$24 \cdot \dfrac{10^5}{10^3}$ Simplify.

$24 \cdot 10^2$

$(2.4 \times 10^1) \cdot 10^2$ Express the first factor using scientific notation.

2.4×10^3 Simplify.

$2.4 \times 10^3 = 2,400$

Calculator

$4\ \boxed{\text{EXP}}\ 3\ \boxed{+/-}\ \boxed{\times}\ 6\ \boxed{\text{EXP}}\ 5\ \boxed{=}$

Display	**Written on Paper**
2.400^{03}	2.4×10^3
2.4×10^3	$= 2,400$

or

$4\ \boxed{\text{EE}}\ \boxed{+/-}\ 3\ \boxed{\times}\ 6\ \boxed{\text{EE}}\ 5\ \boxed{=}$

Display	**Written on Paper**
2400	$= 2.4 \times 10^3$

Note: The process of division of exponents is modeled in Example 2. The power of ten, 10^{-3}, is expressed as $\dfrac{1}{10^3}$. Then $\dfrac{10^5}{10^3}$ is divided. The division is $\dfrac{10^5}{10^3} = 10^{5-3}$ *or* 10^2. The product 24 is expressed using scientific notation. This leads to the final answer of 2.4×10^3 *or* 2,400.

EXAMPLE 3

 Express the product of $40 \cdot 0.0000008$ using scientific notation.

$$40 \cdot 0.0000008 = (4 \times 10^1) \cdot (8 \times 10^{-7})$$

Pencil and Paper

$(4 \times 10^1) \cdot (8 \times 10^{-7})$ Multiply whole numbers.

$(4 \cdot 8) \cdot (10^1 \cdot 10^{-7})$ Multiply powers of ten.

$32 \cdot \dfrac{10^1}{10^7}$ Simplify.

$(3.2 \times 10^1) \cdot \dfrac{1}{10^6}$ Express the first factor using scientific notation.

$3.2 \cdot \dfrac{10^1}{10^6}$ Multiply.

$3.2 \cdot \dfrac{1}{10^5}$ Simplify.

$3.2 \times 10^{-5} = 0.000032$

Calculator

$4\ \boxed{\text{EXP}}\ 1\ \boxed{\times}\ 8\ \boxed{\text{EXP}}\ 7\ \boxed{+/-}\ \boxed{=}$

Display	Written on Paper
3.200^{-05}	3.2×10^{-5}
3.2×10^{-5}	$= 0.000032$

or

$4\ \boxed{\text{EE}}\ 1\ \boxed{\times}\ 8\ \boxed{\text{EE}}\ \boxed{+/-}\ 7\ \boxed{=}$

Display	Written on Paper
0.000032	$= 3.2 \times 10^{-5}$

EXAMPLE 4

 Express the product of $0.0003 \cdot 0.000007$ using scientific notation.

$$0.0003 \cdot 0.000007 = (3 \times 10^{-4}) \cdot (7 \times 10^{-6})$$

Pencil and Paper

$(3 \times 10^{-4}) \cdot (7 \times 10^{-6})$ Multiply whole numbers.

$(3 \cdot 7) \cdot (10^{-4} \cdot 10^{-6})$ Multiply powers of ten.

$21 \cdot \dfrac{1}{10^4 \cdot 10^6}$

$(2.1 \times 10^1) \cdot \dfrac{1}{10^{10}}$ Express the first factor using scientific notation.

$2.1 \cdot \dfrac{10^1}{10^{10}}$ Multiply.

$2.1 \cdot \dfrac{1}{10^9}$ Simplify.

$2.1 \times 10^{-9} = 0.0000000021$

Calculator

$3\ \boxed{\text{EXP}}\ 4\ \boxed{+/-}\ \boxed{\times}\ 7\ \boxed{\text{EXP}}\ 6\ \boxed{+/-}\ \boxed{=}$

Display	Written on Paper
2.100^{-09}	2.1×10^{-9}
2.1×10^{-9}	$= 0.0000000021$

or

$3\ \boxed{\text{EE}}\ \boxed{+/-}\ 4\ \boxed{\times}\ 7\ \boxed{\text{EE}}\ \boxed{+/-}\ 6\ \boxed{=}$

Display	Written on Paper
0.000000002	$= 2.1 \times 10^{-9}$

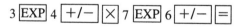

Note: The calculator answers are slightly different due to the way they round and truncate answers.

9.4 **Exercises**

In 1–12, find each product. Express the answer using both scientific notation and standard decimal notation.

1. $(4 \times 10^2)(5 \times 10^3)$

2. $(8 \times 10^2)(7 \times 10^0)$

3. $9{,}000 \cdot 0.00007$

4. $(4.28 \times 10^0)(1600)$

5. $(3.14 \times 10^{-1})(2.05 \times 10^{-2})$

6. $0.008 \cdot 0.00357$

7. $0.009\,(8 \times 10^5)$

8. $60{,}000 \cdot 0.000008$

9. $(6.179 \times 10^4)\,0.87$

10. $17\,(4.9 \times 10^{-3})$

11. $(5.08 \times 10^{-1})^3$

12. $(8.973 \times 10^1)^4$

Most scientific calculators have the capability of performing operations on numbers entered in scientific notation. Consider the following examples. Be sure your calculator is in scientific mode.

EXAMPLE 5

Solve each equation using a calculator. Express each answer in both scientific notation and standard decimal notation. Where appropriate, round the answer to the nearer thousandth.

	Calculator Display	Answer Expressed Using Scientific Notation	Answer Expressed in Standard Decimal Notation
$a = 7825 + 22709$	3.0534^{04}	3.0534×10^4	30534
$b = 6.0075 - 0.00574$	6.00176^{00}	6.00176×10^0	6.0018 (rounded)
$c = 36.902 \cdot 15{,}716$	5.79951832^{05}	5.79951832×10^5	579951.8 (rounded)
$d = 0.00786 \div 21$	3.742857143^{-04}	$3.742857143 \times 10^{-4}$	0.0004 (rounded)

In 13–24, solve each equation using a calculator. Express each answer in both scientific notation and standard decimal notation. Where appropriate, round the answer to the nearer thousandth.

	Answer Expressed Using Scientific Notation	Answer Expressed in Standard Decimal Notation
13. $a = 257.89 + 0.07002$		
14. $a = 807{,}000 + 253{,}581{,}923$		
15. $a = 0.0000238 + 0.000100927$		
16. $b = 531.317 - 89.3975$		
17. $b = 1.7 - 0.89326$		
18. $b = 0.00091 - 0.000234$		
19. $c = 78.2 \cdot 145{,}719.87$		
20. $c = 0.007128 \cdot 2925.10096$		
21. $c = 2{,}590{,}827 \cdot 10{,}891{,}762$		
22. $d = 87.91278 \div 0.012345$		
23. $d = 0.0025 \div 2719$		
24. $d = 0.24687531 \div 2.4687531$		

25. In 1991 the total expenditure for personal health care in the U.S. was approximately 661 billion dollars. In the year 2000 this is projected to be approximately one trillion five hundred billion dollars. Express each of these expenditures using both standard decimal notation and scientific notation.

APPLICATIONS

The prevention and the control of disease are a major function of every health care person. Diseases are caused by microorganisms. The study of microorganisms is microbiology. The microbiologist studies the cause and effect of microorganisms and disease and applies this knowledge to medicine.

By definition, *microbiology* means the study of simple forms of living matter which cannot be seen by the naked eye. This "smallness" has brought about new units of measure. These units are expressed by using powers of ten and exponents.

meter	1×10^0 meter
centimeter	1×10^{-2} meter
millimeter	1×10^{-3} meter
micrometer	1×10^{-6} meter
nanometer	1×10^{-9} meter
picometer	1×10^{-12} meter

Microorganisms are usually measured in micrometers. For example, a bacteria may be 0.002 micrometer long. Using scientific notation, this measure would be 2×10^{-3} micrometer or 2×10^{-9} meter ($2 \times 10^{-3} \cdot 1 \times 10^{-6}$).

9.5 Exercises

1. *Protozoa* are the cause of various types of malaria. Other species are responsible for amoebic dysentery. Protozoa range from 3 to 1,000 micrometers.

   a. Express this range in scientific notation using micrometers as the unit of measure.

   b. Express this range in scientific notation using meters as the unit of measure.

2. *Viruses* — the smallest infectious agents — invade the cell, increase in number, and then break through the cell wall and discharge more virus particles into the bloodstream. Smallpox, chickenpox, infectious hepatitis and German measles are among the diseases caused by viruses. Viruses range from $\dfrac{1}{2,500}$ micrometer to $\dfrac{1}{50,000}$ micrometer.

   a. Express this range in decimals using micrometers as the unit of measure.

   b. Express this range in scientific notation using micrometers as the unit of measure.

3. *Yeast* is a spherical-shaped or oval-shaped plant cell. Yeast is a rich source of vitamin B and is used in making bread and beer. By fermentation, it is also responsible for the spoilage of fruits, syrups, and jellies. The average dimensions of a yeast are 3 to 5 micrometers wide and 5 to 10 micrometers long.

   a. Express these dimensions in scientific notation using micrometers as the unit of measure.

   b. Express these dimensions in decimals using meters as the unit of measure.

4. *Bacteria* are the smallest living things that can be called "living." Nonpathogenic bacteria are helpful and useful bacteria. The curing of tobacco, tea, coffee, cocoa, and leather as well as the making of sauerkraut, vinegar, and cheese are among the accomplishments of bacteria. Pathogenic bacteria, commonly called *germs,* invade plant and animal tissue and are the cause of a multitude of diseases.

   The average length of a bacteria is $\dfrac{1}{1,000}$ micrometer.

   a. Express this length in scientific notation using micrometers as the unit of measure.

   b. Using division, determine how many times larger a bacteria is than the smallest virus.

   Viruses range from $\dfrac{1}{2,500}$ micrometer to $\dfrac{1}{50,000}$ micrometer.

5. Bacterial growth is the greatest cause of infectious disease. The greatest defense against bacteria is aseptic procedures, sterilization, disinfection, and isolation techniques. One bacteria cell can become 256,000 cells within 8 generations. Express this new number of cells using scientific notation.

6. Several microorganisms have been measured. Express each of the following values in scientific notations.

   a. 0.0000136

   b. 0.0002893

   c. 0.004682

   d. 0.000000368

   e. 0.0003010

   f. 0.009000

7. Give the decimal value of the following scientific notations.

   a. 1.963×10^{-8}

   b. 4.6×10^{5}

   c. 9.3×10^{1}

   d. 9.3×10^{-1}

8. The population of four bacterial samples were determined. Complete the following chart expressing the populations in scientific notation.

	Bacteria per mL	Bacteria per 5 mL	Bacteria per 10 mL	Bacteria per 100 mL
Sample A	16,992			
Sample B	2,689,000			
Sample C	26,890			
Sample D	268			
Sample E	12			

9. Chemical analysis revealed the following concentrations per mL of blood. Give the values of these scientific notations.

   Chemical A 8.35×10^{-12} _____ per mL

   Chemical B 12.8×10^{-4} _____ per mL

   Chemical C 1.89×10^{-6} _____ per mL

   Chemical D 0.60×10^{-3} _____ per mL

Unit **10**

Estimation and Significant Digits

Objectives

After studying this unit the student should be able to:

- **Estimate sums and differences.**
- **Estimate products and quotients.**
- **Round numbers to the indicated significant digit.**
- **Perform operations with numbers using the process of retention of significant digits.**

ESTIMATION

Mentally estimating sums, differences, products, and quotients is an important skill in today's calculator world. The following table reviews rounding numbers to various places.

Round to the Nearer						
Number	**Hundred**	**Ten**	**Whole**	**Tenth**	**Hundredth**	**Thousandth**
635.7218	600	640	636	635.7	635.72	635.713
0.9706	0	0	1	1.0	0.97	0.971
7,251.0519	7,300	7,250	5,251	7,251.1	7,251.05	7,251.052
0.0872	0	0	0	0.1	0.09	0.087

Rounding can be used to help estimate answers.

EXAMPLE 1

Approximate the sum of $17.892 + 20.021 + 34.1$.

From the data, it makes sense to round each number to the nearer whole number.

$$
\begin{array}{ll}
17.892 \longrightarrow & 18 \\
20.021 \longrightarrow & 20 \\
\underline{+\ 34.1} \longrightarrow & \underline{+\ 34} \\
& \text{Sum} \approx 72
\end{array}
$$
$\approx$ means "approximately equal to."

Rounding is also useful when finding differences.

EXAMPLE 2

Approximate the difference of $8,150 - 3,827$.

$$
\begin{array}{ll}
8150 \longrightarrow & 8200 \\
\underline{-\ 3827} \longrightarrow & \underline{-\ 3800} \\
& \text{Difference} \approx 4400
\end{array}
$$
Note that each number was rounded to the nearer hundred.

When estimating a product it is desirable to have at least one factor a multiple of 10.

EXAMPLE 3

Approximate the product of 471.9×22.3.

$$
\begin{array}{ccc}
471.9 \longrightarrow & 500 \\
\times\ 22.3 \longrightarrow & \times\ 22 \\
\hline
\text{Product} \approx 11{,}000
\end{array}
\qquad or \qquad
\begin{array}{ccc}
471.9 \longrightarrow & 500 \\
\times\ 22.3 \longrightarrow & \times\ 20 \\
\hline
\text{Product} \approx 10{,}000
\end{array}
$$

Notice that the actual product is 10,523.37.

Compatible numbers are useful when estimating quotients. Compatible numbers are numbers that can easily be mentally divided.

EXAMPLE 4

Approximate the quotient of $4{,}732 \div 7$.

$$
4{,}732 \div 7 = \frac{4732}{7} \rightarrow \frac{4900}{5} \approx 700
$$

Note that 4900 and 7 are compatible numbers.

10.1 Exercises

In 1–24, estimate each sum, difference, product, or quotient. Use appropriate rounding.

1.
$$
\begin{array}{r}
318 \\
679 \\
+\ 107 \\
\end{array}
$$

2.
$$
\begin{array}{r}
16{,}372 \\
+\ 3{,}681 \\
\end{array}
$$

3.
$$
\begin{array}{r}
17.82 \\
25.097 \\
+\ 139.2064 \\
\end{array}
$$

4.
$$
\begin{array}{r}
822 \\
-\ 695 \\
\end{array}
$$

5.
$$
\begin{array}{r}
14.709 \\
-\ 3.912 \\
\end{array}
$$

6.
$$
\begin{array}{r}
1.0092 \\
-\ 0.015 \\
\end{array}
$$

7.
$$
\begin{array}{r}
7.3 \\
\times\ 9.6 \\
\end{array}
$$

8.
$$
\begin{array}{r}
12.302 \\
\times\ 0.98 \\
\end{array}
$$

9.
$$
\begin{array}{r}
319.76 \\
\times\ 0.067 \\
\end{array}
$$

10. $32.1 \div 15.8$

11. $191.7 \div 64.1$

12. $3.12 \div 4.3$

13. $713.035 + 99.09$

14. $7{,}601 + 349 + 9{,}022$

15. $19.0019 + 49.911 + 101.19$

16. $397.7 - 19.81$

17. $6.18102 - 3.00157$

18. $4979.782 - 25.41$

19. 453.12×0.87

20. 47.0765×4.61

21. 2.19×3.782

22. $\dfrac{51.68}{2.62}$

23. $\dfrac{176.7318}{10.004}$

24. $\dfrac{2{,}174.8962}{0.0479}$

In 25–36, estimate the sum, difference, product, or quotient for each equation.

25. $n = 7{,}649{,}614 \div 3{,}905$

26. $1.79 + 3.02 + 7.31 = c$

27. $a = 403 + 579$

28. $1.0715 - 0.889 = t$

29. $r = 14{,}891 - 6{,}208$

30. $Z = (2{,}951 - 871) - 312$

31. $y = (0.0672)(6.73)$

32. $3.17 \times 14.5 = m$

33. $c = 100.3 \cdot 0.82$

34. $937.82 \div 6.39 = b$

35. $q = 146.9067 \div 0.074$

36. $r = 76.61 \div 7.02$

37. Explain why it is a good idea to use estimation when computing even though a calculator is being used.

SIGNIFICANT DIGITS

Often, results of measurements are estimated and rounded to different degrees of accuracy. In measuring, there are some distances which are certain and some that are estimated.

EXAMPLE 1

What is the indicated distance between the arrows?

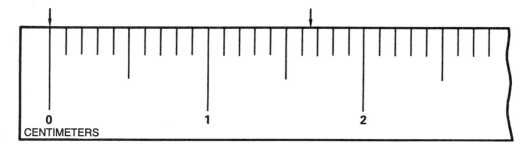

It is certain that the distance is between 1 and 2 centimeters.

It is certain that the distance is between 1.6 and 1.7 centimeters.

It is uncertain what the next reading is. It may be 1.65 centimeters, 1.66 centimeters, or even 1.67 centimeters. In this measurement two digits are certain and one is uncertain. The three digits are called significant digits. *Significant digits* are digits which indicate the number of units that are reasonably sure of having been counted in making a measurement.

When taking a measurement, the significant digits include:

- all the digits which are sure, plus
- one digit which is estimated.

Significant digits are important for finding the product or quotient of approximations. Certain rules are followed when determining how many significant digits a number contains.

- All non-zero digits are significant.
  These numbers have 4, 4, 2, and 5 significant digits, respectively.
  9,763   97.63   4.5   5.4889

- All zeros between significant digits are significant.
  Each of these numbers has 5 significant digits.
  59,002   350.67   99.009   3.0007

- All final zeros to the right of the decimal point are significant.
  These numbers have 4, 4, 5, and 7 significant digits, respectively.
  523.0   28.00   5.5000   11.80000

- If a numeral names a whole number greater than 1 and a decimal point follows the final zero, all zeros are significant.
  These numbers have 2, 3, 4, and 5 significant digits, respectively.
  30.   560.   6,000.   82,000.

- If a numeral names a whole number greater than 1 and no decimal point follows the final zero, the final zeros are not significant.
  Each of these numbers has 2 significant digits. The significant digits are underlined.
  <u>76</u>0   <u>98</u>00   <u>88</u>,000   <u>12</u>0,000

- If a numeral names a number between 0 and 1, the initial zeros are not significant. These numbers have 1, 2, 3, and 1 significant digits, respectively. The significant digits are underlined.
  0.0<u>8</u>   0.00<u>81</u>   0.00<u>760</u>   0.0000<u>4</u>

- If a number is written in proper scientific notation, all digits are significant except those in the power. These numbers have 3, 2, and 4 significant digits, respectively. The significant digits are underlined.
  $\underline{1.64} \times 10^{-19}$ $\underline{9.7} \times 10^{3}$ $\underline{6.495} \times 10^{7}$

Note: Zeros representing nonsignificant digits are placeholders stating how large or small the number is.

RETENTION OF SIGNIFICANT DIGITS

- Retain as many significant digits in data and results as will give only one uncertain digit.

  On reading a burette, the number 15.34 is recorded. All figures are significant since 15.3 represents actual scale divisions and the 4 represents an estimation between two scale divisions and is the only digit which is uncertain. Another analyst may read this same setting as 15.33 or 15.35.

- To have all of its digits significant, every digit of a number except the last must be certain.

EXAMPLE 1

Give the number of significant digits for 62.73.

The number 62.73 has four significant digits. The 6, 2, and 7 are correct and the 3 states that the number is closer to 3 hundredths. If the numbers 62.725 or 62.734 are rounded to four significant digits, the result would be 62.73.

- In retaining significant digits, all numbers are rounded to the number which has the least number of significant digits.

EXAMPLE 2

Round 2.78, 6,282, and 0.5096 to 3 significant digits.

Number	Rounded to 3 Significant Digits
2.78	2.78
6,282	6,280
0.5096	0.510

EXAMPLE 3

Round 3.091, 4,120 and 591 to 2 significant digits.

Number	Rounded to 2 Significant Digits
3.091	3.1
4,120	4,100
591	590

EXAMPLE 4

Round 2,457, 0.057, and 378 to 1 significant digit.

Number	Rounded to 1 Significant Digit
2,457	2,000
0.057	0.06
378	400

10.2 Exercises

In 1–12, indicate the number of significant digits for each number.

1.	634	2.	50.037	3.	730.	4.	3.14
5.	0.2001	6.	207	7.	0.0278	8.	61.359
9.	0.0319	10.	50.0	11.	1,402.7	12.	0.0039

In 13–48, round to the indicated number of significant digits.

Three Significant Digits

13.	5,293	14.	13.489	15.	31.708	16.	309.09
17.	3.7800	18.	16.8	19.	2.0007	20.	1.4365
21.	7,969	22.	14.55	23.	5,201.39	24.	4.201

Two Significant Digits

25.	12.947	26.	3.718	27.	45.601	28.	410.08
29.	2.9100	30.	15.2	31.	5.0006	32.	0.0212
33.	0.998	34.	4,271.380	35.	0.00042	36.	3.71

One Significant Digit

37.	217.8	38.	93.82	39.	2.50	40.	4,500,781
41.	0.07681	42.	0.3709	43.	0.000432	44.	0.667
45.	37,562	46.	0.00051	47.	2,109.5	48.	0.01090

RETENTION OF SIGNIFICANT DIGITS IN ADDITION AND SUBTRACTION

- In adding or subtracting numbers, the accuracy of the answer is in the column where the uncertain digit has the greatest place value.

EXAMPLE 1

Add 0.0457 + 37.82 + 9.36942.

    0.0457   The uncertain digits are underlined.
    37.82   ◄————   The uncertain digit 2 in the number 37.82 has
  + 9.369642   the greatest place value of the uncertain digits.

0.05   The digit 2 in 37.82 has a place value of hundredths.
37.82   Round the numbers to the nearer hundredth.
+ 9.37   Add the numbers.
―――――
47.24

The result, 47.24, is accurate to the nearer hundredth.

EXAMPLE 2

Subtract 96 − 35.8.

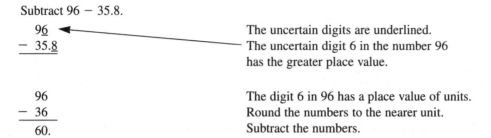

$$\underline{96}$$
$$-\ 35.\underline{8}$$

The uncertain digits are underlined.
The uncertain digit 6 in the number 96
has the greater place value.

$$96$$
$$-\ 36$$
―――――
$$60.$$

The digit 6 in 96 has a place value of units.
Round the numbers to the nearer unit.
Subtract the numbers.

Placing a decimal point in the answer 60 shows that there are two significant digits in the result.

RETENTION OF SIGNIFICANT DIGITS IN MULTIPLICATION AND DIVISION

- For practical purposes, in multiplying and dividing numbers with differing numbers of digits, retain as many significant figures as are found in the number having the least number of significant digits.

EXAMPLE 3

Multiply 35.19 × 0.0451 × 4.0787.

35.19	0.0451	4.0787	
4	3	5	**Significant Digits**

Since 0.0451 has the least number of significant digits,
round all numbers to 3 significant digits.
35.2 × 0.0451 × 4.08 = 6.4770816; rounds to 6.48.
The product, 6.48, is also rounded to 3 significant digits.

Note: When computing with a calculator, *retain all digits and round only the final result.* In manual computation, retain as many digits as are plausible, then round the final result. This procedure will ensure the proper accuracy.

EXAMPLE 4

Multiply using a calculator. 35.19 × 0.0451 × 4.0787.

The product 6.4731783 rounded to 3 significant digits is 6.47.

10.3 Exercises

In 1–16, solve each equation using the process of retaining significant digits.

1. $b = 3.073 + 4.28 + 9.7654$

2. $1.8972 + 0.913 = x$

3. $p = 89.7 + 14.000 + 9.302$

4. $c = 325.7 + 141.72 + 0.983$

5. $a = 29.753 - 13.84$ 6. $d = 42.5 - 27$

7. $18.203 - 7.6 = c$ 8. $1.7392 - 0.706 = t$

9. $R = 30 \times 0.049$ 10. $(71.32)(4.814) = b$

11. $0.712 \cdot 0.014 = d$ 12. $a = 15.0 \times 7.53$

13. $y = 0.4200 \div 179$ 14. $0.1492 \div 300 = a$

15. $1.542 \div 71.9 = x$ 16. $p = 0.0004 \div 0.0018$

APPLICATIONS

Estimation should never be used when calculating the amount of medication to be administered, the amount of a drug to be added to a liquid, or the amount of liquid to be used with a drug. Estimation does aid the health care worker in judging how much drug, solution, or other substances will be needed. It will prevent unneeded preparations of solutions and incomplete procedures due to a lack of a needed ingredient. Estimation is a "foresight." It also serves as a check on the calculations that may be necessary to prepare a solution.

The amount of accuracy needed is dependent upon the purpose of the procedure and the measuring instruments that are to be used. For this reason, numbers cannot simply be "dropped." Even a small quantity such as 0.00008 gram may be meaningful. Expressing measurements using scientific notation aids in determining significant digits and accuracy. This practice may save a life or help to avoid a death.

10.4 Exercises

1. A group of researchers each conducted independent studies on determining the size of a pathogenic bacterium. The average length from each researcher is:

   Researcher **A**   0.00163 micrometer

   Researcher **B**   0.001542 1 micrometer

   Researcher **C**   0.001491 micrometer

   Researcher **D**   0.0015 micrometer

   Since Researcher **D** provided a measurement only accurate to the nearer ten-thousandth, this becomes the level of significance for all the measurements.

   a. Round the measurements to the nearer ten-thousandth micrometer.

   b. Find the average measurement by adding the four measurements and dividing by four. Round the answer to the nearer ten-thousandth micrometer.

2. A 2.2-liter flask of thiamine assay medium contains 40 grams of dextrose; 0.0000008 gram of biotin; 0.02 gram of manganese sulfate; plus other substances. To find the number of grams of each substance present in 1.5 liters of the medium perform the indicated operations. Round each answer to three significant digits.

   a. dextrose: $\dfrac{1.5}{2.2} \times 40$

   b. biotin $\dfrac{1.5}{2.2} \times 0.0000008$

   c. manganese sulfate: $\dfrac{1.5}{2.2} \times 0.02$

3. A laboratory technician is asked to weigh a specimen provided from the surgical team. The mass is determined to be 12.3426 grams. A second technician finds the mass to be 12.3424 grams. A third technician weighs the specimen and records the mass as 12.3429 grams. Find the average of the recorded masses by adding the three masses and dividing by three. Round the answer to the nearer ten-thousandth gram.

4. The acceptable adult dose of a specific medication is determined to be 0.4 milliliter. A technician is requested to divide a 25-milliliter volume into equal adult doses. Estimate the number of adult doses available.

5. Using a scale divided into millimeters, a student measured the width of the thumb nails of five people. The following results were recorded: 13.750 mm, 12.525 mm, 15.125 mm, 14.450 mm, 16.025 mm.

   a. Explain the problem with expressing the measurements to this detail considering the limits of the scale being used.

   b. Explain how you would correctly measure and record this data.

6. Six patients were weighed, each on a different scale. The weights recorded were: 97.52 kg, 79.3842 kg, 52.163 kg, 83.92 kg, 65.7 kg, 87.09 kg.

   a. Convert the numbers to obtain equal levels of significant numbers for the purpose of further comparison. It is assumed that each scale used in this study was fully capable of measuring to the degree of accuracy recorded.

   b. What is the total weight of all six patients?

   c. What is the average weight of the six patients?

Unit 11

Section Two
Applications to Health Work

Objectives

After studying this unit the student should be able to:
• **Use the basic principles of decimal numbers to solve health work problems.**

Every person who is involved in the health care field is dedicated to a common goal—the patient. Moreover, the patient is the whole reason and the very fact for having health care workers. Technicians, technologists, physicians, nurses, and assistants are involved with the well-being of the patient.

Laboratory technicians and technologists, in particular, have a significant responsibility to the prevention, diagnosis, and treatment of diseases and illness. The medical laboratory data constitute the bulk of quantitative and objective information about the patient. Errors in computation or in measurement may be fatal. Remember the cause of death could be: MISPLACED DECIMAL. For this reason, it is essential for all health care workers to concentrate on quality control.

DEFINITIONS

• *Quality control* means applying all possible means to guarantee that laboratory findings, administration of medications, and all other health care procedures are reliable and valid. The elements of quality control in the health care field include:

• *Having a positive and helpful attitude.* Health care workers should not assume that they "know it all." Rather, checking results and seeking advice or assistance when in doubt is the route to follow.

• *Having a thorough knowledge of health care work principles.* Health care workers should have sound judgment about the proficiency in their particular area.

• *Checking the equipment for accuracy.* Health care workers should realize that equipment does not maintain its standardization or calibration forever. The time taken to check the equipment may save a life.

• *Setting standards.* Health care workers should concentrate on labeling solutions, medications, and other items and tools. They should also concentrate on not destroying these standards by being unconcerned or careless. A mislabeled bottle may have grave results.

• *Calculating with care.* Careless arithmetical errors are one of the most common sources of error. These errors are needless and the carelessness may be fatal.

11.1 Exercises

1. A laboratory assistant is asked to prepare dextrose agar. The substances will be dissolved in distilled water to make 500 milliliters of solution. The substances are: beef extract, 1.5 grams; dextrose, 5 grams; tryptose, 5 grams; sodium chloride, 2.5 grams; agar, 7.5 grams. What is the total mass of the ingredients that are to be dissolved?

2. A laboratory assistant prepares a 0.25-preparation of Czapek solution agar. A 1.0-preparation contains: saccharose, 30 grams; sodium nitrate, 2 grams; dipotassium phosphate, 1 gram; magnesium sulfate, 0.5 gram; potassium chloride, 0.5 gram; ferrous sulfate, 0.01 gram; agar, 15 grams. This is dissolved in distilled water to make a 1 000-milliliter solution. Find the mass of each substance in a 0.25-preparation.

   a. saccharose

   b. sodium nitrate

   c. dipotassium phosphate

   d. magnesium sulfate

   e. potassium chloride

   f. ferrous sulfate

   g. agar

3. A 5-gram container of potassium chloride is purchased for bacteriological work. Each culture medium requires 0.25 gram of potassium chloride. How many culture media will the 5-gram container supply?

4. The liquid output for an individual is 2.679 liters on Monday and 3.168 liters on Tuesday. Round each answer to the nearer thousandth.

   a. What is the total output for the two days?

   b. What was the average output for the two days?

   Note: average $= \dfrac{\text{total output}}{\text{two days}}$

   c. How many times greater is the output on Tuesday than on Monday?

5. A patient is prescribed 2.5 grams of medication. One gram is equal to 15.432 4 grains. How many grains of medication does the patient receive?

6. For each meal, a patient eats 85.05 grams of meat. How many grams of meat are eaten in six meals?

7. To determine the average length of microorganisms, several microorganisms are measured. The measurements are:

0.018 millimicron	0.193 millimicron
0.020 millimicron	0.191 millimicron
0.017 millimicron	0.179 millimicron
0.018 millimicron	0.183 millimicron
0.018 millimicron	0.188 millimicron

   Using the formula: average = total ÷ number of measurements, find the average length. Round the answer to the nearer thousandth.

8. A *manometer* is used to determine the rate at which oxygen is used by a bacteria culture. This rate is measured at 30-minute intervals and recorded in chart form.

Interval	Amount of Oxygen Used	Increase or Decrease of Oxygen
First 30-minute interval	1.275 625 milliliters	— — —
Second 30-minute interval	1.373 75 milliliters	?
Third 30-minute interval	1.413 milliliters	?
Fourth 30-minute interval	1.334 5 milliliters	?
Fifth 30-minute interval	1.775 milliliters	?

a. What is the total oxygen used during all the intervals?

b. Which interval shows the greatest increase in the use of oxygen?

c. What is the greatest increase in the use of oxygen?

d. Which interval shows the least change in the use of oxygen?

e. What is the smallest increase in the use of oxygen?

f. In which interval was there a decrease in the amount of oxygen used?

g. What is the amount of decrease?

9. A storage container holds 9.750 liters of ethyl alcohol.

a. If 150 milliliters, 200 milliliters, 100 milliliters, 1 000 milliliters, and 2 500 milliliters are removed from the container, how many liters of alcohol remain?

Note: 1 liter = 1 000 milliliters

b. In the morning, the 9.750-liter container has 4.250 liters in it. If an additional 3.125 liters are poured in, what is the total?

c. How much alcohol could be stored in 7 of these storage containers?

d. How much alcohol can be stored in a container that is 0.65 the size of the 9.750-liter container?

10. At a certain hospital, eight babies are born during a one-month period. Their masses are: 4.348 kilograms; 4.892 kilograms; 4.621 kilograms; 4.485 kilograms; 4.213 kilograms; 4.077 kilograms; 3.397 kilograms; and 3.533 kilograms.

a. What is the mass of the largest baby born during this month?

b. What is the mass of the smallest baby born during this month?

c. What is the average mass of the babies born during the one-month period? Round the answer to the nearer thousandth.
Note: average = total mass ÷ number of babies.

11. It takes 79 milliliters of wax to cover 1 square meter of floor tile. How many milliliters of wax are needed to wax a room 18.183 meters long and 12.89 meters wide?
Note: area = length × width; round the answer to the nearer thousandth square meter.

12. When operating at full capacity, a heating unit consumes 25.693 liters of fuel per hour.

a. If operated at this level, how many liters will be burned in a 24-hour period?

b. Assuming that the heating unit must continue to operate at full capacity, what size storage tank must be used to ensure a 10-day supply?

13. The excreted urine is monitored for a patient over a 12-hour period. During the same period of time, liquid intake is also measured. The following are the results of this monitoring process.

	Liquid Intake	**Urine Excreted**
0 hours	250 milliliters	0 milliliters
2 hours	100 milliliters	100 milliliters
4 hours	50 milliliters	75 milliliters
6 hours	125 milliliters	50 milliliters
8 hours	75 milliliters	50 milliliters
10 hours	75 milliliters	25 milliliters
12 hours	0 milliliters	25 milliliters

a. Find, in milliliters, the total liquid intake.

b. Find, in milliliters, the total urine excreted.

c. What can be concluded about the patient?

14. A weight-loss clinic has several patients on a weight-loss program. In one week the following losses are recorded for four patients:

Patient 1	**Patient 2**	**Patient 3**	**Patient 4**
2.938 kilograms	3.891 kilograms	1.216 kilograms	0.112 kilograms

a. What is the average loss for these four patients?

b. What is the total mass lost by the four patients?

SI METRICS STYLE GUIDE

SI metrics is derived from the French name Le Systeme International d'Unités. The metric unit names are already in accepted practice. SI metrics attempts to standardize the names and usages so that students of metrics will have a universal knowledge of the application of terms, symbols, and units.

The English system of mathematics (used in the United States) has always had many units in its weights and measures tables which were not applied to everyday use. For example, the pole, perch, furlong, peck, and scruple are not used often. The measurements, however, are used to form other measurements and it has been necessary to include the measurements in the tables. Including these measurements aids in the understanding of the orderly sequence of measurements greater or smaller than the less frequently used units.

The metric system also has units that are not used in everyday application. Only by learning the lesser-used units is it possible to understand the order of the metric system. SI metrics, however, places an emphasis on the most frequently used units.

In using the metric system and writing its symbols, certain guidelines are followed. For the student's reference, some of the guidelines are listed.

1. In using the symbols for metric units, the first letter is capitalized only if it is derived from the name of a person.

   SAMPLE:

Unit	Symbol	Unit	Symbol
meter	m	newton	N (named after Sir Isaac Newton)
gram	g	degree Celsius	°C (named after Anders Celsius)

   EXCEPTION: The symbol for liter is L. This is used to distinguish it from the number one (1).

2. Prefixes are written with lowercase letters.

   SAMPLE:

Prefix	Unit	Symbol
centi	meter	cm
milli	gram	mg

   EXCEPTIONS:

Prefix	Unit	Symbol
tera	meter	Tm (used to distinguish it from the metric tonne, t)
giga	meter	Gm (used to distinguish it from gram, g)
mega	gram	Mg (used to distinguish it from milli, m)

3. Periods are not used in the symbols. Symbols for units are the same in the singular and the plural (no "s" is added to indicate a plural).

   SAMPLE: 1 mm *not* 1 mm.   3 mm *not* 3 mms

4. When referring to a unit of measurement, symbols are not used. The symbol is used only when a number is associated with it.

   SAMPLE:

   The length of the room   *not*   The length of the room is expressed in m.
   is expressed in meters.   (*The length of the room is 25 m* is correct.)

5. When writing measurements that are less than one, a zero is written before the decimal point.

   SAMPLE: 0.25 m *not* .25 m

6. Separate the digits in groups of three, counting from the decimal point to the left and to the right. A space is left between the groups of digits.

   SAMPLE:

   5 179 232 mm *not* 5,179,232 mm   0.566 23 mg *not* 0.56623 mg   1 346.098 7 L *not* 1,346.097 L

   A space is also left between the digits and the unit of measure.

   SAMPLE: 5 179 232 mm *not* 5 179 232mm

7. Symbols for area measure and volume measure are written with exponents.

   SAMPLE: 3 cm^2 *not* 3 sq. cm   4 km^3 *not* 4 cu. km

8. Metric words with prefixes are accented on the first syllable. In particular, kilometer is pronounced "kill'-o-meter." This avoids confusion with words for measuring devices which are generally accented on the second syllable, such as thermometer (ther-mom'-e-ter).

SECTION 2: Summary, Review and Study Guide

VOCABULARY

base

compatible numbers

estimation

exponent

expression

evaluating an expression

period

power

repeating decimal

repetend

rounding

scientific notation

significant digits

terminating decimal

truncate

CONCEPTS, SKILLS AND APPLICATIONS

Objectives With Study Guide

Upon completion of Section 2, you should be able to:

- **Express a common fraction with a denominator which is a power of ten as a decimal fraction.**

By observation, express each quantity as a decimal.

$$\frac{7}{8} = \frac{875}{1000} = 0.875$$

$$5 + \frac{4}{20} = 5 + \frac{20}{100} = 5.20$$

- **Express a decimal fraction as a mixed number with a denominator which is a power of ten.**

Express 5.239 as a mixed number. Do not express in lowest terms.

$$5.239 = 5 + \frac{2}{10} + \frac{3}{100} + \frac{9}{1000}$$

$$= 5 + \frac{200}{1000} + \frac{30}{1000} + \frac{9}{1000}$$

$$= 5\frac{239}{1000}$$

- **Compare decimal fractions.**

Which number is smaller: 5.2748 or 5.2738?

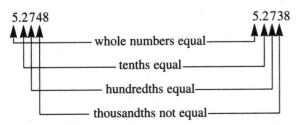

Since, in the thousandths place 3 < 4, 5.2738 < 5.2748

- **Round decimal numbers to any required number of places.**

  Round each number to the place-value position that is underlined.

  10.0̲719   Since 7 > 5, 10.0719 ≈ 10.1 (tenths)

  157.8̲413   Since 1 < 5, 157.8413 ≈ 157.84 (hundredths)

- **Express common fractions as decimal fractions.**

  Using division (numerator ÷ denominator), express each fraction as a decimal.

  $\dfrac{4}{16} = 0.25$ $\dfrac{11}{6} = 1.8333333 = 1.8\overline{3}$

- **Add decimal fractions.**

  Find the sum. 2.035 + 18.9652

  ```
     2.035
  +18.9652
   21.0002
  ```

- **Subtract decimal fractions.**

  Find the difference. 1,507.2 − 31.829

  ```
  1507.200
  −  31.829
  1475.371
  ```

- **Multiply decimal fractions.**

  Find the product. 3.5 • 6.21

  ```
     6.21          2
   × 3.5         +1
     3105          3
   18630
   21.735
  ```

- **Divide decimal fractions.**

  Find the quotient. 15.29 ÷ 0.55

  ```
             27.8
  0.55)15.290
         110
         429
         385
         440
         440
           0
  ```

- **Use Order of Operations to evaluate expressions.**

Compute. $[4(10.8 - 7.3)^3 - 5(14 - 9.7)] \div 3$

$[4(10.8 - 7.3)^3 - 5(14 - 9.7)] \div 3$

$[4(3.5)^3 - 5(4.3)] \div 3$

$[4(42.875) - 5(4.3)] \div 3$

$[171.5 - 21.5] \div 3$

$150 \div 3$

$= 50$

- **Multiply and divide powers.**

Find the product.
$(3^2)(3^4) = 3^{2+4} = 3^6 = 729$

Find the quotient.
$10^5 \div 10^2 = 10^{5-2} = 10^3 = 1000$

Find the power.
$(2^3)^2 = 2^{3(2)} = 2^6 = 64$

- **Evaluate expressions.**

Showing substitution evaluate the expression $3x^2 - 4xy + Z^3$ for $x = 3$, $y = 2$ and $Z = 5$.

$3x^2 - 4xy + Z^3$

$3(3^2) - 4(3)(2) + 5^3$

$= 128$

- **Express numbers using scientific notation.**

Express 0.007108 using scientific notation.

$0.007108 = 7.108 \times 10^{-3}$

Express 5.168×10^4 using standard decimal form.

$5.168 \times 10^4 = 51,680$

- **Perform arithmetic operations using scientific notation.**

Solve each equation. Express each answer using both scientific notation and standard decimal form.

$a = 257 + 37,102$
$a = 37359 \ or \ a = 3.7359 \times 10^4$

$b = 5.003 - 0.00718$
$b = 4.99582 \ or \ a = 4.99582 \times 10^0$

$c = (2.75 \times 10^3) \, (4.0 \times 10^2)$
$\quad = (2.75 \bullet 4.0) \, (10^3 \bullet 10^2)$
$\quad = 110 \bullet 10^5$
$\quad = 1.1 \times 10^7 \ or \ 11,000,000$

$d = 0.036 \div 91$
$d = 0.0003956044 \ or \ 3.956044 \times 10^{-4}$

- **Estimate sums and differences.**

  Estimate the sum or difference.

  $$27.90 \longrightarrow 28$$
  $$18.75 \longrightarrow 19$$
  $$\underline{+46.5 \longrightarrow 47}$$
  $$Sum \approx 94$$

  $$508 \longrightarrow 510$$
  $$\underline{392 \longrightarrow 390}$$
  $$Difference \approx 120$$

- **Estimate products and quotients.**

  Estimate the product or quotient.

  $$32 \longrightarrow 30$$
  $$\underline{\times 59 \longrightarrow 60}$$
  $$Product \approx 1800$$

  $$\frac{432}{6} \longrightarrow \frac{420}{6} \approx 70$$

- **Round numbers to the indicated significant digit.**

  Round to

Number	3 Significant Digits	2 Significant Digits	1 Significant Digit
3742	3740	3700	4000
0.0830	0.0830	0.083	0.08

  Round to the indicated number of significant digits.

- **Perform operations with numbers using the process of retention of significant digits.**

  Solve each equation using the process of retaining significant digits.

  $t = 0.0932 + 8.93617 + 45.82$

  $t = 54.85$ *(rounded to nearer hundredth)*

  $a = 57 - 14.69$

  $a = 42$ *(rounded to nearer whole number)*

  $b = 3.005 \times 2.8$

  $b = 3.0$ *(rounded to nearer tenth)*

  $y = 1.739 \div 6.23$

  $y = 0.28$ *(rounded to nearer hundredth)*

Review

Major concepts and skills to be mastered for the Section 2 Test.

In 1–10, complete each sentence. Choose terms from the vocabulary list.

1. ____?____ is a way to express numbers using powers of 10.

2. When the last remainder is zero the decimal is a ____?____ .

3. A ____?____ refers to an expression with exponents.

4. The process of finding a number close to the exact amount is called ____?____ .

5. The ____?____ indicates the number of times the base is used as a factor.

6. To ____?____ a number means to cut it off at a particular place value position, ignoring any digits that follow.

7. An ____?____ is a symbol or combination of symbols representing a mathematical relationship.

8. Finding an approximation for a number is referred to as ____?____ .

9. ____?____ are digits which indicate the number of units that are reasonably sure of having been counted in making a measurement.

10. A ____?____ is a number used as a factor.

In 11–14, express each quantity as a decimal.

11. $\dfrac{2}{10}$ 12. $\dfrac{45}{100}$ 13. $5 + \dfrac{87}{100}$ 14. $3 + \dfrac{14}{10{,}000}$

In 15–18, express each decimal as a mixed number. Do not express in lowest terms.

15. 4.7   16. 21.85   17. 1.0084   18. 105.0909

In 19 and 20, compare each pair of decimals. Use the symbols <, >, or =.

19. 0.0909 _____ 0.1010   20. $5.\overline{6}$ _____ 5.67

In 21, order from smallest to largest.

21. 6.99998, 7.0001, 6.9999, 7.0090, 7.10006

In 22–25, round each number to the place-value position that is underlined.

22. 15.8$\underline{7}$19   23. 0.$\underline{0}$079   24. 30$\underline{8}$.497   25. $\underline{0}$.42

In 26–29, express each fraction as a decimal.

26. $\dfrac{7}{8}$ 27. $\dfrac{5}{3}$ 28. $\dfrac{3}{5}$ 29. $\dfrac{21}{11}$

In 30 and 31, add.

30.   72.09
   +18.9091

31.   3.172
   0.8913
  + 6.0719

In 32 and 33, solve each equation.

32. $n = 15.6 + 7.609 + 0.7012$ 33. $2.1514 + 5.7280 + 21.081 = x$

In 34 and 35, subtract.

34. 17.3   35. $0.667 - 0.6666$

In 36 and 37, solve each equation.

36. $t = 0.987 - 0.789$ 37. $9 - 0.0045 = b$

In 38 and 39, multiply.

38.   3.005
      $\times$ 4.2

39.   2.395 • 0.81

In 40 and 41, solve each equation.

40. $d = (0.001)(20.001)$

41. $3.05 • 2.009 = a$

In 42 and 43, divide. Round to the nearer tenth.

42. $13.7\overline{)311.812}$

43. $0.00922\overline{)1.576052}$

In 44 and 45, solve each equation.

44. $y = 16.48 \div 2.2$

45. $144.4 \div 36 = b$

In 46–49, compute.

46. $8(2 + 3) - 15 • 2$

47. $4(2^5) - 56 \div 2$

48. $50 \div (7^2 + 1) + 2(5 + 10)$

49. $(8 + 12) \div 10(5) + 3(5^2 + 50)$

In 50–53, find each product, quotient, or power. Write each answer using exponents.

50. $(3^2)(3^4)$ 51. $a^5 \div a^2$ 52. $(2r)^3$ 53. $(4a^2b)^2$

In 54 and 55, evaluate. Show the substitution steps. Use $a = 4.5$, $b = 3$, and $c = 2$.

54. $2a - bc + b^2$

55. $a(bc)^2 - a \div c$

In 56 and 57, express using scientific notation.

56.   31,458,203

57.   0.003178

In 58 and 59, express using standard decimal form.

58. 3.2718×10^{-4}

59. 21.905×10^5

In 60–64, solve each equation. Express each answer using both scientific notation and standard decimal form.

60. $a = 1,285.71 + 0.90103$

61. $b = 791.02 - 5338.7153$

62. $c = (2.5 \times 10^2)(2.0 \times 10^{-3})$

63. $c = 8500 • 0.00006$

64. $d = 0.4715 \div 365$

In 65 and 66, estimate the sum or difference.

65.   25.718
      34.019
      +10.2

66.   9250
      $-$ 5172

In 67 and 68, for each equation estimate the sum or difference.

67. $5.33 + 12.667 + 10.75 = y$

68. $b = 12,071 - 92$

In 69 and 70, estimate the product or quotient.

69.   24.692
         $\times$ 4.091

70.   911.3 $\div$ 44.7

In 71 and 72, for each equation estimate the product or quotient.

71. $n = (39.65)(18.712)$

72. $732.8 \div 36.64 = t$

In 73–76, round to the indicated number of significant digits.

73. 3,912   1 significant digit

74.   732   2 significant digits

75. 4.0512   2 significant digits

76.   0.6087   3 significant digits

In 77–80, solve each equation using the process of retaining significant digits.

77. $t = 7.2091 + 4.05 + 0.9$

78. $257.301 - 57.301 = a$

79. $b = 12.5 \times 7.002$

80. $2.6840 \div 4.0 = y$

SECTION 2 TEST

In 1–4, express each quantity as a decimal.

1. $\dfrac{5}{8}$

2. $\dfrac{37}{100}$

3. $4\dfrac{27}{1000}$

4. $3 + \dfrac{27}{10,000}$

In 5–8, express each decimal as a mixed number. Do not express in lowest terms.

5. 8.2

6. 17.13

7. 4.002

8. 91.1705

In 9 and 10, compare each pair of decimals. Use the symbols <, >, or =.

9. 4.33 _____ $4\overline{3}$

10. 19.009 _____ 20.0001

In 11 and 12, order from smallest to largest.

11. $3.\overline{9}$, 4,001, 3.9989, 4,10001, $3.8\overline{9}$

12. 11.0092, 11.10007, 10.99998, 11.0001, 10.9990

In 13–16, round each number to the place-value position that is underlined.

13. 10<u>2</u>.09

14. 3.7<u>1</u>4

15. 0.<u>0</u>48

16. <u>2</u>6.192

In 17–20, express each fraction as a decimal.

17. $\dfrac{9}{4}$

18. $\dfrac{4}{12}$

19. $\dfrac{7}{3}$

20. $\dfrac{6}{8}$

In 21 and 22, add.

21. $\begin{array}{r} 15.0712 \\ +\,21.9028 \\ \hline \end{array}$

22. $\begin{array}{r} 0.8190 \\ 10.237 \\ +\,13.15 \\ \hline \end{array}$

In 23 and 24, solve each equation.

23. $y = 8.25 + 9.0732 + 4.571$

24. $3.14720 + 15.3 + 10.002 = n$

In 25 and 26, subtract. Round to the nearer hundredth.

25. $\begin{array}{r} 3 \\ -\,1.78 \\ \hline \end{array}$

26. $\begin{array}{r} 42.\overline{6} \\ -\,12.7 \\ \hline \end{array}$

In 27 and 28, solve each equation.

27. $p = 8.125 - 7.8752$

28. $2.0975 - 2 = t$

In 29 and 30, multiply.

29. $\begin{array}{r} 3.75 \\ \times\,5.001 \\ \hline \end{array}$

30. $(25.34)\,(10.7)$

In 31 and 32, solve each equation.

31. $r = 0.32 \cdot 0.47$

32. $(7.5)\,(100.09) = b$

In 33 and 34, divide. Round to the nearer tenth.

33. $216.5\overline{)7014.68}$

34. $0.35\overline{)0.5862115}$

In 35 and 36, solve each equation. Round to the nearer hundredth.

35. $c = 52.8083 \div 0.02$

36. $39.9 \div 6.105 = t$

In 37–40, compute.

37. $7(5.3 - 2.1) \div 4 \cdot 5$

38. $5(4^3) + (320 \div 4)$

39. $10,000 \div (7^2 + 1) - 2(10^2)$

40. $3(492.4 - 15.2(2)) \div 2^3 + 26.75$

In 41–44, find each product, quotient, or power. Write each answer using exponents.

41. $(5^2)\,(5^4)$

42. $7^5 \div 7^3$

43. $(3t)^4$

44. $(6ab^3)^3$

In 45 and 46, evaluate. Show the substitution step. Use $a = 6$, $b = 2.5$ and $c = 4$.

45. $4(ab)^2 - c^3$

46. $5a - 4b + c^4$

In 47 and 48, express using scientific notation.

47. 2,713,408   48.   0.03714

In 49 and 50, express using standard decimal form.

49. 5.8213×10^{-3} 50. 6.281×10^{4}

In 51–54, solve each equation. Express each answer using both scientific notation and standard decimal form.

51. $r = 32.381 + 1.602$ 52. $s = 28.001 - 16.975$

53. $t = (216)(0.0012)$ 54. $u = 0.388 \div 200$

In 55 and 56, estimate each sum or difference.

55.   227   56.   14,495
      4798.9   − 10,199.28
   + 364.875

In 57 and 58, for each equation estimate the sum or difference.

57. $q = 27.91 + 15.3 + 54.8$ 58. $z = 109.75 - 49.6$

In 59 and 60, estimate the product or quotient.

59.   9.25   60. $5,315 \div 99$
  $\times$ 19.8

In 61 and 62, for each equation estimate the product or quotient.

61. $t = (87.9)(60.2)$ 62. $402 \div 0.75 = a$

In 63 and 64, round to the indicated number of significant digits.

63. 981   2 significant digits   64.   14,692   1 significant digit

In 65–68, solve each equation using the process of retaining significant digits.

65. $p = 42.0 + 19.38 + 22.981$ 66. $c = 471.625 - 14.3$

67. $(15.1)(6) = r$ 68. $17.309 \div 6.2 = q$

Section Three

Metric Measure

Unit 12

Introduction to Metric Measure

Objectives

After studying this unit the student should be able to:

- **Identify metric measures in terms of a standard unit.**
- **Express equivalence for metric measures.**

THE MEASURING PROCESS

To *measure an object,* compare it to a predetermined unit of measure and assign a number in relationship to this standard unit.

The process of measuring an object consists of:
- Selecting a unit.
- Dividing the object into units.
- Counting the number of units in the object.
- The number of units is the measure of the object.

Two health care students measure a piece of gauze. Chad measures the width of the gauze and finds the width to be 6 index fingers.

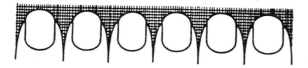

Sara measures the same piece of gauze and finds the width to be 9 index fingers.

Since very few index fingers are exactly the same width, using the index finger as a unit of measure is unsatisfactory. To have a useful system, there must be uniform or standard units of measure. Among the many systems of measure is the International System of Units — the metric system.

METRIC MEASURE AND THE PLACE-VALUE SYSTEM

The metric system is a system of measure which uses the base 10, or the decimal system of numeration. The decimal system uses grouping by tens. Since there are only ten digits in the decimal system, 0, 1, 2, 3, 4, 5, 6, 7, 8, 9 — two or more digits are used when expressing a number that is larger than nine.

Each digit in a number has place value associated with it. The place values are powers of ten.

Place-Value	Thousands	Hundreds	Tens	Units	Tenths	Hundredths	Thousandths
Meaning	1 000	100	10	1	0.1	0.01	0.001
Power of 10	10^3	10^2	10^1	10^0	10^{-1}	10^{-2}	10^{-3}

EXAMPLE 1

Illustrate each place value for 5,237.806.

$$\begin{array}{cccccccc} \text{THOUSANDS} & \text{HUNDREDS} & \text{TENS} & \text{UNITS} & & \text{TENTHS} & \text{HUNDREDTHS} & \text{THOUSANDTHS} \\ 5 & 2 & 3 & 7 & . & 8 & 0 & 6 \end{array}$$

Since the metric system is based upon powers of ten, it is important to be able to mentally multiply by a power of ten.

EXAMPLE 2

Mentally multiply 23.57 • 100.

Count the number of zeros in the power of ten.

1̲0̲0̲

2 zeros

Move the decimal point of the number to the right this same number of zeros.

23.57.

$23.57 \times 100 = 2{,}357$

Note: The number $100 = 10^2$. The exponent, 2, determines the number of places to the right that the decimal point is moved.

EXAMPLE 3

Mentally multiply 23.57 • 0.001.

Count the number of places to the right of the decimal point in the power of ten.

0.001

3 decimal places

Move the decimal point of the number to the left this same number of decimal places.

0.023ₓ57

23.57 × 0.001 = 0.02357

Note: The decimal $0.001 = 10^{-3}$. The exponent, -3, determines the number of places the decimal point is moved to the left.

12.1 Exercises

In 1–9, mentally multiply the numbers by the power of ten.

	1 000	100	10	Number	0.1	0.01	0.001
1.				6.28			
2.				0.913			
3.				57.194			
4.				0.07			
5.				371.721⁵			

	10^3	10^2	10^1	Number	10^{-1}	10^{-2}	10^{-3}
6.				37.594			
7.				0.816³			
8.				407.38			
9.				0.003⁵⁷			

UNITS OF MEASURE AND PREFIXES

In the SI metric system, the two kinds of units are:

• base units

• derived units

Two of the base units are the quantity of length measure and the quantity of mass measure. The base unit for length measure is the meter (m). The base unit for mass measure is the kilogram (kg). A kilogram is equal to 1,000 grams. For purposes of comparison, the gram (g) will be used.

One of the derived units is the quantity of volume measure. The derived unit for volume measure for fluids is the liter (L) and for solids it is the cubic centimeter (cm^3).

Other units of measure are multiples and sub-multiples of the base or derived units. Every metric measure is a power of ten of the base or derived units. Thus, the metric system is sometimes defined as the decimal system of weights and measures.

Different Greek and Latin prefixes are used as names for the powers of ten. This expansion of the place-value chart shows these prefixes and the meanings.

Place-Value	Thousands	Hundreds	Tens	Units	Tenths	Hundredths	Thousandths
Meaning	1 000	100	10	1	0.1	0.01	0.001
Power of 10	10^3	10^2	10^1	10^0	10^{-1}	10^{-2}	10^{-3}
Metric Prefix	kilo	hecto	deka	unit of measurement	deci	centi	milli

To write a measure in the metric system, a prefix and a basic metric unit are combined.

Using the meter and the appropriate prefix, equivalences can be formed.

1 000 meters	=	1 kilometer
100 meters	=	1 hectometer
10 meters	=	1 dekameter
0.1 meter	=	1 decimeter
0.01 meter	=	1 centimeter
0.001 meter	=	1 millimeter

12.2 Exercises

In 1–12, find each equivalent.

1. 1 thousand = ___?___ hundreds

2. 1 hundred = ___?___ thousand

3. 1 hundred = ___?___ tens

4. 1 ten = ___?___ hundred

5. 1 ten = ___?___ units

6. 1 unit = ___?___ ten

7. 1 unit = ___?___ tenths

8. 1 tenth = ___?___ unit

9. 1 tenth = ___?___ hundredths

10. 1 hundredth = ___?___ tenth

11. 1 hundred = ___?___ thousandths

12. 1 thousandth = ___?___ hundredth

13. To change from tenths to tens, multiply by __a__ or move the decimal point __b__ places to the __c__ .

14. To change from units to hundredths, multiply by __a__ or move the decimal point __b__ places to the __c__ .

15. To change from hundreds to units, multiply by __a__ or move the decimal point __b__ places to the __c__ .

16. To change from thousandths to hundredths, multiply by __a__ or move the decimal point __b__ places to the __c__ .

17. To change from tenths to thousandths, multiply by __a__ or move the decimal point __b__ places to the __c__ .

18. To change from hundredths to hundreds, multiply by __a__ or move the decimal point __b__ places to the __c__ .

19. To change from thousands to units, multiply by __a__ or move the decimal point __b__ places to the __c__ .

20. To change from thousandths to units, multiply by __a__ or move the decimal point __b__ places to the __c__ .

21. To change from hundredths to thousandths, multiply by __a__ or move the decimal point __b__ places to the __c__ .

In 22–28, complete the chart to show the ranking of 1 centiliter, 1 deciliter, 1 dekaliter, 1 hecto-liter, 1 kiloliter, 1 liter, and 1 milliliter from largest to smallest. Then indicate the equivalent liter measure.

	Volume Measure from Smallest to Largest	Equivalence
22.		___?___ = 1 000 liter
23.		
24.		
25.	1 liter	1 liter = 1 liter
26.		
27.	1 centiliter	
28.		1 milliliter = ___?___

In 29–35, complete the chart to show the ranking of 1 centigram, 1 decigram, 1 dekagram, 1 gram, 1 hectogram, 1 kilogram, and 1 milligram from smallest to largest. Then indicate the equivalent gram measure.

	Mass Measure from Smallest to Largest	Equivalence
29.		
30.		
31.		
32.	1 gram	1 gram = 1 gram
33.		
34.		
35.		

In 36–41, list the metric prefix.

36. thousands   37. hundreds   38. tens

39. tenths   40. hundredths   41. thousandths

42. Explain why it is necessary to have a standard set of units of measurement.

APPLICATIONS

The metric system is a universal system of measure. In recent years, countries such as Canada and the United States have been making the change from other standard units of measure to the metric system. Change is not always easy. Changing the dials on a machine is relatively easy, compared to changing what a person has become accustomed to using.

The best way to learn a system of measure is to use it. Equivalent measures from one system of measure to another system of measure are discussed in later units. This skill, while useful, is not essential to the effective use of the metric system.

The metric system has standard units which allow the measure of length, volume or capacity, and mass. The standard units of measure are: LENGTH: meter (m); VOLUME or CAPACITY: liter (L); and MASS: kilogram (kg).

12.3 Exercises

1. The mass of a patient is determined to be 88.694 kilograms. When recorded it is indicated as 88,694 grams. Is the recorded mass correct?

2. At sea level the mass of 1 milliliter of pure water is equal to 1 gram. Using the equivalent of 1 milliliter equals 1 gram, find the number of grams in a solution containing 925 milliliters of distilled water, 4 grams of NaCl, and 165 grams of KCl.

3. Central supply has 38.654 meters of gauze. A technician is requested to divide the gauze into 150-centimeter lengths. Find the number of 150-centimeter lengths obtained.

4. A graduated pipette is used to measure the following samples.

   Sample *A:* 0.04 mL   Sample *D:* 1.00 mL

   Sample *B:* 0.08 mL   Sample *E:* 1.02 mL

   Sample *C:* 0.10 mL

   a. Which sample is the smallest?

   b. Which sample is the largest?

5. A hospital blood bank has 5 units of O-type blood on hand. A trauma patient receives a transfusion of 1.75 units. How many units of O-type blood remain on hand?

6. An ocular micrometer, in a microscope, is used to measure the width of several cells. Arrange the following measurements from smallest to largest.

   0.0025 mm, 0.00100 mm, 0.0009 mm, 0.00089 mm, 0.00505 mm

7. A cell is found to be 0.009 mm wide. How many cells of this size could be placed side-by-side and fit in a space not exceeding 1 mm in width?

Unit 13

Metric Length Measure

Objectives

After studying this unit the student should be able to:
- **Express metric length measure in smaller or larger metric units.**
- **Determine the radius, diameter, and circumference of a circle.**

UNITS OF LENGTH MEASURE

The base unit of metric length measure is the *meter.* Length measure can also be expressed using other prefixes such as kilometers, hectometers, dekameters, decimeters, centimeters, and millimeters. The kilometer, meter, centimeter and millimeter are the most commonly used units. A given metric length measure can be expressed in larger or smaller metric units.

- To express a metric length unit as a smaller metric length unit, multiply by a positive power of ten such as 10, 100, 1 000, 10 000, or 100 000.
- To express a metric length unit as a larger metric length unit, multiply by a negative power of ten such as 0.1, 0.01, 0.001, 0.000 1, or 0.000 01.

This chart summarizes the units of metric length measure, the symbols, the equivalences, and the relationship between the units.

Unit	Kilometer	Hectometer	Dekameter	Meter	Decimeter	Centimeter	Millimeter
Symbol	km	hm	dam	m	dm	cm	mm
Equivalence	1 000 meters	100 meters	10 meters	1 meter	0.1 meter	0.01 meter	0.001 meter
Relationship Between Units	1 km is 10 hm	1 hm is 10 dam	1 dam is 10 m	1 m is 10 dm	1 dm is 10 cm	1 cm is 10 mm	

The steel rule also shows the relationship between the units.

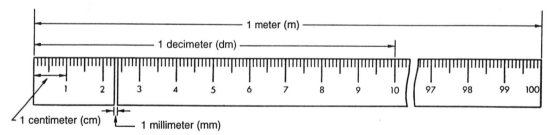

13.1 Exercises

In 1–6, use the chart to find the relationship between the units.

1. 1 km = ___?___ hm   2. 1 hm = ___?___ dam   3. 1 dam = ___?___ m

4. 1 m = ___?___ dm   5. 1 dm = ___?___ cm   6. 1 cm = ___?___ mm

7. 1 hm = ___?___ km   8. 1 dam = ___?___ hm   9. 1 m = ___?___ dam

10. 1 dm = ___?___ m   11. 1 cm = ___?___ dm   12. 1 mm = ___?___ cm

13. To change from mm to m, multiply by ___a___ or move the decimal point ___b___ places
     to the ___c___ .

14. To change from cm to m, multiply by ___a___ or move the decimal point ___b___ places
     to the ___c___ .

15. To change from dm to mm, multiply by ___a___ or move the decimal point ___b___
     places to the ___c___ .

16. To change from dam to dm, multiply by ___a___ or move the decimal point ___b___
     places to the ___c___ .

17. To change from m to cm, multiply by ___a___ or move the decimal point ___b___ places
     to the ___c___ .

18. To change from mm to cm, multiply by ___a___ or move the decimal point ___b___
     places to the ___c___ .

19. To change from cm to dm, multiply by ___a___ or move the decimal point ___b___ places
     to the ___c___ .

20. To change from m to dam, multiply by ___a___ or move the decimal point ___b___ places
     to the ___c___ .

In 21–26, give the meaning of each metric prefix.

21. milli   22. deci   23. hecto

24. deka   25. centi   26. kilo

Expressing metric length measure in larger or smaller units uses the principle of multiplying numbers by powers of ten.

EXAMPLE 1

Tim is 1.82 meters tall. How many centimeters is this?

1.82 m = ___?___ cm   *Note:* A meter is a longer unit of length than is a centimeter.
1 m = 100 cm   Thus, multiply by a positive power of 10. Since there
1.82 • 100 = 182   are 10^2 centimeters in a meter, multiply by 100.
1.82 m = 182 cm

Tim is 182 centimeters tall.

EXAMPLE 2

Tina is 1 423 mm tall. How many meters is this?

1 423 mm = ___?___ m   *Note:* A millimeter is a shorter unit of length than is a meter.
1 mm = 0.001 m   Thus, multiply by a negative power of 10. Since a milli-
1 423 • 0.001 = 1.423   meter is $\frac{1}{1\,000}$ of a meter, multiply by 10^{-3} or 0.001.
1 423 mm = 1.423 m

Tina is 1.423 meters tall.

13.2 Exercises

In 1–39, express each measure in a larger equivalent metric unit or a smaller equivalent metric unit.

1. 1 m = __?__ cm
2. 1 dm = __?__ cm
3. 1 m = __?__ mm

4. 6.5 m = __?__ cm
5. 12 cm = __?__ mm
6. 3.2 km = __?__ m

7. 2.68 hm = __?__ m
8. 4.75 dam = __?__ dm
9. 10 km = __?__ m

10. 0.142 cm = __?__ mm
11. 42 mm = __?__ cm
12. 617 cm = __?__ m

13. 3 275 m = __?__ km
14. 0.753 mm = __?__ dm
15. 15.7 dm = __?__ m

16. 22 dam = __?__ hm
17. 0.32 mm = __?__ m
18. 0.457 cm = __?__ m

19. 823.75 mm = __?__ cm
20. 47.8 dam = __?__ km
21. 712 m = __?__ km

22. 6.2 km = __?__ m
23. 85 cm = __?__ mm
24. 35 cm = __?__ m

25. 25 m = __?__ km
26. 4.8 m = __?__ cm
27. 47 dam = __?__ hm

28. 312 mm = __?__ m
29. 0.625 cm = __?__ mm
30. 48.9 mm = __?__ cm

31. 320 m = 0.32 __?__
32. 3 725 mm = 37.25 __?__
33. 8.71 hm = 8 710 __?__

34. 4 975 cm = 4.975 __?__
35. 18 dm = 1.8 __?__
36. 2 500 __?__ = 2.5 km

37. 17.37 __?__ = 1.737 dam
38. 425 __?__ = 42.5 cm
39. 3.25 __?__ = 3 250 mm

In 40–43, complete the chart.

	Kilometers	Hectometers	Dekameters	Meters	Decimeters	Centimeters	Millimeters
40.				37.594			
41.				0.815 3			
42.				407.38			
43.				0.003 57			

In 44–49, arrange each group of measures from largest to smallest.

44. 14 cm; 14 mm; 14 m
45. 1.62 m; 170 cm; 1 600 mm

46. 0.005 km; 17 m; 2 000 cm
47. 12 m; 1 211 cm; 0.123 km

48. 40 cm; 45 mm; 5 dm
49. 0.25 km; 255 m; 2 560 dm

In 50–54, solve.

50. If 245 centimeters of cloth are required for a lab coat, how many lab coats can be cut from a piece of cloth 13 meters in length?

51. Jean walks an average of 1 500 meters each day she works at the clinic. During one month she works 20 days. How many kilometers does she walk?

52. Craig takes a medical research trip. He keeps the following record of the distances traveled.

    1st day: 368 km   4th day: 419 km
    2nd day: 279 km   5th day: 317 km
    3rd day: 520 km   6th day: 193 km

    a. Find the total distance driven.

    b. Find the average distance covered per day. Round the answer to the nearer kilometer.

    Note: average = total distance ÷ number of days.

53. A rectangular research area, 60 m by 150 m, is to be enclosed by railing that costs $12.60 per meter. Find the cost of the railing.

54. Posts for the rectangular research area, 60 m by 150 m, are to be placed at each corner and 2.5 m apart.

   a. Find how many posts are needed.

   b. If each post cost $7.20, find the cost of all the posts.

LENGTH MEASURE OF A CIRCLE

Length measure of a circle can be found. Before introducing the formula some special terms should be noted.

DEFINITIONS

- The *center* is the point from which all points on the circle are the same distance.
- The *diameter (d)* is the length of the line segment through the center that has both its endpoints on the circle.
- The *radius (r)* is the length of the line segment connecting the center to a point on the circle. The radius is one-half the diameter or $r = \frac{1}{2}d$.
- The *circumference (C)* is the distance around the circle.
- *Pi (π)* is a constant value which compares the circumference and the diameter. The value to seven decimal places is 3.1415927. When calculating by hand use 3.14 for π. When using a calculator, use the π key.

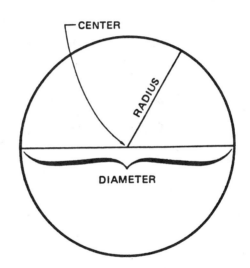

The formula for the circumference of a circle is:
$C = \pi d$ *or*
$C = 2\pi r$

EXAMPLE 1

Find the circumference of a circle with a diameter of 4 cm.

$C = \pi d$	Write the formula.
$C = 3.14\ (4\ cm)$	Substitute the specific value for the unknown.
$C = 12.56\ cm$	Compute.

EXAMPLE 2

Find the circumference of a circle with a radius of 3 cm.

$C = 2\pi r$	Calculator
$C = 2(3.14)\ (3\ cm)$	$C = 2(\pi)\ (3\ cm)$
$C = 6.28\ (3\ cm)$	$C = 18.849556\ cm$
$C = 18.84\ cm$	

EXAMPLE 3

Find the diameter of a circle whose circumference is 12.57 cm. (Round to the nearer hundredth.)

$C = \pi d$
$12.57 = \pi d$ Substitute the specific value for the unknown.

$\dfrac{12.57}{\pi} = d$ To solve for d, divide each side by π.

$4.0011553 = d$ Use the π key on your calculator.

$$12.57 \boxed{\div}\, \boxed{\pi}\, \boxed{=}$$

$d = 4.00$ cm (nearer hundredth)

EXAMPLE 4

Find the radius of a circle whose circumference is 31.42 mm. (Round to the nearer hundredth.)

$C = 2\pi r$
$31.42 = 2\pi r$ Substitute the specific value for the unknown.

$\dfrac{31.42}{2\pi} = r$ To solve for r, divide each side by 2π.

$10.001297 = r$ Use the π key on your calculator.

$$31.42 \boxed{\div}\, \boxed{(}\, 2\, \boxed{\times}\, \boxed{\pi}\, \boxed{)}\, \boxed{=}$$

$r = 10.00$ mm (nearer hundredth)

13.3 Exercises

In 1–4, find the circumference of each circle. Use the π key on your calculator.

  1. $r = 4$ m   2. $d = 6$ mm   3. $d = 10$ m   4. $r = 2.5$ cm

In 5–8, find the diameter of each circle. Use the π key on your calculator. Round answers to the nearer hundredth.

  5. $C = 18.85$ cm   6. $C = 36$ mm   7. $C = 1.5$ m   8. $C = 9.37$ cm

In 9–12, find the radius of each circle. Use the π key on your calculator. Round answers to the nearer hundredth.

  9. $C = 87.3$ cm   10. $C = 3.2$ m   11. $C = 125$ mm   12. $C = 10$ cm

13. The circumference of a patient's bicep is 41 cm. Find the diameter and radius.

14. If you know the radius of a circle how can you find the diameter?

In 15–20, give the abbreviation for each metric measure.

15. decimeter   16. kilometer   17. hectometer

18. meter   19. centimeter   20. millimeter

MEASURING LENGTH

No measurement is exact. The precision of a measurement depends upon the unit of measure that is used. The diameter (distance across the center) of a half-dollar can be measured in centimeters or millimeters.

Unit of Measure — centimeter   Unit of Measure — millimeter

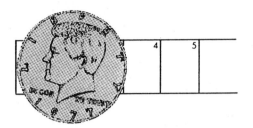   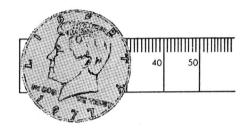

The diameter of the half-dollar, to the nearer centimeter, is 3 centimeters.

The diameter of the half-dollar, to the nearer millimeter is 31 millimeters.

In each case there is some error in the measurement. The error of measurement is the difference between the exact length and the measured length. The error of measurement is less when a smaller unit is used. Since the millimeter is a smaller unit of measure, it is more precise than the centimeter. In general, the smaller the unit of measure, the more precise the measurement. Thus, 31 mm is more precise than 3 cm.

13.4 Exercises

In 1–4, use a meter stick to measure each item using the given units of measure. Indicate which measure is more precise.

	Item	Unit of Measure	Measurement	More Precise Measurement
1.	length of a dollar bill	centimeter		
		millimeter		
2.	width of a classroom door	meter		
		centimeter		
3.	height of a classroom door	meter		
		centimeter		
4.	diameter of a quarter	decimeter		
		millimeter		

13.5 Exercises

In 1–10, choose the best unit of length (km, m, cm, or mm) for measuring each of these items.

1. The distance between your eyes.

2. The length of a car.

3. The diameter of the earth.

4. The height of a mountain.

5. The length of your arm.

6. The length of an eyelash.

7. The length of a shoelace.

8. The distance around a person's head.

9. The thickness of a hair.

10. Rainfall.

In 11–15, place these measures in order from the largest measure to the smallest measure. **Hint:** First, convert each measure into the same prefix.

11. 15 mm; 16 km; 6 000 m; 50 cm; 6 000 mm

12. 5 km; 25 mm; 2 000 cm; 150 m; 42 km

13. 2 750 cm; 10 km; 5 000 mm; 35 m; 3 025 cm

14. 1.75 m; 200 cm; 0.01 km; 450 mm; 2.75 cm

15. 3.15 km; 52.5 m; 0.31 m; 1.3 mm; 7.38 m

In 16–20, use the given unit to estimate and then measure the length of each object. Express the measurements in the indicated units.

Object	Estimate (a)	Measurement (b)	Expressing the measurement using other units	
			(c)	(d)
16. A person's right foot	__?__ mm	__?__ mm	__?__ cm	__?__ m
17. A person's height	__?__ m	__?__ m	__?__ dm	__?__ cm
18. The width of a ballpoint pen tip	__?__ mm	__?__ mm	__?__ cm	__?__ dm
19. A person's waist	__?__ cm	__?__ cm	__?__ mm	__?__ m
20. The length of this room	__?__ m	__?__ m	__?__ dam	__?__ hm

APPLICATIONS

Length measures have wide applications in health occupations. The height of patients, lengths of cloth supplies, and size of paper supplies are only a few of the situations in which length measure becomes important. Until recent years most health services used the English units of length measure with occasional use of the metric units. Presently, the metric system is progressively replacing other units of measure.

When purchasing glassware or other supplies for the laboratory, a knowledge of the metric length measure is valuable. Test tubes are available in various lengths and diameters. Filter paper, chromatography paper, and recording paper are sold in sizes expressed in metric values. When comparing quantities and cost, the ability to determine per unit cost is important for good fiscal management.

13.6 Exercises

1. During each 15-minute test, a recording kymograph is geared to use 3.275 m of 12-cm wide paper.

   a. How much paper should be purchased for 400 15-minute tests?

   b. How many tests can be made with 45 m of paper?

   c. How many meters of paper are required to run a 27-minute test?

   d. How many centimeters of paper are required to run a 3-minute test?

   e. How many millimeters of paper are required to run a 5-minute test?

2. A baby is 45.750 cm long when born. To determine the growth rate, the baby is measured each week for a five-week period. Calculate the weekly growth rate and the total growth.

Week	Length	Growth
birth	45.750 cm	
a. 1	46.025 cm	___?___
b. 2	46.125 cm	___?___
c. 3	46.125 cm	___?___
d. 4	46.650 cm	___?___
e. 5	46.950 cm	___?___
f. **TOTAL GROWTH**		___?___ cm

3. A team of students measure the height of five patients. When the data is compiled the results show the following:

Patient *A:* 1.896 m

Patient *B:* 1 783 mm

Patient *C:* 174.6 cm

Patient *D:* 2.006 m

Patient *E:* 164.8 cm

Remember: The units of measure must be the same before they can be added.

a. Find the average height of the patients. Average equals total heights divided by the number of patients. Round the answer to the nearer thousandth.

b. Which patient is the tallest?

c. Which patient is the shortest?

d. What steps did you have to take to be able to average and compare the measurements?

4. Brand *A* of gauze is found to cost $23.89 per 15-meter roll. Brand *B* comes in 25-meter rolls costing $42.67 each.

a. Find the cost per meter of Brand *A*.

b. Find the cost per meter of Brand *B*.

c. Assuming the quality is equal, which is the better buy?

5. Chromatography paper is available in 50-meter rolls, 12.5 centimeters wide, at a cost of $59.85. This roll is then cut in lengths 875 centimeters long before using.

a. How many pieces can be cut from the 50-meter roll?

b. What is the cost per piece?

6. A patient's room measures 8 meters by 3 meters. Each room is to be fitted with three 2-meter baseboard heaters.

a. How many meters of baseboard heaters will be present in each room?

b. If there are 12 rooms in the ward, each of the same size and equally equipped, how many total meters of baseboard heaters will be required?

c. If the cost is $75 per meter of baseboard heater, what is the total cost of the heaters in the ward?

7. A patient's waist measures 105 centimeters. Suppose a roll of gauze contains 1.6 meters. Approximately how many times could the gauze be completely wrapped around the patient's waist?

Unit **14**

Metric Area Measure

Objectives

After studying this unit the student should be able to:
- **Express metric area measures in smaller or larger metric units.**
- **Determine the area of figures using the appropriate formulas.**

MEASURING AREA

The area of a region is the number of square units that it takes to cover the region. In the metric system, area measure is a derived unit. It is found by using the base unit of length measure — the meter. One unit of square measure is the square centimeter. To find the area, the number of squares can be counted.

The area of each region is found by counting the number of square centimeters.

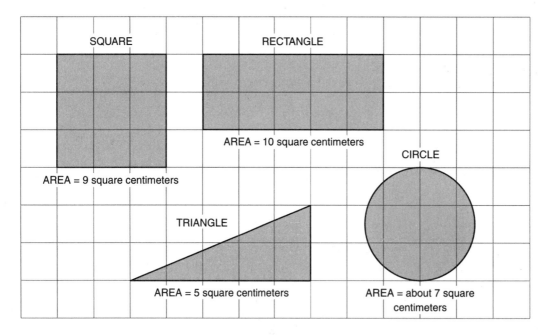

Note: Metric area measures, such as 9 square centimeters, 10 square centimeters and 5 square centimeters can be written 9 cm², 10 cm² and 5 cm². The exponent, 2, indicates that the measure consists of 2 dimensions.

In real life, counting squares is impractical. Examining the areas and dimensions more closely leads to the development of formulas for calculating the areas of regions.

FINDING THE AREA OF A SQUARE

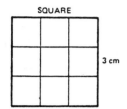

SQUARE

3 cm

3 cm

The area of this square is 9 cm². Each side is 3 cm.
Since (3 cm) (3 cm) = 9 cm²,
the area = (3 cm) (3 cm)
or area = side • side.

- The formula for the area of a square with sides *s* units long is:

$A = s • s$

   or

$A = s^2$

Note: The symbol for the side, *s*, includes the unit of measure.

EXAMPLE 1

Find the area of a square with sides of 4 meters.

$A = s • s \ or \ s^2$	Write the formula.
$A = (4 \ m) \ (4 \ m) \ or \ (4 \ m)^2$	Substitute the specific values.
$A = 16 \ m^2$	Compute.

FINDING THE AREA OF A RECTANGLE

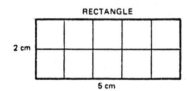

RECTANGLE

2 cm

5 cm

The area of this rectangle is 10 cm².
The length is 5 cm; the width is 2 cm.
Since (5 cm) (2 cm) = 10 cm²,
area = length • width.

- The formula for the area of a rectangle with length *l* units and width *w* units is:

$A = l • w$

   or

$A = lw$

Note: The symbols *l* and *w* include the unit of measure.

EXAMPLE 2

Find the area of a rectangle 3 m by 4.5 m.

$A = l • w$	Write the formula.
$A = (3 \ m) \ (4.5 \ m)$	Substitute the specific values.
$A = 13.5 \ m^2$	Compute.

FINDING THE AREA OF A TRIANGLE

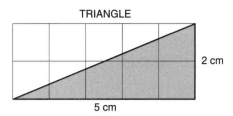

TRIANGLE

5 cm

2 cm

The area of this triangle is 5 cm^2.
The area of the rectangle is 10 cm^2.

Since 5 cm^2. is $\frac{1}{2}$ (10 cm^2) the area of the

triangle $= \frac{1}{2}$ (10 cm^2) or

area $= \frac{1}{2}$ (area of rectangle).

The parts of a triangle have special names:

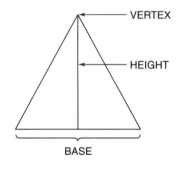

VERTEX

HEIGHT

BASE

- The *vertex* is the common endpoint of 2 line segments.
- The *base* is the length of the side on which the triangle sets.
- The *height* is the line segment from the opposite vertex intersecting the base at a right angle.

- The formula for the area of a triangle with base *b* and height *h* units is:

$$A = \frac{1}{2} \cdot b \cdot h$$

or

$$A = \frac{1}{2}bh$$

Note: The symbols *b* and *h* include the unit of measure. The height is sometimes referred to as the altitude.

EXAMPLE 3

Find the area of a triangle with a base of 12 mm and a height (altitude) of 8 mm.

$A = \frac{1}{2} \cdot b \cdot h$ Write the formula.

$A = \frac{1}{2}(12 \text{ mm}) (8 \text{ mm})$ Substitute the specific values.

$A = 48 \text{ mm}^2$ Compute.

FINDING THE AREA OF A CIRCLE

A rectangular region can be used to obtain a very close approximation for the area of a circle. To obtain a "near rectangle," the circle is cut into 16 pie-shaped sections.

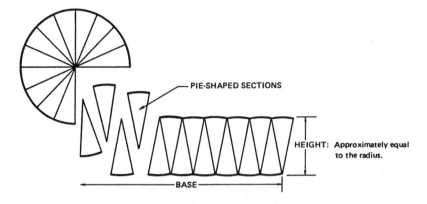

To finish constructing this "near rectangle," the right end pie-shaped section is cut in half. This wedge-shaped piece is then placed on the left side.

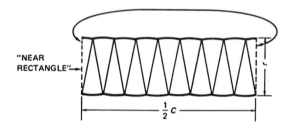

The area of this "near rectangle" is base times the height.

$A = b \cdot h$

$A = \dfrac{1}{2}(C)(r)$ $\qquad\qquad$ base $= \dfrac{1}{2}C$; height $= r$

$A = \dfrac{1}{2}\pi(d)(r)$ $\qquad\qquad$ circumference $= \pi d$

$A = \dfrac{1}{2}(\pi)(2)(r)(r)$ $\qquad\qquad$ diameter $= 2r$

$A = \dfrac{1}{2}(2)(\pi)(r)(r)$ $\qquad\qquad$ Rearrange factors.

$A = \pi r^2$ $\qquad\qquad$ Simplify.

- The formula for the area of a circle is:

$A = \pi r^2$

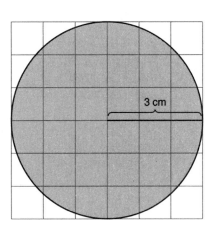

The area of this circle is approximately 28 square centimeters. The radius is 3 cm. Since (3 cm) (3 cm) = 9 cm², the *area* = (9 cm²) (3.1416) *or* 28.274 4 cm².

EXAMPLE 4

Find the area of a circle with a radius of 4 m. Round to nearer hundredth.

$A = \pi r^2$ Write the formula

$A = (\pi)\,(4\ \text{m})^2$ Substitute the specific values.

$A = (\pi)\,(16\ \text{m}^2)$

$A = (\pi)\,(16)\ \text{m}^2$ Use the $\boxed{\pi}$ key on your calculator.

$A = 50.265482\ \text{m}^2$ $\boxed{\pi}\ \boxed{\times}\ 16\ \boxed{=}$

$A = 50.27\ \text{m}^2$

14.1 Exercises

In 1–4, use the grid to estimate the number of square centimeters in each region.

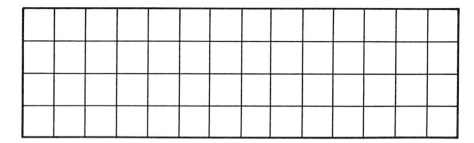

1.    2.

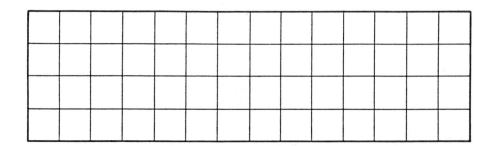

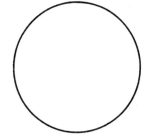

3.

4.

In 5–16, use the appropriate formula to calculate the area of each region. Round answers to the nearer hundredth.

5. Square: $s = 25$ cm

6. Rectangle: $l = 10$ cm
   $w = 2.5$ cm

7. Triangle: $b = 6$ mm
   $h = 3$ mm

8. Circle: $r = 8$ cm

9. Square: $s = 6.2$ mm

10. Rectangle: $l = 6$ mm
    $w = 8$ mm

11. Triangle: $b = 5$ m
    $h = 12$ m

12. Circle: $d = 8$ mm

13. Rectangle: $l = 2.5$ cm
    $w = 2.5$ cm

14. Circle: $r = 5$ m

15. Square: $s = 4$ m

16. Triangle: $b = 8$ cm
    $h = 5$ cm

In 17–24, choose the most appropriate unit of area measure to be used in finding the areas of these objects. Possible choices are: square kilometers, square meters, square centimeters, and square millimeters.

17. A parking lot.

18. This page.

19. A state or province.

20. A fingernail.

21. A floor tile.

22. The laboratory floor.

23. The pupil of an eye.

24. The skin in the palm of the right hand.

UNITS OF AREA MEASURE

The derived unit of metric area measure is the *square meter.* Area measure can also be expressed using other prefixes such as square kilometers, square hectometers, square dekameters, square decimeters, square centimeters, and square millimeters. The square kilometer, square hectometer, square meter, square centimeter, and square millimeter are the most commonly used units. A given metric area measure can be expressed in larger or smaller metric units.

- To express a metric area unit as a smaller metric area unit, multiply by 100, 10 000, 1 000 000, etc.
- To express a metric area unit as a larger metric area unit, multiply by 0.01, 0.000 1, 0.000 001, etc.

Note: Since metric area measure means 10 • 10, all numbers are powers of one hundred such as 100^1, 100^2, 100^3; *and* 100^{-1}, 100^{-2}, 100^{-3}.

This chart summarizes the unit of metric area measure, the symbols, the equivalences, and the relationships between the units.

Unit	Square Kilometer	Square Hectometer	Square Dekameter	Square Meter	Square Decimeter	Square Centimeter	Square Millimeter
Symbol	km^2	hm^2	dam^2	m^2	dm^2	cm^2	mm^2
Equivalence	1 000 000 m^2	10 000 m^2	100 m^2	1 m^2	0.01 m^2	0.000 1 m^2	0.000 001 m^2
Relationship Between Units	1 km^2 is 100 hm^2	1 hm^2 is 100 dam^2	1 dam^2 is 100 m^2	1 m^2 is 100 dm^2	1 dm^2 is 100 cm^2	1 cm^2 is 100 mm^2	

14.2 Exercises

In 1–12, use the chart to find the relationship between the units.

1. 1 km^2 = ___?___ hm^2 2. 1 hm^2 = ___?___ dam^2

3. 1 dam^2 = ___?___ m^2 4. 1 m^2 = ___?___ dm^2

5. 1 dm^2 = ___?___ cm^2 6. 1 cm^2 = ___?___ mm^2

7. 1 hm^2 = ___?___ km^2 8. 1 dam^2 = ___?___ hm^2

9. 1 m^2 = ___?___ dam^2 10. 1 dm^2 = ___?___ m^2

11. 1 cm^2 = ___?___ dm^2 12. 1 mm^2 = ___?___ cm^2

13. To change from dam^2 to dm^2, multiply by___a___ or move the decimal point ___b___ places to the ___c___ .

14. To change from m^2 to hm^2, multiply by___a___ or move the decimal point ___b___ places to the ___c___ .

15. To change from cm^2 to mm^2, multiply by___a___ or move the decimal point ___b___ places to the ___c___ .

16. To change from mm^2 to dm^2, multiply by___a___ or move the decimal point ___b___ places to the ___c___ .

17. To change from m^2 to cm^2, multiply by___a___ or move the decimal point ___b___ places to the ___c___ .

18. To change from dm^2 to m^2, multiply by___a___ or move the decimal point ___b___ places to the ___c___ .

19. To change from dm² to cm², multiply by___a___ or move the decimal point ___b___
    places to the ___c___ .

20. To change from m² to mm², multiply by___a___ or move the decimal point ___b___
    places to the ___c___ .

In 21–26, give the appropriate symbol for each unit.

21. square meter   22. centimeter   23. square decimeter

24. square millimeter   25. hectometer   26. square dekameter

Expressing area measure in larger or smaller units uses the principle of multiplying numbers
by powers of one hundred.

EXAMPLE 1

A square is 3 cm by 3 cm. Find the area in square millimeters.

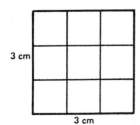 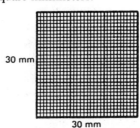

$A = s^2$ $A = s^2$

$A = (3\ \text{cm})\ (3\ \text{cm})$ $A = (30\ \text{mm})\ (30\ \text{mm})$

$A = 9\ \text{cm}^2$ $A = 900\ \text{mm}^2$

$$1\ \text{cm}^2 = 100\ \text{mm}^2$$

$$9 \cdot 100 = 900$$

$$9\ \text{cm}^2 = 900\ \text{mm}^2$$

EXAMPLE 2

A rectangle is 20 mm wide and 50 mm long. Find the area in square centimeters.

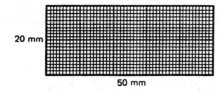 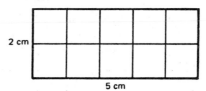

$A = lw$ $A = lw$

$A = (20\ \text{mm})\ (50\ \text{mm})$ $A = (2\ \text{cm})\ (5\ \text{cm})$

$A = 1\ 000\ \text{mm}^2$ $A = 10\ \text{cm}^2$

$$1\ \text{mm}^2 = 0.01\ \text{cm}^2$$

$$1\ 000 \cdot 0.01 = 10$$

$$1\ 000\ \text{mm}^2 = 10\ \text{cm}^2$$

EXAMPLE 3

Express 2.25 hm^2 as square meters.

2.25 hm^2 = ____?____ m^2

1 hm^2 = 10 000 m^2

2.25 • 10 000 = 22 500

2.25 hm^2 = 22 500 m^2

Note: One hm^2 is a larger unit of area than is 1 m^2. Thus, multiply by a positive power of 10. Since there are 10^4 km^2 in 1 m^2 multiply by 10^4 (10 000) or move the decimal point 4 places to the right.

EXAMPLE 4

Express 314 278 mm^2 as square meters.

314 278 mm^2 = ____?____ m^2

1 mm^2 = 0.000 001 m^2

314 278 • 0.000 001 = 0.314 278

314 278 mm^2 = 0.314 278 m^2

Note: One mm^2 is a smaller unit of area than is 1 m^2. Thus, multiply by a negative power of 10. Since 1 mm^2 is $\dfrac{1}{1\ 000\ 000}$ of a m^2 multiply by 10^{-6} (0.000 001) or move the decimal point 6 places to the left.

14.3 Exercises

In 1–30, convert each measure into a larger equivalent metric unit or a smaller equivalent metric unit.

1. 1 km^2 = ____?____ m^2

2. 1 m^2 = ____?____ cm^2

3. 1 cm^2 = ____?____ mm^2

4. 1 m^2 = ____?____ mm^2

5. 2.31 cm^2 = ____?____ mm^2

6. 5 m^2 = ____?____ mm^2

7. 5.25 km = ____?____ km^2

8. 5.92 dam^2 = ____?____ dm^2

9. 8.73 hm^2 = ____?____ m^2

10. 0.318 cm^2 = ____?____ mm^2

11. 1 375 mm^2 = ____?____ cm^2

12. 12 750 cm^2 = ____?____ m^2

13. 3 215 382 m^2 = ____?____ km^2

14. 13 309 mm^2 = ____?____ dm^2

15. 450 dm^2 = ____?____ m^2

16. 2 500 dam^2 = ____?____ km^2

17. 2 250 000 mm^2 = ____?____ m^2

18. 9 272 cm^2 = ____?____ m^2

19. 876.2 mm^2 = ____?____ cm^2

20. 1 075.73 dam^2 = ____?____ km^2

21. 184.73 m^2 = ____?____ km^2

22. 285 719 m^2 = ____?____ km^2

23. 5 314.62 cm^2 = ____?____ m^2

24. 13.48 cm^2 = ____?____ mm^2

25. 1 975 km^2 = ____?____ m^2

26. 3.17 m^2 = ____?____ cm^2

27. 497.2 dam^2 = ____?____ km^2

28. 341 562 mm^2 = ____?____ m^2

29. 5 219 cm^2 = ____?____ mm^2

30. 25 mm^2 = ____?____ cm^2

In 31–35, use the given unit to estimate and then calculate the area of each object. Express the calculated areas in the indicated units.

Object	Estimate (a)	Area (b)	Expressing the Area Using Other Units (c)
31. The classroom floor	___?___ m^2	___?___ m^2	___?___ dam^2
32. This book cover	___?___ cm^2	___?___ cm^2	___?___ mm^2
33. The face of a watch	___?___ mm^2	___?___ mm^2	___?___ cm^2
34. A gum wrapper	___?___ cm^2	___?___ cm^2	___?___ mm^2
35. A fingernail	___?___ mm^2	___?___ mm^2	___?___ cm^2

In 36–38, arrange each group of areas from largest to smallest.

36. 0.029 dam^2; 31 m^2; 3 000 dm^2

37. 3.15 dm^2; 31 800 mm^2; 317 cm^2

38. 53.6 km^2; 540 000 dam^2; 5 350 hm^2

In 39–50, choose the appropriate equivalent measure from Column II and write the letter in the blank under Column I. Answers may be used more than once.

Column I		Column II
39. 408 cm^2	_____	a. 530 dm^2
40. 0.53 km^2	_____	b. 0.53 dam^2
41. 40 800 mm^2	_____	c. 53 000 m^2
42. 5.3 m^2	_____	d. 530 000 dm^2
43. 40.8 hm^2	_____	e. 5.3 dm^2
44. 5 300 dm^2	_____	f. 5 300 m^2
45. 408 m^2	_____	g. 4.08 km^2
46. 53.0 dam^2	_____	h. 0.408 0 m^2
47. 4.08 km^2	_____	i. 408 cm^2
48. 530 cm^2	_____	j. 0.000 408 mm^2
49. 4 080 cm^2	_____	k. 4 080 000 m^2
50. 53 000 mm^2	_____	l. 0.004 08 m^2
		m. none of the above

APPLICATIONS

The ability to calculate area of a room, laboratory, or a piece of filter paper may be important in a hospital or clinical assignment. The design of these objects usually allow easy calculation. Irregularly shaped objects create special problems that will not be covered in this text.

The purchasing agent for a clinic calculates the area of a floor when ordering waxes. Laboratory technicians are using metric area measure when determining the number of microorganisms per square millimeter, centimeter, or meter of surface area. The number of viable and nonviable microorganisms in an area requires the knowledge of area measure.

14.4 Exercises

1. The surface area of a medical library is determined by the medical data statistician. The measurements show the room to be 6.983 meters long, 5.179 meters wide, and 2.438 meters high.

   a. What is the surface area of the floor to the nearer thousandth square meter?

   b. What is the surface area of the four walls to the nearer thousandth square meter?

2. A study reveals that in a public restroom there are approximately 96 human pinworm eggs present per square meter of floor surface area. At least $\frac{1}{8}$ of the eggs are viable. The restroom is 7.6 m × 14.9 m in size.

   a. If 9.7 square meters are deducted for various fixtures in the room, find the area of the floor.

   b. Find the number of viable human pinworm eggs on the floor of this restroom.

3. A glass plate is coated with a thin layer of vaseline. The plate is then placed in an open area outside the clinic for six hours. At the end of this time the plate is returned to the laboratory for microscopic examination. It is discovered that 54 pollen grains are found per square centimeter of surface area. Calculate the surface area for each plate size and determine the number of pollen grains present.

	Total Plate Size	**Area**	**Total Polen Grains**
a.	6 cm × 6 cm		
b.	10 cm × 15 cm		
c.	150 mm × 200 mm		
d.	75 mm × 150 mm		
e.	120 mm diameter		
f.	80 mm diameter		
g.	9 cm diameter		

Unit 15

Metric Volume/Capacity Measure

Objectives

After studying this unit the student should be able to:

• **Determine the volume of solids and fluids.**

• **Express metric volume or capacity measures in smaller or larger metric units.**

MEASURING VOLUMES OF SOLIDS

The volume of a three-dimensional solid is the number of unit cubes that are required to fill the inside of the figure. In the metric system, volume measure for solids is a derived unit. It is found by using the base unit of length measure — the meter. One common unit of volume measure for solids is the cubic centimeter.

EXAMPLE 1

Find the volume of this rectangular solid.

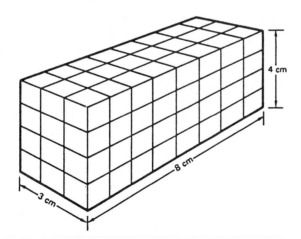

There are 96 cubes in this rectangular solid;
each is a one-centimeter cube.
VOLUME = 96 cubic centimeters.

EXAMPLE 2

This cylindrical solid has 4 π cubes per layer. To find the volume, the number of layers is multiplied by the number of cubes per layer.

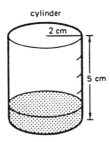

cylinder

2 cm

5 cm

VOLUME = 62.83 cubic centimeters (rounded to hundredths).

Note: Metric volume measures for solids, such as 96 cubic centimeters and 62.832 cubic centimeters, can be written 96 cm^3 and 62.832 cm^3. The exponent, 3, indicates that the measure consists of 3 dimensions.

It may be impractical to determine the number of cubes to find volume. Examining the volume and dimensions more closely leads to the development of a formula for calculating the volume of solids.

FINDING THE VOLUME OF A RECTANGULAR SOLID

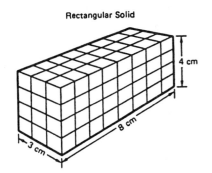

Rectangular Solid

4 cm

8 cm

3 cm

The volume of this rectangular solid is 96 cm^3.
The length is 8 cm; the width is 3 cm; the height is 4 cm.
Since (8 cm) (3 cm) (4 cm) = 96 cm^3, the volume = (8 cm) (3 cm) (4 cm), or volume = length • width • height.

• The formula for the volume of a rectangular solid is:

$V = l \cdot w \cdot h$
  or
$V = lwh$

Note: The symbols for *l*, *w*, and *h* include the unit of measure. In order to find the volume, the unit of measure *must* be the same.

EXAMPLE 3

Find the volume of a rectangular solid 3 m by 6 m by 4 m.

$V = lwh$	Write the formula.
$V = (3\text{ m}) (6\text{ m}) (4\text{ m})$	Substitute the specific values.
$V = (18\text{ m}^2) (4\text{ m})$	Compute.
$V = 72\text{ m}^3$	

FINDING THE VOLUME OF A CYLINDRICAL SOLID

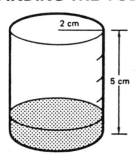

The volume of this cylindrical solid is 62.831853 cm³.
The height is 5 cm; the radius is 2 cm; the base area is 4π.
Since $4\pi = \pi r^2$ and $(4\pi)(5) = 62.831853$ cm³,
the volume $= (4\pi)(5)$ or volume $= \pi$ radius² • height.

• The formula for the volume of a cylindrical solid is:

$$V = \pi r^2 \cdot h$$

or

$$V = \pi r^2 h$$

Note: The symbols r and h include the unit of measure. In order to find the volume, the unit of measure *must* be the same.

EXAMPLE 4

Find the volume of a cylindrical solid with a radius of 3 cm and a height of 6 cm.

$V = \pi r^2 h$	Write the formula.
$V = \pi (3 \text{ cm})^2 (6 \text{ cm})$	Substitute the specific values.
$V = \pi (9 \text{ cm}^2) (6 \text{ cm})$	Compute.
$V = 169.656 \text{ cm}^3$	

Note: When calculating volumes, it is necessary that all the dimensions be expressed using the same metric unit. When the dimensions have different units, express all the dimensions in the same unit, then find the volume.

15.1 Exercises

In 1–10, find the volume of each rectangular or cylindrical solid.

	Length	Width	Height	Volume
1.	2 cm	2 cm	2 cm	
2.	3 cm	2 cm	2 cm	
3.	3 m	2 m	4 m	
4.	10 mm	20 mm	6 mm	
5.	6 dm	5 dm	4 dm	

	Radius	Height	Volume
6.	2 mm	50 mm	
7.	4 m	10 m	
8.	2 cm	10 cm	
9.	3 cm	4 cm	
10.	1 dm	0.5 dm	

11. Two cylindrical solids have the same height. The radius of the first solid is 2 cm and the radius of the second solid is 4 cm. Does the second solid have a volume exactly twice as much as the first solid?

12. Find the volume in cubic meters.

$$V = \underline{\ \ ?\ \ } \ m^3$$

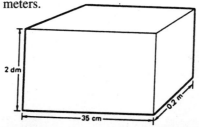

13. Find the volume in cubic decimeters.

    V = __?__ dm³

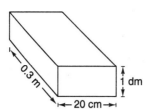

14. Find the volume in cubic centimeters.

    V = __?__ cm³

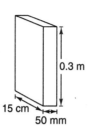

15. Find the volume in cubic centimeters.

    V = __?__ cm³

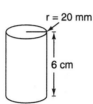

16. Find the volume in cubic millimeters.

    V = __?__ mm³

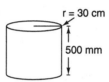

17. Find the volume in cubic decimeters.

    V = __?__ dm³

In 18–25, the dimensions of a rectangular solid are: length, 5 cm; width, 6 cm; height, 2 cm. Adjust the dimensions as stated, find the new volume, and compare it to the original volume.

18. The volume of the solid is __?__ cm³.

19. Double one dimension. The new volume is __?__ times the original volume.

20. Double two dimensions. The new volume is __?__ times the original volume.

21. Double three dimensions. The new volume is __?__ times the original volume.

22. Halve one dimension. The new volume is __?__ times the original volume.

23. Halve two dimensions. The new volume is __?__ times the original volume.

24. Halve all three dimensions. The new volume is __?__ times the original volume.

25. Explain why it is impossible to find the volume of a rectangle.

UNITS OF VOLUME FOR SOLIDS

The derived unit of metric volume measure for solids is the *cubic meter*. Cubic measure for solids can also be expressed cubic kilometers, cubic hectometers, cubic dekameters, cubic decimeters, cubic centimeters and cubic millimeters. The cubic meter, cubic decimeter, and cubic centimeter are the most commonly used units. A given metric volume measure can be expressed in larger or smaller metric units.

- To express a metric volume unit for solids as a smaller metric volume unit, multiply by 1 000, 1 000 000, 1 000 000 000, etc.
- To express a metric volume unit for solids as a larger metric volume unit, multiply by 0.001, 0.000 001, 0.000 000 001, etc.

Note: Since metric volume measurement means $10 \cdot 10 \cdot 10$, all numbers are powers of one thousand such as $1\,000^1$, $1\,000^2$, $1\,000^3$; and $1\,000^{-1}$, $1\,000^{-2}$, $1\,000^{-3}$.

This chart summarizes the units of metric volume measure for solids, the symbols, the equivalences, and the relationships between the units.

Unit	Cubic Kilometer	Cubic Hectometer	Cubic Dekameter	Cubic Meter	Cubic Decimeter	Cubic Centimeter	Cubic Millimeter
Symbol	km^3	hm^3	dam^3	m^3	dm^3	cm^3	mm^3
Equivalence	1 000 000 000 m³	1 000 000 m³	1 000 m³	1 m³	0.001 m³	0.000 001 m³	0.000 000 001 m³

Relationship Between Units	1 km³ is 1 000 hm³	1 hm³ is 1 000 dam³	1 dam³ is 1 000 m³	1 m³ is 1 000 dm³	1 dm³ is 1 000 cm³	1 cm³ is 1 000 mm³

15.2 Exercises

In 1–12, use the chart to find the relationship between the units.

1. $1 \, km^3 =$ ___?___ hm^3

2. $1 \, hm^3 =$ ___?___ dam^3

3. $1 \, dam^3 =$ ___?___ m^3

4. $1 \, m^3 =$ ___?___ dm^3

5. $1 \, dm^3 =$ ___?___ cm^3

6. $1 \, cm^3 =$ ___?___ mm^3

7. $1 \, hm^3 =$ ___?___ km^3

8. $1 \, dam^3 =$ ___?___ hm^3

9. $1 \, m^3 =$ ___?___ dam^3

10. $1 \, dm^3 =$ ___?___ m^3

11. $1 \, cm^3 =$ ___?___ dm^3

12. $1 \, mm^3 =$ ___?___ cm^3

13. To change from m^3 to dm^3, multiply by __a__ or move the decimal point __b__ places to the __c__ .

14. To change from cm^3 to dm^3, multiply by __a__ or move the decimal point __b__ places to the __c__ .

15. To change from dm^3 to m^3, multiply by __a__ or move the decimal point __b__ places to the __c__ .

16. To change from hm^3 to dam^3, multiply by __a__ or move the decimal point __b__ places to __c__ .

17. To change from m^3 to hm^3, multiply by __a__ or move the decimal point __b__ places to the __c__ .

18. To change from mm^3 to cm^3, multiply by __a__ or move the decimal point __b__ places to the __c__ .

19. To change from dm³ to cm³, multiply by ___a___ or move the decimal point ___b___ places to the ___c___ .

20. To change from cm³ to mm³, multiply by ___a___ or move the decimal point ___b___ places to the ___c___ .

Expressing volume measurements in larger or smaller units uses the principle of multiplying numbers by powers of one thousand.

EXAMPLE 1

The volume of a rectangular solid is 30 cm³. Find the volume in cubic centimeters.

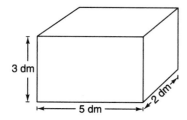 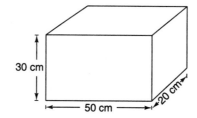

$V = lwh$

$V = 5 \text{ dm} (2 \text{ dm}) (3 \text{ dm})$

$V = 30 \text{ dm}^3$

$V = lwh$

$V = 50 \text{ cm} (20 \text{ cm}) (30 \text{ cm})$

$V = 30\,000 \text{ cm}^3$

$$30 \text{ dm}^3 = \underline{\ ?\ } \text{ cm}^3$$

$$1 \text{ dm}^3 = 1\,000 \text{ cm}^3$$

$$30 \cdot 1\,000 = 30\,000$$

$$30 \text{ dm}^3 = 30\,000 \text{ cm}^3$$

EXAMPLE 2

The volume of a rectangular solid is 72 000 mm³. Find the volume in cubic centimeters.

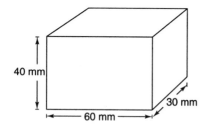 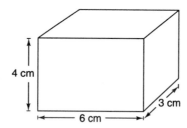

$V = lwh$

$V = 60 \text{ mm} (30 \text{ mm}) (40 \text{ mm})$

$V = 72\,000 \text{ mm}^3$

$V = lwh$

$V = 6 \text{ cm} (3 \text{ cm}) (4 \text{ cm})$

$V = 72 \text{ cm}^3$

$$72\,000 \text{ mm}^3 = \underline{\ ?\ } \text{ cm}^3$$

$$1 \text{ mm}^3 = 0.001 \text{ cm}^3$$

$$72\,000 \cdot 0.001 = 72$$

$$72\,000 \text{ mm}^3 = 72 \text{ cm}^3$$

15.3 Exercises

In 1–20, express each measure in a larger equivalent metric unit or a smaller equivalent metric unit.

1. 108 dm^3 = ___?___ m^3 2. 286 cm^3 = ___?___ mm^3

3. 2.785 km^3 = ___?___ m^3 4. 3.4 m^3 = ___?___ km^3

5. 0.001 m^3 = ___?___ dm^3 6. 9 km^3 = ___?___ cm^3

7. 7 dm^3 = ___?___ cm^3 8. 858 000 mm^3 = ___?___ cm^3

9. 650 000 mm^3 = ___?___ m^3 10. 380 m^3 = ___?___ mm^3

11. 518 mm^3 = ___?___ cm^3 12. 0.04 dm^3 = ___?___ m^3

13. 12.3 cm^3 = ___?___ mm^3 14. 730 mm^3 = ___?___ cm^3

15. 10 dam^3 = ___?___ km^3 16. 280 000 cm^3 = ___?___ m^3

17. 0.25 m^3 = ___?___ cm^3 18. 132 m^3 = ___?___ km^3

19. 3.58 m^3 = ___?___ dm^3 20. 23 cm^3 = ___?___ dm^3

In 21–30, choose the appropriate equivalent measure from Column II and write the letter in the blank under Column I, Column II measures may be used more than once.

Column I	Column II
21. 7 132 cm^3 _____	a. 504.9 dm^3
22. 5.049 m^3 _____	b. 50.4 hm^3
23. 713.2 m^3 _____	c. 0.504 9 hm^3
24. 504.9 dam^3 _____	d. 50 490 000 mm^3
25. 713 200 mm^3 _____	e. 50 490 000 cm^3
26. 50.49 m^3 _____	f. 71.32 dm^3
27. 71.32 dam^3 _____	g. 0.713 2 dam^3
28. 50 490 cm^3 _____	h. 0.007 132 km^3
29. 7.132 hm^3 _____	i. 7.132 dm^3
30. 0.050 4 km^3 _____	j. 71 320 m^3
	k. none of the above

UNITS OF VOLUME FOR FLUIDS

In the metric system, the volume of a figure that is not a solid is measured by using a derived unit. This derived unit is the *liter*. The derived unit is used to measure the volume of liquids or gases such as the volume that a test tube will hold or the amount of solution in a container. Volumes of fluids can also be expressed using other prefixes, such as kiloliters, hectoliters, dekaliters, deciliters, centiliters, and milliliters. The liter, milliliter, and the kiloliter are the most commonly used units. Basic fact: 1 mL = 1 cm^3 or 1 cc.

A given metric volume measure for fluids can be expressed in larger or smaller metric units.

- To express a metric volume unit for fluids as a smaller metric volume unit, multiply by a positive power of ten such as 10, 100, 1 000, 10 000, or 100 000.
- To express a metric volume unit for fluids as a larger metric volume unit, multiply by a negative power of ten such as 0.1, 0.01, 0.001, 0.000 1, or 0.000 01.

This chart summarizes the units of metric volume measure for fluids, the symbols, the equivalences, and the relationships between the units.

Unit	Kiloliter	Hectoliter	Dekaliter	Liter	Deciliter	Centiliter	Milliliter
Symbol	kL	hL	daL	L	dL	cL	mL
Equivalence	1 000 liters	100 liters	10 liters	1 liter	0.1 liter	0.01 liter	0.001 liter
Relationship Between Units	1 kL is 10 hL	1 hL is 10 daL	1 daL is 10 L	1 L is 10 dL	1 dL is 10 cL	1 cL is 10 mL	

15.4 Exercises

In 1–12, use the chart to find the relationship between the units.

1. 1 kL = __?__ hL

2. 1 hL = __?__ daL

3. 1 daL = __?__ L

4. 1 L = __?__ dL

5. 1 dL = __?__ cL

6. 1 cL = __?__ mL

7. 1 hL = __?__ kL

8. 1 daL = __?__ hL

9. 1 L = __?__ daL

10. 1 dL = __?__ L

11. 1 cL = __?__ dL

12. 1 mL = __?__ cL

13. To change from L to hL, multiply by __a__ or move the decimal point __b__ places to the __c__ .

14. To change from dL to kL, multiply by __a__ or move the decimal point __b__ places to the __c__ .

15. To change from mL to L, multiply by __a__ or move the decimal point __b__ places to the __c__ .

16. To change from cL to mL, multiply by __a__ or move the decimal point __b__ places to the __c__ .

17. To change from daL to cL, multiply by __a__ or move the decimal point __b__ places to the __c__ .

18. To change from kL to daL, multiply by __a__ or move the decimal point __b__ places to the __c__ .

19. To change from hL to dL, multiply by __a__ or move the decimal point __b__ places to the __c__ .

20. To change from mL to dL, multiply by __a__ or move the decimal point __b__ places to the __c__ .

In 21–26, give the unit each symbol represents.

21. mL

22. hL

23. kL

24. dL

25. cL

26. daL

Expressing volume measures for fluids in larger or smaller units uses the principle of multiplying numbers by powers of ten.

EXAMPLE 1

Express 0.328 L as milliliters.

0.328 L = __?__ mL

1 L = 1 000 mL

0.328 • 1 000 = 328

0.328 L = 328 mL

Note: One L is a larger unit of volume than is 1 mL. Thus, multiply by a positive power of 10. Since there are 10^3 mL in 1 L multiply by 10^3 (1 000) or move the decimal point 3 places to the right.

EXAMPLE 2

Express 2 200 cL as kL.

2 200 cL = __?__ kL

1 cL = 0.000 01 kL

2 200 • 0.000 01 = 0.022 00

2 200 cL = 0.022 00 kL

Note: One cL is a smaller unit of volume than is 1 kL. Thus, multiply by a negative power of 10. Since 1 cL is $\dfrac{1}{100\ 000}$ of 1 kL multiply by 10^{-5} (0.000 01) or move the decimal point 5 places to the left.

15.5 Exercises

In 1–10, express each measure using a larger equivalent metric unit or a smaller equivalent metric unit.

1. 2 500 mL = __?__ L

2. 1 530 mL = __?__ L

3. 8 000 mL = __?__ L

4. 3.2 L = __?__ mL

5. 0.5 L = __?__ mL

6. 0.75 L = __?__ mL

7. 250 mL = __?__ L

8. 5 L = __?__ mL

9. 1 L = __?__ mL

10. 10 mL = __?__ L

In 11–15, common objects are listed in Column I. Choose an appropriate measure from Column II and write the letter in the blank. *Remember:* 1 mL = 1 cm^3.

Column I		Column II
11. A 3-cm^3 syringe	a.	32 000 L _____
12. A swimming pool	b.	82 L _____
13. A small water glass	c.	200 mL _____
14. An automobile gas tank	d.	8 mL _____
15. A tablespoon	e.	3 mL _____

In 16–29, choose the most appropriate measure of volume for fluids for each container.

16. Graduated Flask
    40 mL; 1 000 mL; 20 L

17. Test Tube
    50 mL; 15 L; 15 mL

18. Plastic Storage Jug
    75 L; 7.5 L; 2L

19. Metal Storage Can
    5 L; 5 mL; 50 mL

20. Ungraduated Pipette
    400 mL; 4 mL; 4 L

21. Graduated Pipette
    500 mL; 15 L; 15 mL

22. Cup
    30 L; 300 mL; 3 L

23. Liquid Medicine Dispenser
    675 mL; 6.75 L; 6.75 mL

24. Erlenmeyer Flask
    500 mL; 500 L; 5.00 L

25. Syringe
    10 mL; 10 L; 1.0 L

26. Bowl

2.00 L; 2.00 mL; 200 mL

27. Spoon

5 mL; 100 mL; 10 L

28. Graduated Cylinder

250 mL; 2.5 L; 5.00 mL

29. Saline Solution Bag

50 mL; 500 mL; 5.0 L

MEASURING VOLUMES OF FLUIDS

The liter and the cubic meter are both derived units for volume measure. Volume measure for fluids cannot be found directly. In order to find the volume measure for fluids, the volume is calculated using units of volume measure for solids. Using equivalences, the volume for fluids can then be found.

- The volume of a container which has a measurement of 1 000 cm^3 is 1 L. It is written:
  1 000 cm^3 = 1 L.

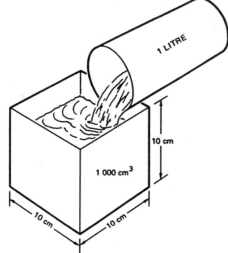

- The volume or a container which has a measurement of 1 cm^3 is 0.001 L. It is written:
  1 cm^3 = 0.001 L

      or

  1 cm^3 = 1 mL.

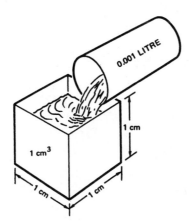

EXAMPLE 1

Find the volume, in liters, of this rectangular container.

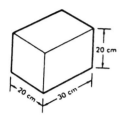

$V = lwh$ Find the volume in cubic centimeters.

$V = 20 \text{ cm} (30 \text{ cm}) (20 \text{ cm})$

$V = 12\,000 \text{ cm}^3$

$12\,000 \text{ cm}^3 = \underline{\ \ ?\ \ } \text{ L}$ Express the volume in liters.

$1 \text{ cm}^3 = 0.001 \text{ L}$

$12\,000 \cdot 0.001 = 12$

$12\,000 \text{ cm}^3 = 12 \text{ L}$

EXAMPLE 2

Find the volume, in milliliters, of the rectangular container.

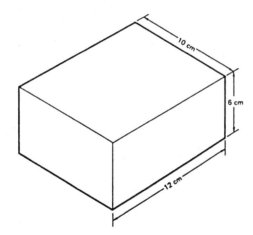

$V = lwh$ Find the volume.

$V = 10 \text{ cm} (12 \text{ cm}) (6 \text{ cm})$

$V = 720 \text{ cm}^3 = \underline{\ \ ?\ \ } \text{ mL}$ Express the volume in milliliters.

$1 \text{ cm}^3 = 1 \text{ mL}$

$720 \cdot 1 = 720$

$720 \text{ cm}^3 = 720 \text{ mL}$

EXAMPLE 3

For each container:
a. Find the volume in liters.
b. Express the volume in milliliters.

$r = 3.5$ cm
$h = 13.0$ cm

DEXTROSE

$r = 25$ cm
$h = 75$ cm

$V = \pi r^2 h$

$V = \pi r^2 h$

$V = \pi r^2 h$
$V = (3.1415927)\,(3.5)^2\,(13)$
$V = 500.30$ cm^3 = _____ mL
1 cm$^3 = 1$ mL
$500.30\,(1) = 500.30$
500.30 cm$^3 = 500.30$ mL

Find the volume.

Express in milliliters.

$V = \pi r^2 h$
$V = (3.1415927)\,(25)^2\,(75)$
$V = 147262.16$ cm^3 = _____ mL
1 cm$^3 = 1$ mL
$147262.16\,(1) = 147262.16$
147262.16 cm$^3 = 147262.16$ mL

15.6 Exercises

In 1–10, for each container:
a. Find the volume in milliliters.
b. Express the volume in liters.

1.

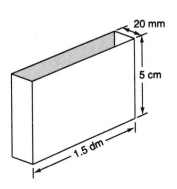

12 cm
10 cm
8 cm

2.

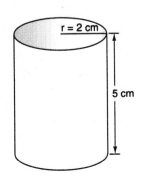

2 dm
2 dm
2 dm

3.

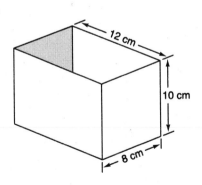

20 mm
5 cm
1.5 dm

4.

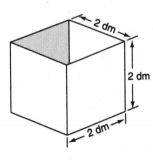

r = 2 cm
5 cm

5.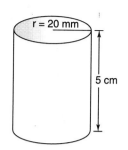

6.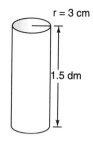

7.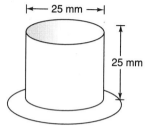

8.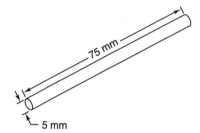

9.

10.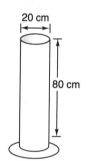

In 11–15, arrange each group of measurements from largest to smallest.

11. 21 L; 0.2 hL; 19 000 mL; 2 000 cm³; 25 000 cm³

12. 628 mL; 0.5 L; 0.3 cL; 1 cm³; 24 m³

13. 250 daL; 50 kL; 49 000 L; 250 dam³; 50 cm³

14. 9 L; 12 000 mL; $\frac{1}{10}$ hL; $\frac{1}{100}$ mm³; 90 hm³

15. 1 hL; 200 L; 1 500 mL; 150 cm³; 1 m³

APPLICATIONS

The air in a room and liquid in a container both involve volume measure. These volumes can be calculated. The heating engineer makes calculations to determine the rate at which air is replaced in any given area of the hospital. The laboratory technician must calculate volume when determining the amount of solution required to fill petri dishes. The skill of volume calculation is widely applied in many health areas. Mastering this skill is vital to an effective worker.

15.7 Exercises

1. During a specific period of time, the liquid intake and output for a patient are recorded. The intake is 0.250 L, 0.100 L, 0.200 L, 0.350 L, and 0.075 L. The output for the same period of time is 0.465 L.

   a. What is the total intake for this period of time?

   b. How does the output compare with the intake (greater than, equal to, less than)?

2. A graduated cylinder is used to measure the amounts of the same solution to be placed in a beaker and a flask. Determine the total amount of solution needed to fill each beaker and flask with the indicated amounts.

Beaker	Flask	TOTAL
a. 0.750 L	125 mL	__?__
b. 0.010 L	25 mL	__?__
c. 0.005 L	10 mL	__?__

3. Liquid medication is administered to a patient every four hours.

   a. If the dose is 15 mL, how many milliliters are administered in 24 hours?

   b. If the dose is 5 mL, how many milliliters are administered in 16 hours?

4. A 50 mL graduated pipette is used to measure 46 samples, each containing 30 mL.

   a. What is the total volume measured?

   b. How many samples can be taken from a flask containing 750 mL?

5. A patient's room measures 5.750 m by 6.125 m by 3.000 m. A circulating fan can exchange 12.750 m^3 of air each hour. How much time is required to exchange all the air in the room?

6. In one section of the hospital, 127 L of dextrose solution are used each day of the week.

   a. How many liters of solution are used in one week?

   b. How many liters of solution are used during the month of June?

Unit 16

Metric Mass and Temperature Measures

Objectives

After studying this unit the student should be able to:
- **Express metric mass in smaller or larger metric units.**
- **Determine the relationships between volume, capacity, and mass.**
- **Express Fahrenheit temperature readings as Celsius temperature readings.**
- **Express Celsius temperature readings as Fahrenheit temperature readings.**

UNITS OF MASS

In order to move an object, it is necessary to push, pull, or lift it. When moving the object, the mass or quantity of the object is actually being moved. The weight, or earth's gravitational pull, is how much the object pushes down on a scale. Two objects containing the same mass have equal weights when weighed at the same place. The kilogram is the base unit of mass in the metric system. Mass measure may also be expressed using other prefixes, such as hectograms, dekagrams, grams, decigrams, centigrams, and milligrams. The kilogram, gram, and milligram are the most commonly used units. A given metric mass unit can be expressed in larger or smaller metric units.

- To express a metric mass unit as a smaller metric mass unit, multiply by a positive power of ten such as 10, 100, 1 000.

- To express a metric mass unit as a larger metric mass unit, multiply by a negative power of ten such as 0.1, 0.01, 0.001.

This chart summarizes the units of metric mass measure, the symbols, the equivalences, and the relationships between the units.

Unit	Kilogram	Hectogram	Dekagram	Gram	Decigram	Centigram	Milligram
Symbol	kg	hg	dag	g	dg	cg	mg
Equivalence	1 000 grams	100 grams	10 grams	1 gram	0.1 gram	0.01 gram	0.001 gram
Relationship Between Units	1 kg is 10 hg	1 hg is 10 dag	1 dag is 10 g	1 g is 10 dg	1 dg is 10 cg	1 cg is 10 mg	

16.1 Exercises

In 1–12, use the chart to find the relationship between the units.

1. 1 kg = ____?____ hg

2. 1 hg = ____?____ dag

3. 1 dag = ____?____ g

4. 1 g = ____?____ dg

5. 1 dg = ____?____ cg

6. 1 cg = ____?____ mg

7. 1 hg = ____?____ kg

8. 1 dag = ____?____ hg

9. 1 g = ____?____ dag

10. 1 dg = ____?____ g

11. 1 cg = ____?____ dg

12. 1 mg = ____?____ cg

13. To change from hg to g, multiply by____a____ or move the decimal point ____b____ places to the ___c___ .

14. To change from cg to dg, multiply by____a____ or move the decimal point ____b____ places to the ___c___ .

15. To change from kg to g, multiply by____a____ or move the decimal point ____b____ places to the ___c___ .

16. To change from mg to cg, multiply by____a____ or move the decimal point ____b____ places to the ___c___ .

17. To change from dag to dg, multiply by____a____ or move the decimal point ____b____ places to the ___c___ .

18. To change from g to mg, multiply by____a____ or move the decimal point ____b____ places to the ___c___ .

19. To change from dg to hg, multiply by____a____ or move the decimal point ____b____ places to the   c   .

20. To change from g to kg, multiply by____a____ or move the decimal point ____b____ places to the ___c___ .

In 21–26, give the appropriate abbreviation for each of the following:

21. 100 grams

22. 0.001 gram

23. 0.1 gram

24. 10 grams

25. 1 000 grams

26. 0.01 gram

Expressing metric mass measure in larger or smaller units uses the principle of multiplying numbers by powers of ten.

EXAMPLE 1

Tim has a mass of 86 kg. How many grams is this?

86 kg = ___?___ g

1 kg = 1 000 g

86 • 1 000 = 86 000

86 kg = 86 000 grams

Tim's mass is 86 000 grams.

Note: One kg is a larger unit of mass than is 1 g. Thus, multiply by a positive power of 10. Since there are 10^3 g in 1 kg, multiply by 10^3 (1 000) or move the decimal point 3 places to the right.

EXAMPLE 2

Each day Tony takes a vitamin pill containing 18 milligrams of iron. How many grams of iron does he take in 1 year (365 days)?

18 mg = ___?___ g

1 mg = 0.001 g

18 • 0.001 = 0.018

18 mg = 0.018 g

0.018 g • 365 = 6.570 g

Tony consumes 6.57 grams of iron in 1 year.

Note: One mg is a smaller unit of mass than is 1 g. Thus, multiply by a negative power of 10. Since 1 mg is $\dfrac{1}{1000}$ of a gram, multiply by 10^{-3} (0.001) or move the decimal point 3 places to the left.

16.2 Exercises

In 1–18, convert each measure into a larger equivalent metric unit or a smaller equivalent metric unit.

1. 4.5 kg = ___?___ g

2. 0.32 g = ___?___ mg

3. 652 g = ___?___ kg

4. 10 000 mg = ___?___ kg

5. 0.2 kg = ___?___ g

6. 3 250 mg = ___?___ g

7. 2 kg = ___?___ mg

8. 0.03 g = ___?___ mg

9. 35 mg = ___?___ g

10. 89.5 g = 0.895 ___?___

11. 375 g = 0.375 ___?___

12. 4 750 mg = 4.75 ___?___

13. 0.02 kg = 20 ___?___

14. 0.005 g = 0.000 005 ___?___

15. 3 000 mg = 3 ___?___

16. 45 g = 0.045 ___?___

17. 4 g = 0.004 ___?___

18. 0.002 5 kg = 2.5 ___?___

In 19–24, choose the most appropriate unit of mass to be used in finding the mass of these objects. Possible choices are kilogram, gram, and milligram.

19. A teenager

20. A drop of blood

21. A dime

22. An eyelash

23. A ballpoint pen

24. A stick of gum

In 25–30, choose the most appropriate mass for each object.

25. Shoe

1 kg   1 mg   1 g

26. Glass of Orange Juice

480 mg   480 kg   480 g

27. Nickel

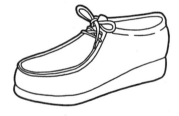

5 mg   5 g   5 kg

28. Syringe

275 mg   275 g   275 kg

29. Drinking Straw

30. Television

57 g   57 mg   57 kg

47 g   47 kg   47 mg

In 31–34, use the given unit to estimate and then find the mass of each object. Express the actual mass in the indicated units.

Object	Estimate (a)	Mass (b)	Expressing The Mass Using Other Units (c)
31. A wallet	___?___ g	___?___ g	___?___ kg
32. A shoe	___?___ kg	___?___ kg	___?___ g
33. The mass of a student	___?___ kg	___?___ kg	___?___ g
34. The change in a wallet or a purse	___?___ g	___?___ g	___?___ mg

RELATIONSHIP BETWEEN METRIC UNITS

In the metric system, a special relationship exists between the measures of volume and mass for water at 4 degrees Celsius.

At 4°C, 1 mL of water has a mass of 1 g. This means that 1 000 mL has a mass of 1 000 g.

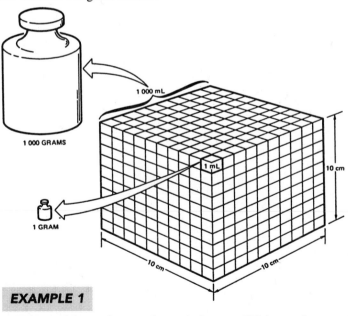

EXAMPLE 1

This container of water is at 4 degrees Celsius and standard pressure. Find the volume and mass of the water.

Volume = 1 dm³   *or*   1 000 cm³

Volume = 1 L   *or*   1 000 mL

Mass = 1 000 g   *or*   1 kg

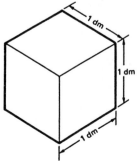

16.3 Exercises

In 1–18, express each measure as an equivalent measure.

1. $5 \text{ dm}^3 = $ ___?___ cm^3

2. $100 \text{ mm} = $ ___?___ cm

3. $1\,500 \text{ mL} = $ ___?___ L

4. $0.20 \text{ dm}^2 = $ ___?___ m^2

5. $625 \text{ g} = $ ___?___ kg

6. $0.175 \text{ m}^3 = $ ___?___ cm^3

7. $400 \text{ m} = $ ___?___ dm

8. $10 \text{ L} = $ ___?___ daL

9. $0.36 \text{ dam}^2 = $ ___?___ m^2

10. $0.4 \text{ g} = $ ___?___ mg

11. $38 \text{ dm}^3 = $ ___?___ m^3

12. $425 \text{ m}^3 = $ ___?___ dam^3

13. $1\,500 \text{ m} = $ ___?___ km

14. $0.5 \text{ L} = $ ___?___ mL

15. $375 \text{ mg} = $ ___?___ g

16. $1 \text{ dam}^2 = $ ___?___ m^2

17. $2 \text{ km} = $ ___?___ m

18. $32 \text{ L} = $ ___?___ mL

In 19–34, complete these relationships for water at 4 degrees Celsius and standard pressure.

	Volumes		**Mass**
19.	12 cm^3	12 mL	___?___ g
20.	0.5 cm^3	___?___ mL	0.5 g
21.	___?___ dm^3	7 L	7 kg
22.	___?___ cm^3	20 mL	___?___ g
23.	___?___ dm^3	___?___ L	32 kg
24.	37 cm^3	___?___ mL	___?___ g
25.	___?___ dm^3	0.5 L	___?___ kg
26.	___?___ cm^3	___?___ mL	454 g
27.	10 cm^3	10 mL	10 ___?___
28.	15 dm^3	15 ___?___	15 kg
29.	0.75 ___?___	0.75 L	0.75 kg
30.	18 ___?___	18 mL	18 ___?___
31.	25 ___?___	25 ___?___	25 g
32.	0.5 cm^3	0.5 ___?___	0.5 ___?___
33.	17.5 dm^3	17.5 ___?___	17.5 ___?___
34.	82 ___?___	82 ___?___	82 kg

In 35–40, find the values which measure the same amount of water at 4 degrees Celsius.

35. 5L   5 g   5 kg

36. 5L 5 dm^3 5 cm^3

37. 30 cm^3 30 L   30 mL

38. 30 cm^3 30 kg   30 g

| 39. | 17 mL | 17 g | 17 kg |
| 40. | 17 mL | 17 dm^3 | 17 cm^3 |

MEASURING METRIC TEMPERATURE

Temperature can be measured in degrees Celsius or in degrees Fahrenheit. On the Celsius scale, water boils at 100 °C and freezes at 0 °C.

On the Fahrenheit scale water boils at 212 °F and freezes at 32 °F.

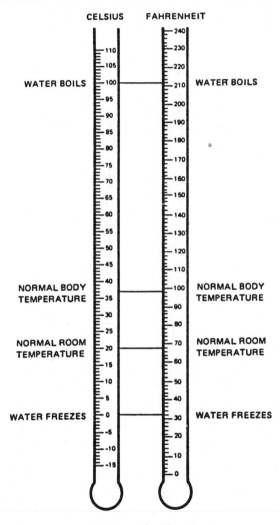

- To express degrees Fahrenheit as degrees Celsius use the relationship:

$$°C = \frac{5}{9}(°F - 32)$$

- To express degrees Celsius as degrees Fahrenheit use the relationship:

$$°F = \frac{9}{5}°C + 32$$

EXAMPLE 1

Express 68° F as degrees Celsius.

$$°C = \frac{5}{9}(°F - 32)$$ Write the formula.

$$°C = \frac{5}{9}(68 - 32)$$ Substitute the specific values.

$$°C = \frac{5}{\cancel{9}_1}(\cancel{36}^4)$$ Calculator:

$$°C = 20$$

$$68\ °F = 20\ °C$$

Calculator:
$5 \boxed{\div} 9 \boxed{\times} \boxed{(}\ 68 \boxed{-} 32 \boxed{)} \boxed{=}$

EXAMPLE 2

Express 35 °C as degrees Fahrenheit.

$$°F = \frac{9}{5}°C + 32$$ Write the formula.

$$°F = \frac{9}{\cancel{5}_1}(\cancel{35}^4) + 32$$ Substitute the specific values.

$$°F = \frac{9}{5}(35) + 32$$ Calculator:

$$°F = 63 + 32$$

$$°F = 95°$$

$$35\ °C = 95\ °F$$

Calculator:
$9 \boxed{\div} 5 \boxed{\times} 35 \boxed{+} 32 \boxed{=}$

16.4 Exercises

In 1–8, express each Celsius temperature as an equivalent Fahrenheit temperature and each Fahrenheit temperature as an equivalent Celsius temperature.

1. 113 °F
2. 50 °F
3. 14 °F
4. −13 °F

5. −10 °C
6. 25 °C
7. 70 °C
8. 110 °C

In 9–14, choose the more appropriate Celsius temperature for each activity.

9.	Having a fever	20 °C	38.5 °C
10.	Air on a summer day	26 °C	45 °C
11.	Sunbathing	10 °C	28 °C
12.	Raining	0 °C	3 °C
13.	Building a snowman	−5 °C	10 °C
14.	Mowing the grass	27 °C	8 °C

APPLICATIONS

All health care workers are involved in measuring mass. The health care assistant, the LPN and the RN may be involved in determining the mass of patients. The pharmacist is involved with measuring medication. Laboratory personnel measure the mass of substances to be used in cultures or in laboratory tests. The dietitian measures the mass of food that is given to each patient. Different health care workers use different types of instruments to measure mass. The instrument that is used must be accurate and the health care worker must read the measurements correctly and record the measurements promptly and correctly.

Health care workers are also involved in measuring temperature. Temperature is measured in units termed degrees Celsius (°C). The freezing temperature of pure water at 760 mm pressure is 0 °C and the boiling temperature is 100 °C. The temperature of the human body serves as an indicator of a person's well-being. Normal body temperature is 36.99 °C or 37 °C. The Celsius scale is and has been the common unit of measure in laboratories for many years.

16.5 Exercises

1. To determine the total daily medication that each patient receives, daily records are kept. This record shows the medication administered to three patients.

Mrs. Larson	Mr. Davonport	Miss Juliet
2.5 g	1.6 g	9.6 g
1.6 g	1.6 g	0.250 g
0.4 g	1.6 g	7.5 g
0.25 g	0.6 g	0.3 g

   a. How much medication does Mrs. Larson receive?

   b. How much medication does Mr. Davonport receive?

   c. How much medication does Miss Juliet receive?

2. The mass loss of a patient is recorded over a five-day period. The initial mass of the patient is 68.500 kg. Determine the patient's new mass for each day and the total mass loss.

	Day	Mass Loss	New Mass
a.	1	250 g	?
b.	2	750 g	?
c.	3	80 g	?
d.	4	500 g	?
e.	5	250 g	?
f.		TOTAL ?	

3. Mrs. Judson receives 250 grams of meat for each meal. Miss Keebler receives 0.56 times as much meat as Mrs. Judson does. How much meat does Miss Keebler receive?

4. To produce 1 000 milliliters of a liquid medium suitable for culturing bacteria, 21 grams of a dehydrated culture medium is needed.

   a. How many liters can be prepared from 975 grams of dehydrated culture medium? Round the answer to the nearer thousandth liter.

   b. How many grams are required to prepare 5.750 L of solution?

5. An analytical balance is used to obtain these measurements: 1.250 grams, 16 milligrams, 46 milligrams, and 0.750 grams.

    a. What is the total mass in grams?

    b. What is the total mass in milligrams?

6. A triple beam balance is used to measure amounts of bacto-dextrose agar used for the culture of bacteria. The amounts measured are: 16.00 g, 4.25 g, 12.75 g, and 22.12 g. A preparation of 10.5 L requires 453 g of bacto-dextrose agar.

    a. What is the total mass of the bacto-dextrose agar?

    b. How many grams of bacto-dextrose agar are required per liter of preparation? Round the answer to the nearer tenth gram.

    c. How many liters can be prepared from the measured amounts of bacto-dextrose agar? Round the answer to the nearer thousandth liter.

    d. How many liters can be prepared from 1 000 grams of bacto-dextrose agar? Round the answer to the nearer thousandth liter.

7. A solution is formed when a granular material is placed in water and heated to 100 °C. A solution condition continues to exist until the temperature drops to 45 °C. A laboratory has only a Fahrenheit thermometer.

    a. Find the Fahrenheit value of 100 °C.

    b. Find the Fahrenheit value of 45 °C.

8. A normal temperature for a human is considered to be 98.7 °F. A fever condition exists for a body temperature that is over 98.6 °F.

    a. Determine the degree Celsius value for normal body temperature.

    b. Determine the degree Celsius value for a fever condition of 102.5 °F. Round the answer to the nearer tenth degree.

    c. Determine the degree Celsius value for a fever condition of 100 °F. Round the answer to the nearer tenth degree.

9. The temperatures of several rooms are recorded with a Fahrenheit thermometer. What are the degree Celsius readings for each temperature measure? Round the answers to the nearer tenth degree.

    a. 75 °F

    b. 78 °F

    c. 81 °F

    d. 86 °F

Unit 17

Section Three
Applications to Health Work

Objectives

After studying this unit the student should be able to:
- **Use the various units of metric measure to solve health work problems.**

Measurement is an important component in all health work. Health care workers are concerned with how much patients eat, the amount of medication they take, the weight they gain or lose, the amount of liquid gained or lost, and the temperature at which the body is functioning. Some form of measurement is taken for each patient every day. The units of measure most commonly used are metric.

- The height of a patient or length of gauze is measured using the meter as the base unit. This unit of measure is convenient for measuring large units such as height, room size, or parking lot dimensions. Centimeters and millimeters are commonly used to measure smaller objects that are still detectable by the human eye. The size of this page, the length of a pin, and the width of a pen may be expressed in centimeters or millimeters. The micrometer is used to measure small objects such a fungi spores and bacterial cells.

- The mass of a patient or the amount of a dry substance is measured using the kilogram as the base unit. For large masses, such as the human body, the kilogram is the most practical unit of measure. The gram is the most common unit of measure in weighing out dry substances in the laboratory. Very small units are measured in milligrams.

- The volume of water taken in by a patient or the volume of water given off may be measured in either liters or milliliters. Large volumes are most commonly measured in liters. The milliliter is the common unit of measure for liquids in the laboratory even though the liter is the base unit.

- The temperature of a patient above or below an accepted norm is a concern of the health care staff. The Celsius scale is the unit of measure for determining this variation. This unit of measure is used in the laboratory as well as in the patient's room.

17.1 Exercises

1. A patient is weighed once a month over a period of five months. Find the change in the patient's mass between each weighing.

	Month	Mass	Change
	Beginning Mass	75.125 kg	— — —
a.	1	75.052 kg	
b.	2	74.983 kg	
c.	3	75.005 kg	
d.	4	77.015 kg	
e.	5	79.250 kg	

2. Several samples are weighed and the average mass is determined. Find the average mass of these samples. Average equals total divided by the number of samples. Round the answer to the nearer thousandth gram.

Sample	Mass
A	0.125 g
B	1.750 g
C	2.015 g
D	6.835 g
E	4.321 g
F	3.895 g
G	7.396 g
H	3.425 g

3. A roll of gauze is 35 meters in length. How many 50-centimeter lengths can be obtained from the roll?

4. A piece of circular filter paper has a diameter of 16 centimeters. Calculate the surface area in square units for one side of the paper.

5. The liquid intake measured for a patient during a specified period of time is 1.275 liters. During the same period the liquid given off is 0.498 liter. Find the difference between these two measures.

6. A storage tank of distilled water has a capacity of 2 000 liters.

   a. If the storage tank contains 1 256.025 liters, how many liters are required to fill the tank?

   b. If 37.850 liters are used from the tank each day, how many liters are left in the tank after seven days of use? The tank is full at the start and not filled during the seven-day period.

7. Agar-agar, a solidifying agent used in bacteriological and mycological study, melts when heated in water to 100 °C. When allowed to cool it solidifies at about 45 °C. What is the temperature difference between melting and solidification?

8. The temperature of a bacteriological culture is measured each hour for six hours. What is the average temperature during the six-hour period?

   1st hour: 23.8 °C   4th hour: 25 °C
   2nd hour: 24.0 °C   5th hour: 25 °C
   3rd hour: 23.7 °C   6th hour: 25.8 °C

9. The diameter of a capillary tube is 0.025 mm. Calculate the area of a circle this size.

10. A Petri dish has a diameter of 9 centimeters. Find the surface area on the floor of the Petri dish.

11. A nutrient agar is inoculated with a pathogenic microorganism. At the end of 24 hours it is discovered that one colony per 9 square millimeters is present. How many colonies are present in a Petri dish, which is –

   a. 9 centimeters in diameter?

   b. 120 millimeters in diameter?

   c. 10 centimeters in diameter?

12. A medication is provided in container quantities of 50 mL. How many times can a 15-mL syringe be filled from this single container?

13. A one-liter graduated cylinder contains 942 mL of distilled water. A total of 127 mL is removed. What is the amount remaining in the cylinder?

14. A hemacytometer is used to measure the number of cells in 0.1 mL of solution. The population is determined to be 15 cells. How many cells can be predicted in these volumes of the same solution?

	Volume	Number of Cells
a.	5 mL	
b.	50 mL	
c.	1 000 mL	
d.	1 L	

15. An analytical balance is used to measure the mass of four different tissue masses. The masses are determined to be:

Specimen **A:** 1.059 grams

Specimen **B:** 0.984 grams

Specimen **C:** 1.001 grams

Specimen **D:** 0.493 grams

a. What is the combined mass of the tissue masses?

b. What is the average mass per tissue mass? Round the answer to the nearer thousandth.

16. A laboratory technician is requested to prepare 25 Petri dishes for a bacteriological study. The dishes are 9 cm in diameter. A nutrient solution is to be prepared and poured to a depth of 0.5 cm. How much nutrient solution is required to fill all 25 dishes?

17. A manometer is used to measure the amount of oxygen used by a culture of micro-organisms.

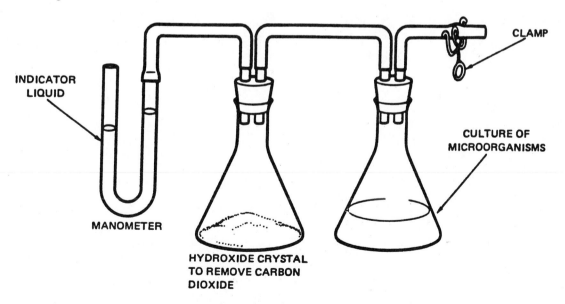

INDICATOR LIQUID

CLAMP

CULTURE OF MICROORGANISMS

MANOMETER

HYDROXIDE CRYSTAL TO REMOVE CARBON DIOXIDE

The liquid in the manometer changes level as the oxygen is used. The diameter of the manometer tube is 0.5 mm. During one ten-minute period the indicator liquid moves 43 mm. Find the volume of gas change in the system.

18. A system requiring a measure of liquid is constructed. One section requires a graduated manometer. It is necessary to calibrate a section of the glass-tubed manometer into 2-cubic millimeter units. If the diameter of the glass tubing is 0.75 mm, what is the distance between each graduation?

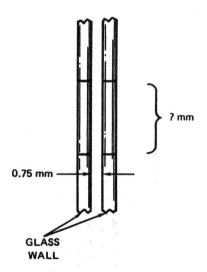

? mm

0.75 mm

GLASS WALL

SECTION 3: Summary, Review, and Study Guide

VOCABULARY

area	derived unit	length
base	diameter	liter
base units	Fahrenheit	mass
Celsius	gram	meter
centi	hecto	milli
circumference	height	pi
deci	kilo	radius
deka	kilogram	vertex
		volume

CONCEPTS, SKILLS AND APPLICATIONS

Objectives With Study Guide

Upon completion of Section 3 you should be able to:

- **Identify metric measures in terms of base units.**

  Order the following volume measure from smallest to largest.

  centiliter, hectoliter, deciliter, kiloliter, liter, milliliter, dekaliter.

  milliliter, centiliter, deciliter, liter, dekaliter, hectoliter, kiloliter.

- **Express equivalences for metric measures.**

  Complete the following chart to show the ranking of 1 centigram, 1 dekagram, 1 decigram, 1 gram, 1 hectogram, 1 kilogram, and 1 milligram from largest to smallest. Then indicate the equivalent gram measure.

Mass Measure from Largest to Smallest	Equivalence
1 kilogram	1 000 grams
1 hectogram	100 grams
1 dekagram	10 grams
1 gram	1 gram
1 decigram	0.1 gram
1 centigram	0.01 gram
1 milligram	0.001 gram

- **Express metric length measure in smaller or larger units.**

  52 m _____ cm
  Since 1 m = 100 cm, multiply by 10^2
  52 m = 5 200 cm

  4 040 mm = _____ m

  Since 1 mm = $\dfrac{1}{1\,000}$ m, multiply by 10^{-3}

  4 040 mm = 4.040 m

- **Determine the radius, diameter, and circumference of a circle.**

  If the circumference of a circle is 37.70 cm, find the diameter and the radius to the nearer whole number.

  $C = \pi d$

  $37.70 = \pi(d)$

  $\dfrac{37.70}{\pi} = d$

  $d = 12$ cm
  Since $d = 12$ cm
  $r = 6$ cm

  Find the circumference of a circle whose radius is 20 mm. Round to the nearer hundredth.

  $C = 2\pi r$
  $C = 2(\pi)(20)$
  $C = 125.66$ mm

- **Express metric area measures in smaller or larger metric units.**

  2.51 hm^2 = _____ dam^2
  Since 1 hm^2 = 100 dam^2, multiply by 10^2
  2.51 hm^2 = 251 dam^2

  $7\ 109$ cm^2 = _____ m^2
  Since 1 cm^2 = $\dfrac{1}{10\ 000}$ m^2, multiply by 10^{-4}
  $7\ 109$ cm^2 = $0.710\ 9$ m^2

- **Determine the area of figures using the appropriate formulas.**

  Find the area of a square whose sides are 10 dm.
  $A = s^2$
  $A = (10\ \text{dm})^2$
  $A = 100\ \text{dm}^2$

  Find the area of a rectangle with sides of length 12 cm and 15 cm.
  $A = lw$
  $A = (12\ \text{cm})(15\ \text{cm})$
  $A = 180\ \text{cm}^2$

  Find the area of a triangle whose base is 3 m and its height is 5 m.
  $A = \dfrac{1}{2}bh$

  $A = \dfrac{1}{2}(3\ \text{m})(5\ \text{m})$

  $A = 7.5\ \text{m}^2$

Find the area of a circle whose radius is 20 mm. Round to the nearer mm^2.

$A = \pi r^2$

$A = \pi (20 \text{ mm})^2$

$A = 1\ 257 \text{ mm}^2$

- **Determine the volume of solids and fluids.**

  Find the volume of a rectangular solid whose dimensions are $l = 3$ cm, $w = 5$ cm, and $h = 10$ cm.

  $V = lwh$

  $V = (3 \text{ cm})\ (5 \text{ cm})\ (10 \text{ cm})$

  $V = 150 \text{ cm}^3$

  Find the volume of a cylindrical solid whose radius is 6 cm and whose height is 15 dm. Round to the nearer dm^3.

  $V = \pi r^2 h$

  $V = \pi (6 \text{ dm})^2\ (15 \text{ dm})$

  $V = 283 \text{ dm}^3$

- **Express metric volume measures in smaller or larger units.**

  $0.418\ 035 \text{ dam}^3 = $ _____ dm^3

  Since 1 $dam^3 = 1\ 000\ 000\ dm^3$, multiply by 10^6

  $0.418\ 035 \text{ dam}^3 = 418\ 035 \text{ dm}^3$

  $3\ 789 \text{ mm}^3 = $ _____ cm^3

  Since 1 $mm^3 = \dfrac{1}{1\ 000}\ cm^3$, multiply by 10^{-3}

  $3\ 789 \text{ mm}^3 = 3.789 \text{ cm}^3$

- **Express metric capacity measures in smaller or larger units.**

  $0.75 \text{ L} = $ _____ mL

  Since 1 L = 1 000 mL, multiply by 10^3

  $0.75 \text{ L} = 750 \text{ mL}$

  $595 \text{ L} = $ _____ kL

  Since 1 kL = $\dfrac{1}{1\ 000}$ L, multiply by 10^{-3}

  $595 \text{ L} = 0.595 \text{ kL}$

- **Express metric mass in smaller or larger metric units.**

  2.78 kg = _____ g

  Since 1 kg = 1 000 g, multiply by 10^3

  2.78 kg = 2 780 g

  419 mg = _____ g

  Since 1 mg = $\dfrac{1}{1\,000}$ g, multiply by 10^{-3}

  419 mg = 0.419 g

- **Determine the relationship between volume, capacity and mass.**

  Complete these relationships for water at 4 degrees Celsius and standard pressure.

Volumes		**Mass**
540 cm^3	__a__ mL	__b__ g
__c__ dm^3	2 L	__d__ kg

  Since 1 000 cm^3 = 1 000 mL = 1 000 g and

  1 dm^3 = 1 L = 1 kg and 1 cm^3 = 1 mL = 1 g

540 cm^3	540 mL	540 g
2 dm^3	2 L	2 kg

- **Express Fahrenheit temperature readings as Celsius temperature readings.**

  Express 77 °F as degrees Celsius.

  $$°C = \frac{5}{9}(F - 32)$$

  $$°C = \frac{5}{9}(77 - 32)$$

  $$°C = 25$$

- **Express Celsius temperature readings as Fahrenheit temperature readings.**

  Express 40 °C as degrees Fahrenheit.

  $$°F = \frac{9}{5}°C + 32$$

  $$°F = \frac{9}{5}(40) + 32$$

  $$°F = 104$$

Review

Major concepts and skills to be mastered for the Section 3 Test.

In 1–10, complete each sentence. Choose terms from the vocabulary list.

1. The prefix for one-thousandth is _____ .

2. _____ represents the ratio between the circumference and diameter of a circle.

3. The _____ is the base unit of mass in the metric system.

4. The _____ of metric volume measure for solids is the cubic meter.

5. The line segment connecting the center of a circle to a point on the circle is named the _____ .

6. The base unit of metric length measure is the _____ .

7. In the metric system, the derived unit of volume for a figure that is not a solid is measured by the _____ .

8. _____ is the prefix representing hundreds.

9. On the _____ scale normal human body temperature is 98.6°.

10. The _____ of an object is the amount of matter that the object contains.

11. Order the following length measures from largest to smallest.
    hectometer, meter, decimeter, dekameter, centimeter, kilometer, millimeter.

In 12–15, complete the following chart to show the ranking of 1 liter, 1 kiloliter, 1 centiliter, 1 dekaliter, 1 milliliter, 1 deciliter, and 1 hectoliter from smallest to largest. Then indicate the equivalent liter measure.

Volume Measure from Smallest to Largest	Equivalence
12.	13.
14.	15.
16.	17.
18.	19.
20.	21.
22.	23.
24.	25.

In 26–29, complete.

26. 250 mm = _____ cm   27. 4.87 dam = _____ m

28. 571 dm = _____ dam   29. 9 132 m = _____ km

30. Find the radius and diameter of a circle whose circumference is 25 dm. Round each to the nearer whole number.

31. If the diameter of a circle is 26 cm, find the circumference. Round to the nearer hundredth.

In 32–41, complete.

32. 417 dam^2 = _____ m^2

33. 5 282 mm^2 = _____ dm^2

34. 5 m^2 = _____ cm^2

35. 0.075 km^2 = _____ dam^2

36. The side of a square is 7.2 m. Find its area.

37. The length of a rectangle is 11 dm. The width is 8 dm. Find the area.

38. A triangle has a base of 5 cm. Its height is 12 cm. Find the area.

39. The radius of a circle is 30.5 dm. Find its area. Round to the nearer dm^2.

40. The length of a rectangular solid is 4.8 dm. Its width is 3 dm and its height is 15 dm. Find the volume.

41. A cylindrical solid has a radius of 20 cm. Its height is 48 cm. Find its volume. Round to the nearer cm^3.

In 42–45, complete.

42. 52 814 cm^3 = _____ m^3

43. 8.071 3 hm^3 = _____ dam^3

44. 61 cm^3 = _____ mm^3

45. 7 534.78 dm^3 = _____dam^3

In 46–49, complete.

46. 813 cL = _____ dL

47. 5.3 hL = _____ mL

48. 3 718 mL = _____ L

49. 10.73 L = _____ kL

In 50–53, complete.

50. 500 g = _____ kg

51. 4.8 dag = _____ g

52. 75 dg = _____ mg

53. 0.217 kg = _____ mg

In 54–57, for water at 4 degrees Celsius and standard pressure complete these relationships.

	Volumes		**Mass**	
54.	_____ cm^3	20 mL	55.	_____ g
	25 dm^3	56. _____L	57.	_____ kg

58. Express as degrees Celsius, a temperature of 73 °F. Round to the nearer tenth.

59. Express as degrees Fahrenheit, a temperature of 37 °C. Round to the nearer tenth.

SECTION 3 TEST

1. To change from a larger metric unit to a smaller metric unit multiply by a _____ power of 10.

2. To change from a smaller metric unit to a larger metric unit multiply by a _____ power of 10.

3. The value of each metric prefix is _____ times the value of the prefix to its immediate right.

4. The value of each metric prefix is _____ times the value of the prefix to its immediate left.

5. The derived unit for volume measure for fluids is the _____ .

6. Explain how the metric system and decimals are similar.

7. List, in order from largest to smallest, the six most common metric prefixes for the medical field.

8. Is it possible to find the capacity and volume of a container if you know the mass of the water that filled the container? Explain.

In 9–22, complete.

9. 4 219 m = _____ km

10. 117 mm = _____ cm

11. 1.82 dam^2 = _____ dm^2

12. 375 cm^2 = _____ m^2

13. 235 168 mm^2 = _____ dm^2

14. 6.875 km^2 = _____ dam^2

15. 7.52 hm = _____ cm

16. 8 176 cm = _____ dam

$C = 6\pi$ cm

17. $d = ?$

18. $r = ?$

19.
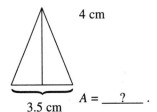

$A = \underline{\quad ? \quad}$.

$s = 3$ cm

20.
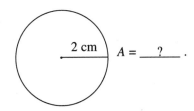

2 cm

$A = \underline{\quad ? \quad}$.

4 cm

21.

4 cm

3.5 cm

$A = \underline{\quad ? \quad}$.

22.

2 cm

$A = \underline{\quad ? \quad}$.

23. Find the volume of a rectangular solid 5 cm by 7 cm by 3 cm.

24. Find the volume of a cylindrical solid with a radius of 4.2 cm and a height of 5 cm.

In 25 and 26, arrange each group of measurements from smallest to largest.

25. 260 daL;   40 kL;   48 000 L;   240 dam^3;   50 cm^3

26. 2 hL;   400 l;   3 000 mL;   300 cm^3;   2 m^3

In 27–34, complete.

27. 18.9 g = _____ mg

28. 32 L = _____ mL

29. 489 mL = _____ L

30. 7.9 kg = _____ g

31. 208 g = _____ hg

32. 219 cL = _____ daL

33. 5.2 dL = _____ mL

34. 0.75 kg = _____ mg

In 35–38, complete these relationships for water at 4 degrees Celsius and standard pressure.

35. _____ dm^3 4L

36. _____ kg

      25 cm^3 37. _____ mL

38. _____ g

In 39–42, circle the correct unit.

39. 1.5   cm^3 or dm^3 1.5 L

40. 1.5   g   or kg

41. 52   dm^3 or cm^3 52 mL

42. 52   kg or g

In 43 and 44, express each Celsius temperature as an equivalent Fahrenheit temperature and each Fahrenheit temperature as an equivalent Celsius temperature. Round to the nearer tenths.

43. 102 °F   44.   −5 °C

Section Four

Ratio, Proportion, and Percents

Unit **18**

Introduction to Ratio and Proportion

Objectives

After studying this unit the student should be able to:

- **Compare quantities using ratios.**
- **Identify mean terms and extreme terms in a proportion.**
- **Determine if a proportion is an equality using the means and extremes.**
- **Determine the unknown component in a simple proportion.**

The ability to accurately compare numbers or quantities is a skill used by many people. Ratios and proportions are one method of comparison.

RATIO

Quantities can be compared in different ways. A comparison between numbers by division is a *ratio*. The numbers that are compared are the *terms*, or *components*, of the ratio.

This diagram can be used to illustrate several comparisons, *or* ratios.

There are 25 shaded squares compared to 75 unshaded squares.
The ratio can be written 25 to 75 *or* 1 to 3.
There are 25 shaded squares compared to 100 total squares.
The ratio can be written 25 to 100 *or* 1 to 4.
The ratio of the unshaded squares compared to the shaded squares is $\frac{75}{25}$ *or* $\frac{3}{1}$.
The ratio of the unshaded squares to the total squares is 75:100 *or* 3:4.

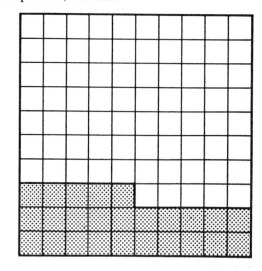

Note: The three notations, "to,":, and — , all identify a ratio.

- In general, the ratio of *a* to *b* can be written as:

a to b; $a:b$; or $\frac{a}{b}$.

Two ratios are equivalent if they have the same value. Two methods can be used to find if the ratios are equivalent.

EXAMPLE 1

Are 10:15 and 12:18 equivalent ratios?

Method 1: Write each ratio as a fraction and express as an equivalent fraction with common denominators.

$$\frac{10}{15} = \frac{2 \cdot \overset{1}{\cancel{5}}}{3 \cdot \cancel{5}}$$
$$\phantom{\frac{10}{15}} = \frac{2}{3}$$

$$\frac{12}{18} = \frac{\overset{1}{\cancel{2}} \cdot 2 \cdot \overset{1}{\cancel{3}}}{\cancel{2} \cdot 3 \cdot \cancel{3}}$$
$$\phantom{\frac{12}{18}} = \frac{2}{3}$$

Note: A ratio is in lowest terms if the components contain no common factors other than 1.

Since the ratios in lowest terms are equal, 10:15 and 12:18 are equivalent ratios.

Method 2: Using a calculator express the ratios as decimals, then compare.

$$\frac{10}{15} = 10 \boxed{\div} 15 \boxed{=}$$
$$\phantom{\frac{10}{15}} = 0.\overline{6}$$

$$\frac{12}{18} = 12 \boxed{\div} 18 \boxed{=}$$
$$\phantom{\frac{12}{18}} = 0.\overline{6}$$

Since the decimals are the same, the ratios 10:15 and 12:18 are equivalent ratios.

A ratio can compare like measurements or unlike measurements. For example, $1.50 per kilogram compares the unlike measurements of dollars and kilograms. Since *per* means *to;* the ratio can be read "$1.50 to 1 kilogram." When the ratio is a comparison of unlike measurements, it is called a *rate*.

PROPORTION

Ratios can be used to form proportions. A *proportion* is an equation which states that two ratios are equal.

• In general, the proportion formed by using the ratios *a:b* and *c:d* is:

$$a{:}b = c{:}d; \quad \text{or} \quad \frac{a}{b} = \frac{c}{d}$$

It is read "*a* is to *b* as *c* is to *d*."

DEFINITIONS

• In a proportion the middle terms or components are the *means* and the end components are the *extremes*.

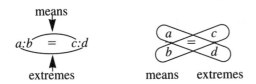

The equality of a proportion may be verified by using the means and extremes. The product of the means is calculated. The product of the extremes is also calculated. If the product of the means equals the product of the extremes, the proportion is an equality. Conversely, if the proportion is an equality, then the product of the means equals the product of the extremes.

- Generally, if $\dfrac{a}{b} = \dfrac{c}{d}$, then $ad = bc$ and

  if $ad = bc$, then $\dfrac{a}{b} = \dfrac{c}{d}$.

EXAMPLE 1

Using the product of the means and the product of the extremes, verify that $\dfrac{8}{12} = \dfrac{2}{3}$.

$\dfrac{8}{12} \overset{?}{=} \dfrac{2}{3}$ $\dfrac{8}{12} \overset{?}{=} \dfrac{2}{3}$

$8 \cdot 3 = 24$ $12 \cdot 2 = 24$

Since $24 = 24$, $\dfrac{8}{12} = \dfrac{2}{3}$.

Since a proportion consists of 2 ratios, its equality can be determined using either cross-products or each ratio's decimal equivalent.

EXAMPLE 2

Determine if $\dfrac{2}{5} = \dfrac{8}{15}$.

Pencil and Paper

$\dfrac{2}{5} = \dfrac{8}{15}$

$2(15) \overset{?}{=} 5(8)$

$30 < 40$, so

$\dfrac{2}{15} \neq \dfrac{8}{15}$

Calculator

$\dfrac{2}{5} = \dfrac{8}{15}$

$2 \boxed{\div} 5 \overset{?}{=} 8 \boxed{\div} 15$

$0.4 \neq 0.5\overline{3}$, so

$\dfrac{2}{15} \neq \dfrac{8}{15}$

As in ratios, proportions may use unlike measurements. The comparison of the unlike measurement is the *rate*. The proportional relationships between ratios of unlike measurements are *rate pairs*.

18.1 Exercises

In 1–5, express each ratio in lowest terms.

1. 8:16   2. $\dfrac{10}{12}$ 3. 9 to 12   4. 30:50   5. $\dfrac{85}{40}$

6. The ratio of the number of men in the class to the number of women in the class.

7. The ratio comparing the number of students in the class to the total number of desks in the room.

In 8–13, identify the mean terms and extreme terms in each proportion.

8. $\dfrac{3}{5} = \dfrac{12}{20}$ 9. $\dfrac{20}{3} = \dfrac{N}{6}$ 10. $\dfrac{1}{x} = \dfrac{33}{100}$

11. $4:9 = 8:18$ 12. $t:5 = 10:20$ 13. $3:6 = 9:a$

In 14–25, for each problem:

a. Using the product of the means equal the product of the extremes. Determine if each proportion is an equality or an inequality, then

b. Using your calculator check your answer by finding and comparing the decimal equivalent for each ratio in the proportion.

14. 3:10 = 30:100   15. 2:3 = 3:4   16. 8:4 = 1:2

17. 1.4:28 = 1:20   18. 20:5 = 10:2   19. 9:12 = 18:24

20. $\dfrac{12.5}{1000} = \dfrac{0.1}{8}$ 21. $\dfrac{3}{2} = \dfrac{30}{18}$ 22. $\dfrac{0.26}{0.52} = \dfrac{1.4}{2.8}$

23. $\dfrac{3}{7.5} = \dfrac{2}{5}$ 24. $\dfrac{15}{4} = \dfrac{37.5}{10}$ 25. $\dfrac{2}{10} = \dfrac{5}{30}$

26. List two ways to determine if two ratios are equivalent.

PROPORTIONAL COMPUTATIONS

A proportion is an equation which states that two ratios are equal. Generally, if $\dfrac{a}{b} = \dfrac{c}{d}$, then $ad = bc$ or if $ad = bc$, then $\dfrac{a}{b} = \dfrac{c}{d}$. Many times one component of a proportion is unknown. By using the product of the means and the product of the extremes, the unknown or variable may be determined or the proportion is "solved."

EXAMPLE 1

Find the unknown, *N*, for which $\dfrac{8}{N} = \dfrac{2}{3}$ will be an equality.

$\dfrac{8}{N} = \dfrac{2}{3}$

$3(8) = 2(N)$ Find the cross-product.

$24 = 2N$ Simplify.
                                                    Solve by dividing both sides of the equation
$\dfrac{24}{2} = \dfrac{2N}{2}$ by 2, the number in front of the variable.

$12 = N$ or $N = 12$

Since $\dfrac{8}{12} = \dfrac{2}{3}$, 12 is the correct answer.

Note: Since 1 times any number is that number, the 1 in front of the *N* need not be shown. In actuality, $1N = N$. The expression 2(*N*), means 2*N*. The parenthesis may be shown or not shown.

EXAMPLE 2

Solve the proportion $\dfrac{2}{5} = \dfrac{C}{20}$.

Pencil and Paper

$20(2) = 5(C)$

$40 = 5C$

$\dfrac{40}{5} = \dfrac{5C}{5}$

$8 = C \quad \text{or} \quad C = 8$

Since $\dfrac{2}{5} = \dfrac{8}{20}$, 8 is the correct answer.

Calculator

$\dfrac{2}{5} = \dfrac{C}{20}$

$C = \dfrac{2(20)}{5}$ calculation-ready form

$C = 2\boxed{\times}\,20\,\boxed{\div}\,5\,\boxed{=}$

$C = 8$

18.2 Exercises

In 1–12, solve each proportion.

1. $\dfrac{1}{2} = \dfrac{3}{b}$

2. $\dfrac{4}{1} = \dfrac{x}{4}$

3. $\dfrac{t}{8} = \dfrac{7}{1}$

4. $\dfrac{2}{C} = \dfrac{1}{10}$

5. $\dfrac{6}{5} = \dfrac{r}{10}$

6. $\dfrac{3}{4} = \dfrac{30}{y}$

7. $P{:}100 = 3{:}10$

8. $2{:}R = 5{:}15$

9. $0.7{:}10 = x{:}100$

10. $a{:}25 = \dfrac{1}{2}{:}5$

11. $8{:}4 = 10{:}n$

12. $10{:}d = 1.5{:}6$

APPLICATIONS

Many health care workers prepare solutions. A solution is a liquid preparation of one or more substances dissolved in a liquid, usually water. Pharmacists prepare solutions of medication to be administered to patients. Registered nurses (RN) may prepare solutions to be administered orally; licensed practical nurses (LPN) may prepare solutions to be administered topically. Medical technologists (4 years of post secondary training), medical laboratory technicians (2 years of post secondary training), and medical laboratory assistants (1 year of post secondary training), prepare solutions to be used in chemical, microscopic, and bacteriological tests.

A solution consists of a solvent and a solute.

DEFINITIONS

- The *solvent* is the liquid in which the substance or substances are dissolved. It is usually water.
- The *solute* is the substance that is dissolved in the liquid. The solute is the drug.

The ratio between the solute (drug) and the solution is the *ratio strength of the solution.* That is:

amount of drug:amount of solution
is the ratio strength of the solution

> *or*

$$\frac{\text{amount of drug}}{\text{amount of solution}} = \text{ratio strength of the solution}$$

A 1:5 solution means $\frac{1}{5}$ or 1 part solute (drug) to 5 parts solution. This may mean that there is 1 milliliter of the drug in 5 milliliters of solution. Since the ratio 1:5 is in lowest terms, it may also be 2 milliliters of drug in 10 milliliters of solution or even 25 milliliters of drug in 125 milliliters of solution.

18.3 Exercises

1. Drugs may be in many forms. One form is crystals. There are 3 parts boric acid in 12 parts of solution. Express the strength of the solution as a ratio.

2. Glycerin is a liquid. There are 8 milliliters of glycerin in 24 milliliters of solution. Express the strength of a solution as a ratio.

3. The ratio strength of a certain solution is 1:5. There are 15 milliliters of the drug. How many milliliters of solution are there? (Use the proportion $\frac{1}{5} = \frac{15}{x}$.)

4. The ratio strength of a Lysol solution is 1:20. If there are 1 000 milliliters of solution, how many milliliters of Lysol are there? (Use the proportion $\frac{1}{20} = \frac{y}{1\ 000}$.)

5. Doses of a 2:50 solution are to be administered to a patient. Each dose is 0.5 milliliter.

   a. How much medication is needed to administer 5 doses?

   b. Using the proportion $\frac{2}{50} = \frac{\text{amount of drug}}{\text{total amount of medication}}$, find the amount of drug.

Unit 19

Computations With Proportions

Objectives

After studying this unit the student should be able to:
• **Determine the unknown in a rate pair.**
• **Determine the unknown in a proportion which contains an implied component.**
• **Determine the unknown in a proportion which contains fractional components.**

RATE PAIR COMPUTATIONS

Rate pairs express proportional relationships between ratios. The ratios are comparisons, in *the same order,* of unlike measurements. Determining the unknown in rate pairs uses the same procedure as in proportions.

EXAMPLE 1

A medical supplier sells 4 tongue depressors for 24 cents. What is the cost of 1 tongue depressor?

The rate is depressors:cost. The rate pair is $\dfrac{\text{depressors}}{\text{cost}} = \dfrac{\text{depressors}}{\text{cost}}$.

$\dfrac{\text{depressors} \rightarrow 4}{\text{cost} \quad \rightarrow 24} = \dfrac{1 \leftarrow \text{depressors}}{N \leftarrow \quad \text{cost}}$

This is read 4 depressors are to 24 cents as 1 depressor is to N cents.

$4(N) = 1(24)$ Cross-multiply.

$4N = 24$ Simplify.

$\dfrac{4N}{4} = \dfrac{24}{4}$ Divide each side by 4, the number in front of the unknown.

$N = 6$

One tongue depressor costs 6 cents.

Note: The answer is a descriptive quantity. It describes what type of measurement the unknown is.

The key to setting up a correct proportional rate pair is to be sure that each side of the proportion compares quantities in the same order. Notice that Example 2 is set up two different ways.

EXAMPLE 2

If your heart pumps 4 liters of blood in 90 seconds, how many liters are pumped in 60 seconds? Round to the nearer tenth of a liter.

$$\frac{\text{liters}}{\text{seconds}} \quad \frac{4}{90} = \frac{x}{60} \quad \frac{\text{liters}}{\text{seconds}}$$

$$90(x) = 4(60)$$

$$x = \frac{4(60)}{90} \qquad \text{calculation-ready form}$$

$$x = 2.7$$

$$\frac{\text{seconds}}{\text{liters}} \quad \frac{90}{4} = \frac{60}{y} \quad \frac{\text{seconds}}{\text{liters}}$$

$$90(y) = 4(60)$$

$$y = \frac{4(60)}{90}$$

$$y = 2.7$$

In 60 seconds, your heart would pump 2.7 liters of blood.

19.1 Exercises

In 1–5, an incorrect rate pair is written in each proportion. Write the correct rate pairs. Determine the value of each unknown component.

Rates	Incorrect Rate Pairs	Correct Rate Pairs	Answer
1. 400 meters in 80 seconds N meters in 10 seconds	$\frac{400}{80} \neq \frac{10}{N}$		
2. 8 kilograms lost in 4 weeks 2 kilograms lost in x weeks	$\frac{8}{4} \neq \frac{x}{2}$		
3. 12 cents per y stirring rods 144 cents for 12 stirring rods	$\frac{y}{0.12} \neq \frac{1.44}{12}$		
4. $6 for 1 hour of work Z for 100 hours of work	$\frac{Z}{100} \neq \frac{1}{6}$		
5. 1 centimeter for 80 kilometers q centimeters for 360 kilometers	$\frac{80}{1} \neq \frac{q}{360}$		

In 6–11, set up two different proportional rate pairs to correctly describe each situation. Remember that the comparisons on each side of the equation must be in the same order. Label the comparisons.

6. 8 kilograms lost in 4 weeks;
    2 kilograms lost in y weeks

7. $12 for 1 hour of work;
    Z for 100 hours of work

8. $0.12 per t stirring rods;
    $26.40 for 220 stirring rods

9. 0.5 teaspoon of salt per 0.75 liter water;
    X teaspoons of salt used in 15 liters of water

10. 4 grams make 10 mL of solution
     H grams make 40 mL of solution

11. C Tagamet tablets for $84.00;
     90 Tagamet tablets for $126.00

In 12–19, set up a rate pair. Determine the unknown. Express the answer as a descriptive quantity.

12. Sally makes a salt solution using 64 mL of water and 10 mg of salt. She wants to make more solution of the same strength. If she uses 456 mg of salt, how much water should be used?

13. If 5 mg will make 30 dL of solution, how many mg will be needed for 50 dL of solution?

14. Emily gained $1\frac{1}{2}$ lbs in 12 weeks. If this growth rate continues how many pounds will she gain in 9 months?

15. Suppose aspirin costs $0.0075 per tablet. How much will a container of 5,000 cost?

16. If 20 cL of a solution contains 5 g of a substance, how many grams would be contained in 60 cL of solution?

17. An automatic washer with a continuous conveyor can wash 100 large flasks in 60 minutes. At this rate, how many flasks can be washed in 15 minutes?

18. At a price of 2 test tubes for $0.99, how many test tubes can be purchased for $9.00?

19. The case load in one ward is determined to be 12 patients for each 2 nurses. To have an equal case load, how many nurses would be required for a 70-patient ward?

In 20 and 21, the mortality rate, as determined by a nationwide survey, for a particular surgical procedure has been stated to be 5 deaths for each 3,800 operations.

20. In one year, a total of 10,500 operations using this surgical procedure are conducted. What is the anticipated number of deaths relating to this type of surgery for this period of time?

21. If the research indicated 75 deaths resulted during a given period of time for this type of surgery, how many operations would have taken place?

In 22–26, complete.

22. $4{:}10 = 20{:}50$ because $4 \cdot \boxed{} = 20 \cdot \boxed{}$

23. $8{:}4 = 16{:}8$ because $\boxed{} \cdot 8 = 16 \cdot \boxed{}$

24. $12{:}8 = 3{:}2$ because $12 \cdot 2 = \boxed{} \cdot \boxed{}$

25. $3{:}5 = 9{:}15$ because $\boxed{} \cdot \boxed{} = \boxed{} \cdot \boxed{}$

26. $a{:}b = c{:}d$ because $\boxed{} \cdot d = \boxed{} \cdot c$

When using implied components form the ratios, then determine the unknown.

EXAMPLE 1

Laura is a lab technician earning $12.25 per hour. How much does she earn for a 40-hour work week?

$$\frac{\text{pay} \rightarrow 12.25}{\text{hours} \rightarrow 1} = \frac{x \leftarrow \text{pay}}{40 \leftarrow \text{hours}}$$

$$1x = 40(12.25)$$

$$x = 490$$

Laura earns $490 per 40-hour work week.

When setting up a proportion it is important to express quantities in the same units before writing the proportion.

EXAMPLE 2

During one period in Michael's life, he grew at the rate of 0.25 inch per month. How many months did it take him to grow 0.5 foot?

Since 0.5 foot equals 6 inches, use 6 in the proportion:

$$\frac{\text{inches} \to}{\text{months} \to} \frac{0.25}{1} = \frac{6}{4} \frac{\leftarrow \text{inches}}{\leftarrow \text{months}}$$

$$0.25y = 1(6)$$

$$y = \frac{1(6)}{0.25}$$

$$y = 24$$

It took 24 months for Michael to grow 0.5 foot.

Proportions with implied components can be used to change from one metric unit to another.

EXAMPLE 3

How many milligrams are there in 0.5 gram?

$$\frac{\text{grams}}{\text{milligrams}} \frac{0.5}{x} = \frac{1}{1\ 000} \frac{\text{grams}}{\text{milligrams}}$$

$$1(x) = 0.5\ (1\ 000)$$

$$x = 500$$

There are 500 milligrams in 0.5 gram.

FRACTIONAL COMPONENTS

Many times, proportions have fractions as components. The method used when working with fractions is the same as the method used when working with whole numbers.

EXAMPLE 1

A prescription calls for $\frac{1}{2}$ tablet every 3 hours. How many tablets will be taken in 15 hours?

Pencil and Paper

$$\frac{\text{tablets}}{\text{hours}} \frac{\frac{1}{2}}{3} = \frac{t}{15} \frac{\text{tablets}}{\text{hours}}$$

$$3(t) = \frac{1}{2}(15)$$

$$3t = \frac{15}{2}$$

$$t = \frac{15}{2} \div \frac{3}{1} = \frac{\overset{5}{\cancel{15}}}{2} \cdot \frac{1}{\underset{1}{\cancel{3}}}$$

$$t = \frac{5}{2} \text{ or } 2\frac{1}{2}$$

Calculator

$$\frac{\frac{1}{2}}{3} = \frac{t}{15}$$

$$3t = \frac{1}{2}(15)$$

$$t = \frac{\frac{1}{2}(15)}{3} \qquad \text{calculation-ready form}$$

$$t = 1 \boxed{\div} 2 \boxed{\times} 15 \boxed{\div} 3 \boxed{=}$$

$$t = 2.5 \text{ or } 2\frac{1}{2}$$

In 15 hours, $2\frac{1}{2}$ tablets will be taken.

Some problems contain fractions in both the numerator and denominator.

EXAMPLE 2

In an experiment, Tom used $\dfrac{1}{2}$ gram of sugar for each $\dfrac{4}{5}$ liter of water. How much sugar will he use for 7 liters of water?

Calculator

$$\dfrac{\text{grams of sugar}}{\text{liters of water}} \; \dfrac{\frac{1}{2}}{\frac{4}{5}} = \dfrac{a}{7} \; \dfrac{\text{grams of sugar}}{\text{liters of water}}$$

$$\dfrac{4}{5}(a) = \dfrac{1}{2}(7)$$

$$\dfrac{4}{5}a = \dfrac{7}{2}$$

$$a = \dfrac{\frac{7}{2}}{\frac{4}{5}} \quad \text{calculation-ready form}$$

$$a = \dfrac{7}{2} \cdot \dfrac{5}{4}$$

$$a = \dfrac{35}{8} = 4\dfrac{3}{8}$$

Calculator with Fraction Key

$$\dfrac{\frac{1}{2}}{\frac{4}{5}} = \dfrac{a}{7}$$

$$\dfrac{4}{5}(a) = \dfrac{1}{2}(7)$$

$$\dfrac{4}{5}a = \dfrac{7}{2}$$

$$a = \dfrac{\frac{7}{2}}{\frac{4}{5}}$$

$$7 \boxed{ab \, / \, c} \div 2 \div 4 \boxed{ab \, / \, c} \div 5 \boxed{=}$$

$a = 4 - 3 \rfloor 8$ which means $4\dfrac{3}{8}$.

Tom will use $4\dfrac{3}{8}$ liters of water.

19.2 Exercises

In 1–6, solve for the unknown.

1. $\dfrac{x}{\frac{1}{3}} = \dfrac{9}{16}$

2. $\dfrac{\frac{1}{3}}{1} = \dfrac{y}{5}$

3. $\dfrac{\frac{1}{2}}{p} = \dfrac{4}{9}$

4. $\dfrac{\frac{1}{5}}{\frac{2}{8}} = \dfrac{2}{a}$

5. $\dfrac{\frac{3}{4}}{b} = \dfrac{1\frac{1}{4}}{5}$

6. $\dfrac{\frac{1}{5}}{\frac{2}{3}} = \dfrac{c}{\frac{1}{3}}$

In 7–20, form each proportion. Solve for the unknown. Express the answer as a descriptive quantity.

7. Human blood normally has a ratio of 1 000:1 between red blood cells and white blood cells. A sample of blood had 2 125 white blood cells and 4,250,000 red blood cells. How does this compare with normal blood?

8. Conventional gall bladder surgery usually requires a 2-month recovery period. With laser surgery, the recovery period is about 2 weeks. How many times as fast is the recovery period for laser surgery versus conventional surgery.

9. Distilled water costs $0.25 per liter. How many liters can be purchased for $10.00?

10. A replacement valve on an autoclave costs 7 times as much today as it did 10 years ago. If the valve cost $58.00 ten years ago, how much does it cost today?

11. The physician prescribes $1\frac{1}{2}$ tablets every 4 hours. How many tablets will be taken in 24 hours?

12. A prescription calls for $\frac{1}{2}$ tablet every 3 hours. How many days will a bottle of 60 tablets last?

13. The per patient bed space in a hospital has been calculated to be 180 square feet. A ward in the hospital has 5,760 square feet of bed space. How many patients can be housed in that ward at any one time?

14. At 20 °C and standard pressure, 18 cm^3 would have a mass of how many kilograms?

15. How many grams equal 1 500 milligrams?

16. The prefix nano (n) represents 10^{-9} in the SI (Systéme International) System. How many milliseconds would be in 2 500 nanoseconds?

17. An X-ray technician's take-home pay is $345 per week. What would be his take-home pay for 12 months (52 weeks)?

18. 8.7 millimeters equals how many meters?

19. The staff in clinic *A* treats 152 patients in one day. The staff in clinic *B* treats 38 patients. How many more patients are treated in clinic *A* than clinic *B?*

20. The prefix micro(μ) or (mc) represents 10^{-6} in the SI (Systéme International) System. How many micrograms would be in 750 milligrams?

APPLICATIONS

Proportions are used in calculating the amount of solvent and solute to be used in a solution. The ratio compares the amount of solute to the amount of solution. The proportion equates two ratios formed when comparing different amounts of a solution. Remember that for calculation purposes, 1 gram is equal to 1 milliliter.

EXAMPLE 1

In 100 milliliters of a solution there are 15 milliliters of solute. How much solute is required to prepare 325 milliliters of this solution?

$$\frac{\text{amount of solute}}{\text{amount of first solution}} = \frac{\text{amount of solute}}{\text{amount of second solution}}$$

$$\frac{15 \text{ milliliters}}{100 \text{ milliliters}} = \frac{x \text{ milliliters}}{325 \text{ milliliters}}$$

$$100\,(x) = 15\,(325)$$

$$100\,(x) = 4\,875$$

$$\frac{100\,(x)}{100} = \frac{4\,875}{100}$$

$$x = 48.75 \text{ milliliters}$$

19.3 Exercises

1. A total of 62 grams of sodium chloride is present in 950 milliliters of solution. How many grams are in 125 milliliters of a solution with the same strength?

2. Solution *A* of Fehling's solution is prepared from copper sulfate and distilled water. There are 69.30 grams of copper sulfate in 1 000 milliliters of solution. How much copper sulfate is used in 125 milliliters of solution to obtain a solution with the same strength?

3. Solution *B* of Fehling's solution contains potassium hydroxide, potassium sodium tartrate and distilled water. There are 250 grams of potassium hydroxide and 346 grams of potassium sodium tartrate in 1 000 milliliters of solution. This means that there are 250 parts potassium hydroxide in 1 000 parts of solution and 346 parts potassium sodium tartrate in 1 000 parts of solution.

   a. How many grams of potassium hydroxide are needed to prepare 125 milliliters of this solution?

   b. How many grams of sodium tartrate are needed to prepare 125 milliliters of this solution?

4. In 115 milliliters of Lugol's solution there are 10 grams of potassium iodide and 5 grams of iodine. How much of each solute is in 750 milliliters of Lugol's solution?

   a. potassium iodide

   b. iodine

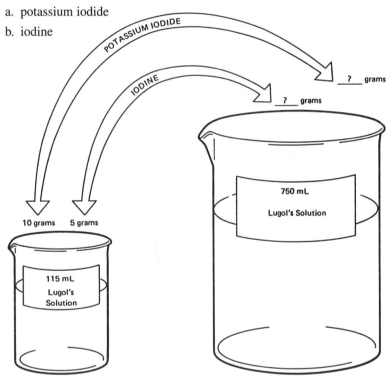

5. A five percent solution is prepared by dissolving five grams of crystals in 100 mL of distilled water. How many grams would be required to prepare a five percent solution with only 30 mL of distilled water?

6. A lab technician can complete 32 lab analyses in an 8-hour shift. The lab contracts with several hospitals to complete a maximum of 500 similar analyses in an 8-hour shift. How many total hours will be required to complete these tests? How many lab technicians would be required if they all worked during the same 8-hour shift?

7. The case load in one ward is determined to be 12 patients for each 16 hours of nursing time in an 8-hour shift. To have an equal case load, how many nursing hours in an 8-hour shift would be required for a 50-patient ward?

8. A local school district uses birth rates in the county hospitals to predict kindergarten enrollment. For example, the number of births in 1988 was used to determine kindergarten enrollment for the 1993-94 school year. Fifteen years of data analysis reveals that 27 children enroll for each 100 births five years earlier. In 1990, a total of 2,347 births were recorded for the county. How many children can be predicted to enroll in kindergarten for the 1995-96 school year?

9. During a one-week period it is found that 21 new patients are admitted to the hospital and 63 previously admitted patients are dismissed.

   a. What is the ratio of those admitted compared to those dismissed during this period of time?

   b. What is the ratio of those admitted compared to all patients admitted and dismissed during this period of time?

   c. What is the ratio of those dismissed compared to all patients admitted and dismissed during this period of time?

10. A national survey reveals that five patients die for every 6,000 that undergo a surgical procedure.

    a. A total of 48,000 similar surgeries were conducted in a six-month period. How many deaths could be anticipated?

    b. Another study found that there were 64 deaths during a three-month period for this type of surgery. How many total surgeries probably occurred during this period of time?

11. Medical service cost for each patient treated at a clinic averaged $53.00 per visit.

    a. If the clinic must recover its monthly operational cost of $132,500, how many patients must be served during the month for the clinic to break even?

    b. What is the ratio of one patient's cost compared to the cost of operating the clinic?

12. Instructions to prepare a disinfectant requires 1 000 mL of distilled water for each 20 mL of stock disinfectant.

    a. What is the ratio of disinfectant to distilled water?

    b. How many 2 500 mL capacity disinfectant containers can be prepared from 1 500 mL of stock disinfectant?

13. The number of patients served in two separate wards of the hospital varies due to the type of service provided. *Ward A* requires three nurses, each working an eight-hour shift, to serve 21 patients. *Ward B* requires three nurses each eight-hour shift to serve 12 patients.

    a. Express the nurse/patient ratio for *Ward A*.

    b. Express the nurse/patient ratio for *Ward B*.

14. Daily processing of patient records requires 20 minutes of computer-operator time per patient.

    a. Using this ratio, calculate the amount of time required to process 125 patient records.

    b. At a cost of $9.50 per hour for record processing, what is the per-patient cost?

    c. Each operator can process three records per hour. How many computer operators would be required to process these records in an eight-hour day?

15. The per-patient bed space in a hospital has been calculated to be 180 square feet.

    a. Express the ratio of patient per square foot.

    b. A ward in the hospital has 5,760 square feet of bed space. Calculate the number of patients that can be housed at any one time.

Unit 20

Introduction to Percents

Objectives

After studying this unit the student should be able to:
- **Express ratios, fractions, and decimals as percents.**
- **Express percents as equivalent fractions or decimals.**
- **Express fractional, decimal, and percent equivalents.**

PERCENT

The constant use of ratio comparisons has led to a special percent notation for ratios. This compares a quantity to one hundred.

Determine the ratio of coverage for each figure compared to the total region.

Figure **A** covers $\dfrac{1}{100}$ of the total region.

Figure **B** covers $\dfrac{25}{100}$ of the total region.

Figure **C** covers $\dfrac{6}{100}$ of the total region.

Figure **D** covers $\dfrac{10}{100}$ of the total region.

Note: Each of these figures is compared to 100.

- A percent is a ratio that compares a number to 100. If n is any number, then $\dfrac{n}{100}$ can be expressed using the symbol $n\%$. Rather than using fractions to express the ratios, equivalent decimal or percent notation can be used.

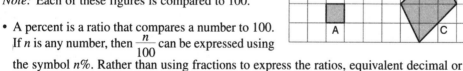

Figure	Ratio	Fraction (hundredths)	Decimal	Percent
A	1:100	$\dfrac{1}{100}$	0.01	1%
B	25:100	$\dfrac{25}{100}$	0.25	25%
C	6:100	$\dfrac{6}{100}$	0.06	6%
D	10:100	$\dfrac{10}{100}$	0.10	10%

Note: The percent symbol, %, is always written to the right and has a decimal and fractional equivalent.

- Generally $n\% = n(0.01) = n\left(\dfrac{1}{100}\right)$ or $\dfrac{n}{100}$.

EXAMPLE 1

Use percent notation to express the ratios 3:20 and 8:5.

$$3{:}20 = \frac{3}{20} \qquad\qquad 8{:}5 = \frac{8}{5}$$

$$\frac{3}{20} = \left(\frac{3}{20}\right)\left(\frac{5}{5}\right) = \frac{15}{100} = 15\% \qquad \frac{8}{5} = \left(\frac{8}{5}\right)\left(\frac{20}{20}\right) = \frac{160}{100} = 160\%$$

EXAMPLE 2

Express the fractions $\dfrac{1}{5}$ and $\dfrac{36}{200}$ using percent notation.

$$\frac{1}{5} = \left(\frac{1}{5}\right)\left(\frac{20}{20}\right) = \frac{20}{100} = 20\% \qquad \frac{36}{200} = \frac{36 \div 2}{200 \div 2} = \frac{18}{100} = 18\%$$

EXAMPLE 3

Express the decimals 0.32 and 2.75 using percent notation.

$$0.32 = \frac{32}{100} = 32\% \qquad\qquad 2.75 = 2\frac{75}{100} = \frac{275}{100} = 275\%$$

Note: In actual use, the percent symbol % is only a reference and is replaced by a ratio, a fraction, or a decimal for computational purposes.

20.1 Exercises

In 1–40, use percent notation to express each ratio, fraction, or decimal.

1. 23 to 100
2. 13:100
3. 40 to 50
4. 3:25

5. 200 to 100
6. 135:100
7. 200 to 200
8. 150:50

9. $\dfrac{1}{4}$ to 100
10. $\dfrac{1}{2}$ to 100
11. 0.5 to 100
12. 0.75:100

13. 120 to 1,000
14. 500 to 1,000
15. 1,200 to 1,000
16. 3 to 10

17. $\dfrac{2}{5}$
18. $\dfrac{23}{25}$
19. $\dfrac{13}{20}$
20. $\dfrac{1}{2}$

21. $\dfrac{3}{4}$
22. $\dfrac{10}{10}$
23. $\dfrac{0}{3}$
24. $\dfrac{25}{10}$

25. $\dfrac{3}{50}$
26. $\dfrac{27}{20}$
27. $\dfrac{80}{200}$
28. $\dfrac{150}{1,000}$

29. 0.230
30. 0.032
31. 0.02
32. 0.25

33. 0.10
34. 0.55
35. 0.09
36. 0.99

37. 1.00
38. 1.25
39. 3.50
40. 4.75

EXPRESSING PERCENTS AS EQUIVALENT COMMON FRACTIONS

Percent means out of one hundred; one hundred is a power of ten. Finding decimal equivalents is simplified since the decimal system is also based on powers of ten. In finding an equivalent fraction, the denominator of one hundred is used and then the fraction is expressed in lowest terms.

EXAMPLE 1

Express 35% as a fraction.

Pencil and Paper

$$35\% = \frac{35}{100}$$

$$\frac{35}{100} = \frac{\cancel{5} \cdot 7}{2 \cdot 2 \cdot \cancel{5} \cdot 5} = \frac{7}{20}$$

$$35\% = \frac{7}{20}$$

Fraction Calculator

$$35\% = \frac{35}{100}$$

$$35 \boxed{ab/c} 100 \boxed{=} 7 \lrcorner 120$$

$$35\% = \frac{7}{20}$$

EXAMPLE 2

Express 146% as a mixed number.

Pencil and Paper

$$146\% = \frac{146}{100}$$

$$\frac{146}{100} = \frac{\overset{1}{\cancel{2}} \cdot 73}{\underset{1}{\cancel{2}} \cdot 2 \cdot 5 \cdot 5}$$

$$146\% = \frac{73}{50} \text{ or } 1\frac{23}{50}$$

Fraction Calculator

$$146\% = \frac{146}{100}$$

$$146 \boxed{ab/c} 100 \boxed{=} 1 \lrcorner 23 \lrcorner 50$$

$$146\% = 1\frac{23}{50}$$

EXAMPLE 3

Express $66\frac{2}{3}\%$ as an equivalent fraction.

Pencil and Paper

$$66\frac{2}{3}\% = \frac{66\frac{2}{3}}{100} = \frac{\frac{200}{3}}{100}$$

$$\frac{\frac{200}{3}}{100} = \frac{200}{3} \div \frac{100}{1} = \frac{200}{3} \cdot \frac{1}{100}$$

$$\frac{\overset{2}{\cancel{200}}}{3} \cdot \frac{1}{\underset{1}{\cancel{100}}} = \frac{2}{3}$$

$$66\frac{2}{3}\% = \frac{2}{3}$$

Fraction Calculator

$$66\frac{2}{3}\% = \frac{66\frac{2}{3}}{100} = \frac{\frac{200}{3}}{100}$$

$$200 \boxed{ab/c} 3 \boxed{\div} 100 \boxed{=} 2 \lrcorner 3$$

$$66\frac{2}{3}\% = \frac{2}{3}$$

EXPRESSING PERCENTS AS EQUIVALENT DECIMALS

To express a percent as a decimal, remove the % symbol and move the decimal point two (2) places to the left.

EXAMPLE 1

Express 8% as a decimal.

Pencil and Paper

$$8\% = \frac{8}{100}$$

$$\frac{8}{100} = 100 \overline{)8.00}^{\,0.08}$$

$$8\% = 0.08$$

Calculator

Note: Some calculators require an arithmetic operation before a percent can be found.

$1 \boxed{\times} 8\% = 0.08$

Other calculators give the decimal equivalent immediately.

$8 \boxed{\%} = 0.08$

EXAMPLE 2

Express 12.5% as a decimal.

Pencil, Paper, and Calculator

$$12.5\% = \frac{12.5}{100}$$

$$\frac{12.5}{100} = 12.5 \div 100 = 0.125$$

$$12.5\% = 0.125$$

Mental Shortcut

Remove % symbol.   12.5

Divide by 100.   0.125

$12.5\% = 0.125$

Note: Dividing by 100 is the same as moving the decimal point two places to the left.

EXAMPLE 3

Express 130.5% as a decimal.

Pencil, Paper, and Calculator

$$130.5\% = \frac{130.5}{100}$$

$$\frac{130.5}{100} = 130.5 \div 100 = 1.305$$

$$130.5\% = 1.305$$

Mental Shortcut

Remove % symbol.   130.5

Move the decimal point   1.305
two places to the left
(divide by 100).

$130.5\% = 1.305$

20.2 Exercises

In 1–20, express each percent as indicated.

Fraction in Lowest Terms

1. 20%   2. 60%   3. $33\frac{1}{3}\%$ 4. 5%   5. 0.2%

6. 7.5%   7. 6.25%   8. 120%   9. 150.5%   10. 210%

Decimal

11. 35%   12. 81%   13. 16.25%   14. $\frac{1}{2}$%   15. 0.05%

16. 0.5%   17. 0.125%   18. 225%   19. 300%   20. 250.65%

In 21–34, complete. Use <, =, or >.

21. 2.5% _____ $2\frac{1}{2}$%

22. $7\frac{1}{4}$% _____ $8\frac{1}{8}$%

23. $3\frac{1}{2}$% _____ $3\frac{1}{3}$%

24. $0.\overline{6}$% _____ 0.67%

25. 14.7% _____ 1.47

26. $3\frac{1}{4}$ _____ 325%

27. 0.075 _____ 0.75%

28. $\frac{1}{5}$ _____ 2.5%

29. 4.5% _____ 0.45

30. $2\frac{1}{5}$ _____ $2\frac{1}{5}$%

31. A % _____ 1% is a number _____ 0.01.

32. A % _____ 100% is a number greater than 1.

33. A number less than 1 is a % _____ 100.

34. A fraction _____ 100% is either a mixed number or an improper fraction.

35. Explain how you know that 25% ≠ $\frac{1}{4}$%.

EXPRESSING DECIMALS AS EQUIVALENT PERCENTS

The basic meaning of percents, decimals, and fractions is adhered to when expressing decimals and fractions as percents. Decimals are expressed as equivalent fractions with a denominator of 100. Then the fraction is expressed as a percent.

EXAMPLE 1

Express 0.75 as a percent.

Mental Shortcut

$$0.75 = 0.75\left(\frac{100}{100}\right)$$

Multiply by 100. $0.75 \cdot 100 = 75$

$$0.75\left(\frac{100}{100}\right) = \frac{75}{100} = 75\%$$

Attach % symbol.   75%

$$0.75 = 75\%$$

$$0.75 = 75\%$$

Note: Multiplying by 100 is the same as moving the decimal point two places to the right.

EXAMPLE 2

Express 0.005 as a percent.

Mental Shortcut

$$0.005 = 0.005 \left(\frac{100}{100} \right)$$ Multiply by 100. $100 \,(0.005) = 0.5$

$$0.005 \left(\frac{100}{100} \right) = \frac{0.5}{100}$$ Attach % symbol. 0.5%

$$\frac{0.5}{100} = 0.5\% \text{ or } \frac{1}{2}\%$$ $0.005 = 0.5\% \text{ or } \frac{1}{2}\%$

EXAMPLE 3

Express 3.61 as a percent.

Mental Shortcut

$$3.61 = 3.61 \left(\frac{100}{100} \right)$$ Multiply by 100. $100 \cdot 3.61 = 361$

$$3.61 \left(\frac{100}{100} \right) = \frac{361}{100}$$ Attach % symbol. 361%

$$\frac{361}{100} = 361\%$$ $3.61 = 361\%$

EXPRESSING FRACTIONS AS EQUIVALENT PERCENTS

Proportions may be used to express a fraction as an equivalent fraction with a denominator of 100.

EXAMPLE 1

Express $\frac{1}{4}$ as a percent.

Pencil and Paper

$$\frac{1}{4} = \frac{n}{100}$$

$$4(n) = (100)$$

$$\frac{4n}{4} = \frac{100}{4}$$

$$n = 25$$

$$\frac{1}{4} = 25\%$$

Calculator

$$\frac{1}{4} = 1 \boxed{\div} 4 \boxed{=} 0.25$$

Multiply by 100. $100(0.25) = 25$

Attach % symbol. 25%

$$\frac{1}{4} = 25\%$$

EXAMPLE 2

Express $\dfrac{3}{2}$ as a percent.

Pencil and Paper

$$\frac{3}{2} = \frac{n}{100}$$

$$2(n) = 3(100)$$

$$\frac{2n}{2} = \frac{300}{2}$$

$$n = 150$$

$$\frac{3}{2} = 150\%$$

Calculator

$$\frac{3}{2} = 3 \boxed{\div} 2 \boxed{=} 1.5$$

Multiply by 100. $100(1.5) = 150$

Attach % symbol.   150%

$$\frac{3}{2} = 150\%$$

EXAMPLE 3

Express $\dfrac{1}{400}$ as a percent.

Pencil and Paper

$$\frac{1}{400} = \frac{n}{100}$$

$$400(n) = 1(100)$$

$$\frac{400n}{400} = \frac{300}{400}$$

$$n = \frac{1}{4} = 0.25$$

$$\frac{1}{400} = \frac{1}{4}\% = 0.25\%$$

Calculator

$$\frac{1}{400} = 1 \boxed{\div} 400 \boxed{=} 0.0025$$

Move the decimal point   0.00 25
two places to the right
(Multiply by 100).

Attach % symbol.   0.25%

$$\frac{1}{4} = 0.25\%$$

20.3 Exercises

In 1–30, express each decimal or fraction as a percent. Round where indicated.

1. 0.47   2. 0.76   3. 0.293   4. 0.926   5. 0.078

6. 0.001   7. 0.0502   8. 0.00064   9. 6.82   10. 51.9

Round to the Nearer Whole Percent

11. $\dfrac{5}{8}$ 12. $\dfrac{1}{6}$ 13. $\dfrac{5}{12}$ 14. $\dfrac{4}{8}$ 15. $\dfrac{95}{100}$

16. $\dfrac{17}{20}$ 17. $\dfrac{4}{7}$ 18. $\dfrac{7}{6}$ 19. $\dfrac{13}{4}$ 20. $\dfrac{11}{9}$

Round to the Nearer Tenth Percent

21. $\dfrac{5}{12}$ 22. $\dfrac{3}{8}$ 23. $\dfrac{5}{6}$ 24. $\dfrac{3}{13}$ 25. $\dfrac{1}{9}$

26. $\dfrac{7}{16}$ 27. $\dfrac{13}{8}$ 28. $\dfrac{28}{7}$ 29. $\dfrac{14}{5}$ 30. $\dfrac{16}{11}$

In 31–52, complete the chart of fractional, decimal, and percent equivalents.

	Mixed Number or Fraction	Decimal	Percent
31.		$0.33\overline{3}$	$33\frac{1}{3}\%$
32.			30%
33.	$\frac{1}{4}$		
34.	$\frac{1}{10}$		
35.			75%
36.	$\frac{1}{5}$		
37.			50%
38.		0.875	
39.			60%
40.	$1\frac{1}{4}$		
41.		0.4	

	Mixed Number or Fraction	Decimal	Percent
42.	$\frac{1}{8}$		$12\frac{1}{2}\%$
43.		2.5	
44.			$\frac{1}{2}\%$
45.	$\frac{2}{1}$		
46.	$\frac{4}{5}$		
47.			$62\frac{1}{2}\%$
48.	$1\frac{1}{2}$		
49.			$66\frac{2}{3}\%$
50.	$\frac{3}{8}$		
51.		0.0025	
52.			85%

APPLICATIONS

The strength of a solution is usually expressed as a percent. A 5% solution of lugol means that there are 5 parts lugol in 100 parts of solution. Since 5% represents the ratio 5:100 or 1:20, it may also mean 1 part lugol in 20 parts of solution or 100 parts lugol in 2,000 parts of solution.

In determining percent of solution strength, the units of measure must be the same for the solution and the solute. A 5% solution of lugol means 1 milliliter lugol in 20 milliliters water or 100 ounces lugol in 2,000 ounces lugol. Ratios such as 1 milliliter lugol in 20 liters water or 100 liters lugol in 2,000 ounces water will not provide a correct percent of solution strength.

For liquid solutes, the strength of the solution is a *percent by volume.* It is the percent of the final solution volume represented by the volume of the solute used to make the solution. It is determined by $\dfrac{\text{volume of solute}}{\text{volume of solution}} \times 100$. A 12% alcohol solution (by volume) would be a solution made from 12 parts (usually milliliters) of alcohol and enough solvent to bring the total volume up to 100 parts (usually milliliters).

For solid solutes, the strength of the solution is a *percent by weight (mass).* It is the percent of the total solution mass contributed by the solute. It is determined by $\dfrac{\text{mass of solute}}{\text{volume of solution}} \times 100$. A 4% boric acid solution (by weight) would be 4 parts (usually grams) of boric acid per 100 parts (usually milliliters) of solution. Remember that for computational purposes, 1 gram is equal to 1 milliliter.

DEFINITIONS

- A *pure solution* is a substance with a strength of 100%. This means that the substance has not been mixed with any other substance.
- A *stock solution* is a solution which is kept on hand. The component has been mixed in solution form and has a strength less than 100%. The strength of stock solutions may greatly exceed that required for safe use. In these instances the stock solution is used to form a new solution with a lower strength. For example, a 15% stock solution of alcohol may be used to form a 5% alcohol solution.

20.4 Exercises

In 1–10, express each solution as a ratio.

1. 4% boric acid solution

2. 50% stock hydrochloric acid solution

3. 75% stock glycerin solution

4. 95% stock lysol solution

5. $\frac{1}{10}$% silver nitrate solution

6. $\frac{2}{5}$% sodium bicarbonate solution

7. 7.5% sodium phosphate solution

8. $\frac{3}{4}$% phenylephrine solution

9. 100% magnesium sulfate

10. 35% ethyl alcohol

In 11–17, the amount of solute in a solution is often expressed as a percent. It may also be expressed as a fraction or a decimal. Express each percent of solute as a fraction and a decimal.

	Percent of Solute	**Fractional Value**	**Decimal Value**
11.	1% sodium chloride		
12.	5% glucose		
13.	20% lactose		
14.	25% sodium chloride		
15.	50% ethyl alcohol		
16.	60% methyl alcohol		
17.	95% ethyl alcohol		

In 18–22, determine the strength of each solution.

18. 7 parts sodium chloride and 50 parts of solution

19. 7 grams sodium chloride and 50 milliliters of solution

20. 30 milliliters pure ethyl alcohol and 75 milliliters of solution

21. 50 grams of glucose in 1 liter (1000 milliliters) of solution

22. 5 tablets each containing 6 grains dissolved in 100 milliliters of solution
     Note: 1 gram = 15 grains

23. What percent solution is obtained when 7 grams of boric acid crystals are in a 500-milliliter solution?

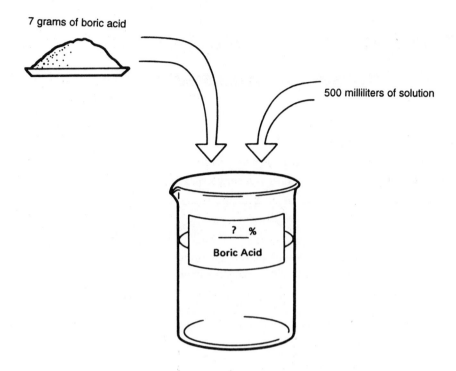

7 grams of boric acid

500 milliliters of solution

___?___ %

Boric Acid

Unit 21

Equations Involving Percents

Objectives

After studying this unit the student should be able to:
- **Determine unknown quantities involving percents by using proportions.**
- **Determine unknown quantities involving percents by using percent equation.**

USING PERCENTS

In real life, three types of percent problems are encountered. Notice that in any problems involving percent, all that is required is setting up a proportion that compares a part to the whole.

EXAMPLE 1

25% of 200 is what number?

$$25\% = \frac{25}{100} \begin{array}{l} \text{part} \\ \text{whole} \end{array} \qquad \frac{x}{200} \begin{array}{l} \text{part} \\ \text{whole} \end{array}$$

$$\frac{25}{100} = \frac{x}{200} \qquad \text{Cross multiply.}$$

$$100(x) = 25(200) \qquad \text{Simplify.}$$

$$100x = 5000$$

$$\frac{100x}{100} = \frac{5000}{100} \qquad \text{Divide each side by 100.}$$

$$x = 50 \qquad \text{Reduce.}$$

25% of 200 is 50.

Remember that proportions have to compare things in the same order.

EXAMPLE 2

40 is what percent of 400?

$$\frac{40}{100} \quad \frac{\text{part}}{\text{whole}} \qquad\qquad \frac{n}{100} \quad \frac{\text{part}}{\text{whole}}$$

$$\frac{40}{400} = \frac{n}{100} \qquad\qquad \text{Cross multiply.}$$

$$400(n) = 40(100) \qquad\qquad \text{Simplify.}$$

$$400n = 4000$$

$$\frac{400n}{400} = \frac{4000}{400} \qquad\qquad \text{Divide each side by 400.}$$

$$n = 10$$

Since $\dfrac{n}{100} = \dfrac{10}{100} = 10\%$, 40 is 10% of 400.

Notice that a whole, rather than a part, is being found in Example 3.

EXAMPLE 3

75 is 50% of what number?

$$50\% = \frac{50}{100} \quad \frac{\text{part}}{\text{whole}} \qquad\qquad \frac{75}{y} \quad \frac{\text{part}}{\text{whole}}$$

$$\frac{50}{100} = \frac{75}{y} \qquad\qquad \text{Cross multiply.}$$

$$50(y) = 100(75) \qquad\qquad \text{Simplify.}$$

$$50y = 7{,}500$$

$$\frac{50y}{50} = \frac{7{,}500}{50} \qquad\qquad \text{Divide each side by 50.}$$

$$y = 50$$

Thus, 75 is 50% of 150.

21.1 Exercises

In 1–40, find each unknown quantity. Round the answer to the nearer tenth if necessary.

1. 50% of 80 is what number?   2. 75% of 120 is what number?

3. 33.3% of 90 is what number?   4. 30% of 43.5 is what number?

5. 5% of 130 is what number?   6. 2.5% of 80 is what number?

7. 0% of 38 is what number?   8. 120% of 50 is what number?

9. 200% of 145 is what number?   10. 500% of 30 is what number?

11. 20 is what percent of 80?   12. 15 is what percent of 75?

13. 15 is what percent of 45?   14. 35 is what percent of 45?

15. 3 is what percent of 42?   16. 28 is what percent of 81?

17. 17 is what percent of 102?
18. 50 is what percent of 25?

19. 300 is what percent of 160?
20. 163 is what percent of 42?

21. 82 is 50% of what number?
22. 76 is 40% of what number?

23. 48 is 15.5% of what number?
24. 4.23 is 1% of what number?

25. 13 is 40% of what number?
26. 121 is 55% of what number?

27. 65 is 10% of what number?
28. 42 is 80% of what number?

29. 50 is 200% of what number?
30. 32 is 350% of what number?

31. 15 is what percent of 300?
32. 27% of 130 is what number?

33. 110 is 30% of what number?
34. 12.5% of 80 is what number?

35. 2.2 is 10% of what number?
36. 22.5 is what percent of 180?

37. 45 is 50% of what number?
38. 12 is what percent of 2,400?

39. 32% of 8 is what number?
40. $1\frac{1}{2}$ is 200% of what number?

ANOTHER METHOD OF CALCULATING WITH PERCENTS

Although proportional rate pairs will solve every type of percent problem, there are quicker methods. In this method the word "of" is expressed by using the times sign (•). Similarly, the equal sign (=) is used to express the word "is."

EXAMPLE 1

25% of 200 is what number?

25% of 200 is what number?

$$25\% \cdot 200 = X$$

$$0.25 \cdot 200 = X$$

$$50 = X$$

25% of 200 is 50.

Calculator

25 $\boxed{\%}$ $\boxed{\times}$ 200 $\boxed{=}$ 50

EXAMPLE 2

40 is what percent of 400?

40 is what percent of 400?

$$40 = n\% \cdot 400$$

$$\frac{40}{400} = \frac{n\% \cdot 400}{400}$$

$$\frac{10}{100} = n\%$$

$$10\% = n$$

40 is 10% of 400.

Calculator

$$\frac{40}{400} = 40 \boxed{\div} 400 \boxed{=} 0.1$$

$$0.1 = 10\%$$

EXAMPLE 3

75 is 50% of what number?

75 is 50% of what number?

$$75 = 50\% \cdot y$$

$$75 = 0.50 \cdot y$$

$$\frac{75}{0.50} = \frac{0.25y}{0.50}$$

$$150 = y$$

75 is 50% of 150

Calculator

$$75 = 50\% \text{ of } y$$

$$\frac{75}{50\%} = y$$

$$75 \boxed{\div} \, 50 \boxed{\%} \boxed{=} 150$$

21.2 Exercises

In 1–40, find each unknown quantity. Round the answer to the nearer tenth if necessary.

1. 20% of 63 is what number?
2. 65% of 90 is what number?
3. 66.6% of 120 is what number?
4. 25% of 39.3 is what number?
5. 14.4% of 100 is what number?
6. $1\frac{1}{2}$% of 40 is what number?
7. 0.5% of 900 is what number?
8. 15% of 80 is what number?
9. 25% of 30 is what number?
10. 400% of 65 is what number?
11. 15 is what percent of 60?
12. 84 is what percent of 400?
13. 16.6 is what percent of 33.2?
14. 22.3 is what percent of 111.5?
15. 2 is what percent of 21?
16. 1.5 is what percent of 90?
17. 15.21 is what percent of 121.68?
18. 100 is what percent of 75?
19. 220 is what percent of 180?
20. 68.5 is what percent of 13.7?
21. 40 is 20% of what number?
22. 92 is 55% of what number?
23. 32 is 28% of what number?
24. 0.5 is 2% of what number?
25. 18 is $\frac{1}{2}$% of what number?
26. 1.5 is 1.5% of what number?
27. 235 is 25% of what number?
28. 16.7 is 0% of what number?
29. 300 is 150% of what number?
30. 85 is 250% of what number?
31. 25 is what percent of 400?
32. 16% of 320 is what number?
33. 52 is 50% of what number?
34. 9.5% of 145 is what number?
35. 1.5 is 20% of what number?
36. 14 is what percent of 70?
37. 150 is 75% of what number?
38. 5 is what percent of 6,000?
39. 85% of 9 is what number?
40. 3.5 is 400% of what number?

In 41–46, complete the chart by determining the value of the unknown quantity. Round the answer to the nearer tenth if necessary.

	Known Quantity	Known Quantity	Unknown Quantity	Answer
41.	World population: 6 billion	Starving: 560,000,000	Percent starving	
42.	Patients admitted: 30	Patients released: 70%	Patients remaining	
43.	25 shots recommended	20 shots given	Percent of shots not given	
44.	Earned: $30,500	Spent: $29,000	Percent spent	
45.	9 hours sleep per night	Total hours in one day	What percent of one day are the non-sleeping hours?	
46.	World population: 6 billion	HIV Positive $\frac{1}{10}\%$	How many people are HIV positive?	

APPLICATIONS

Solutions may be prepared in various strengths. The strength of the solution is indicated by the amount of solute that is in the solution. This amount may be expressed as a percent. A 5% glucose solution means that 5% of the solution is the glucose. Since solid solutes are measured in grams and 1 gram is equal to 1 milliliter, the amount of solute is the product of the percent strength times the amount of solution.

EXAMPLE 1

Determine the number of grams of glucose that are present in 100 milliliters of 5% glucose solution.

5% of 100 milliliters is what number?

$$\frac{5}{100} = \frac{x}{100}$$

$$100\,(x) = 5\,(100)$$

$$100x = 500$$

$$\frac{100x}{100} = \frac{500}{100}$$

$$x = 5 \text{ grams glucose}$$

21.3 Exercises

In 1–7, determine the number of grams of solute present in each solution.

1. 45 milliliters of a 7% procaine solution

2. 125 milliliters of a 13% glucose solution

3. 275 milliliters of a 5% potassium permanganate solution

4. 750 milliliters of a 2% tyrothricin solution

5. 900 milliliters of a 0.9% saline solution

6. 2.750 liters (2 750 milliliters) of a 12% sodium bicarbonate solution

7. Calculate the quantity of glucose required to prepare 1 liter of a 27% solution.

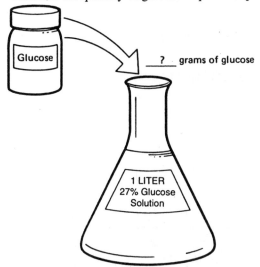

_?_ grams of glucose

1 LITER
27% Glucose
Solution

8. A physiological solution is a solution which matches a person's body chemistry. A 0.9%
    physiological saline solution is prepared. How many grams of sodium chloride are need-
    ed to prepare 500 milliliters of this solution?

In 9 and 10, using tablets, a health care worker prepares 300 milliliters of a $\frac{1}{2}$% solution. Each
tablet weighs 15 milligrams or 0.015 gram.

9. Find the amount of solute needed to prepare the solution.

10. Find the number of tablets needed to prepare the solution.

Note: number of tablets $= \dfrac{\text{amount (mass) of solute}}{\text{mass of one tablet}}$ _or_ $\dfrac{\text{amount (mass) of solute}}{0.015 \text{ gram}}$

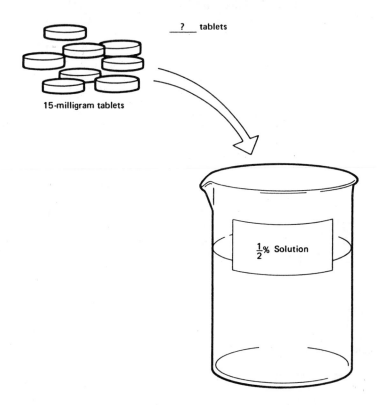

_?_ tablets

15-milligram tablets

$\frac{1}{2}$% Solution

Unit 22

Computations With Percents

Objectives

After studying this unit the student should be able to:
- **Find the percent of increase and the percent of decrease.**
- **Calculate discount or markup.**
- **Estimate answers using percents.**

PERCENT OF INCREASE OR DECREASE

When a measurement increases or decreases it is often convenient to use a percent to compare this increase or decrease to the original measurement.

To find the percent of increase or decrease:
- Find the difference between the original and the new quantity.
- Set up a ratio comparing to amount of increase or decrease and the original quantity.

$$\frac{\text{amount of increase or decrease}}{\text{original quantity}}$$

- Express the ratio as a percent.

$$\% \text{ of increase} = \frac{\text{amount of increase}}{\text{original quantity}} \qquad \% \text{ of decrease} = \frac{\text{amount of decrease}}{\text{original quantity}}$$

EXAMPLE 1

The percent of increase is illustrated by using pulse rates.

Normal pulse rate: (original quantity)	50	Increase: (change)	30
Pulse rate after running: 1.6 kilometers: (new quantity)	80	Percent of increase: $\frac{30}{60} = 0.60$	60%

224

EXAMPLE 2

The percent of decrease is also illustrated by using pulse rates.

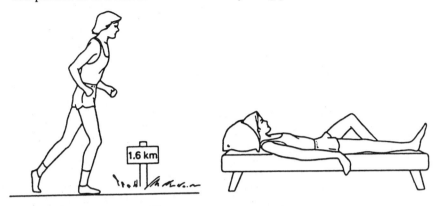

Pulse rate after running:
    1.6 kilometers: 80
    (original quantity)

Decrease:
    (change)   20

Pulse rate after
    two minutes rest: 60
    (new quantity)

Percent of decrease:
$$\frac{20}{80} = 0.25$$ 25%

DISCOUNTS AND MARKUPS

Discounts and markups are usually stated in terms of percents.

DEFINITIONS

- A *discount* is the dollar amount by which the regular price is reduced. The price after the dollar amount of the discount has been subtracted is the *sale price*. Two different methods may be used to calculate the sale price.

EXAMPLE 1

Find the sale price of a $200 item discounted 25%.

Method 1
Compute the dollar amount
of the discount, then subtract.
$200 • 25% = $40
$200 − $40 = $160
The sale price is $160.

Method 2
Find the percent to be paid,
then compute the sale price directly.
100% − 25% = 75%
$200 • 75% = $160
The sale price is $160.

DEFINITION

- A *markup* is the dollar amount by which the original cost is increased to cover expenses and profit in arriving at the charged price.

EXAMPLE 2

Find the price charged for a tablet costing the clinic $.40 if the tablet is marked up 150%.

Tablet Cost to Clinic	Tablet Cost Increase	Tablet Cost to Patient
$.40	150%	?
$.40	150% • .40	?
$.40 + +	$.60	$1.00

Cost + markup = charged price

$.40 + 150% • 40 = charged price

$.40 + $.60 = $1.00

The tablet would cost the patient $1.00.

22.1 Exercises

In 1–10, calculate the percent of increase or decrease by completing the tables.

	Normal Pulse Rate	Pulse Rate After Running	Increase	Increase ÷ Normal Pulse Rate	Percent of Increase (Nearer Tenth)
1.	60	90			
2.	54	78			
3.	80	100			
4.	70	120			
5.	65	88			

	Pulse Rate After Running	Pulse Rate After Two Minutes Rest	Decrease	Decrease ÷ Pulse Rate After Running	Percent of Decrease (Nearer Tenth)
6.	90	75			
7.	78	65			
8.	100	90			
9.	120	100			
10.	88	77			

In 11–20, find each percent of increase or decrease.

11. Pulse rate before running : 65
    Pulse rate after running : 105
    Percent of increase:

12. Present weight: 240 lb
    Weight 20 years ago: 200 lb
    Percent of increase:

13. Normal weight: 98 kg
    Weight after illness: 80 kg
    Percent of decrease:

14. Pulse rate after workout: 120
    Pulse rate normally: 60
    Percent of increase:

15. From 2 to 14

16. From 14 to 2

17. From 500 to 100

18. From 50 to 80

19. From 5 to 15

20. From 350 to 175

In 21–26, find the sale price of each item.

21. Regular price: $50
    Discount: 25%
    Sale price:

22. Regular price: $2.00
    Discount: 5%
    Sale price:

23. Regular price: $80
    Discount: 10%
    Sale price:

24. Regular price: $65.50
    Discount: 15%
    Sale price:

25. Regular price: $3,500
    Discount: 5%
    Sale price:

26. Regular price: $421.78
    Discount: 10%
    Sale price:

In 27–32, find the charged selling price for each item.

27. Cost: $100
    Markup: 25%
    Charged price:

28. Cost: $3.00
    Markup: 100%
    Charged price:

29. Cost: $25
    Markup: 500%
    Charged price:

30. Cost: $3000
    Markup: 40%
    Charged price:

31. Cost: $.85
    Markup: 600%
    Charged price:

32. Cost: $.24
    Markup: 250%
    Charged price:

33. During a recent six month period, the average daily cost of a room in St. Joseph hospital increased from $271.25 to $301.50. If the percent of increase stays constant for the next six months, how much will be the average daily cost of a room in this hospital?

34. Joyce said, "Last year I was in school so I had no salary. This year I am earning $24,500. This is an increase of $24,500 or a 100% increase." Is Joyce's statement correct? Explain.

35. Is an increase of 50% of a number followed by a decrease of 50% equal to the original number? Explain and give an example.

ESTIMATING WITH PERCENTS

Estimating affords the opportunity to perform calculation mentally without the aid of pencils, paper, or a calculator. Mental images and calculations should be utilized in the examples.

EXAMPLE 1

Approximately what percent of the diamond is shaded?

Mentally, "draw in the diagonals" (segments connecting opposite angles). There is 1 out of 4 regions shaded. This is 25%.

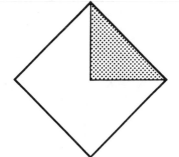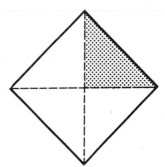

EXAMPLE 2

Mentally calculate 250% of 50.

Think: 100% of 50 = 50 so, 200% of 50 = 100

50% of 50 = 25

250% of 50 is 125.

Using multiples of 10 sometimes helps in estimating percents.

EXAMPLE 3

Estimate 29.7% of 448.02.

Think: 29.7% is rounded to 30%.
448.02 is rounded to 450.
30% of 450 is 135.
29.7% of 448.02 is about 135.

EXAMPLE 4

Mentally determine if 2% of 600 is less than, equal to, or greater than 5% of 300.

Think: 2% of 600 is 12.   2% of 600 <, =, > 5% of 300

    5% of 300 is 15.   12   <   15

                    Thus, 2% of 600 is less than 5% of 600.

22.2 Exercises

In 1–6, estimate each percent.

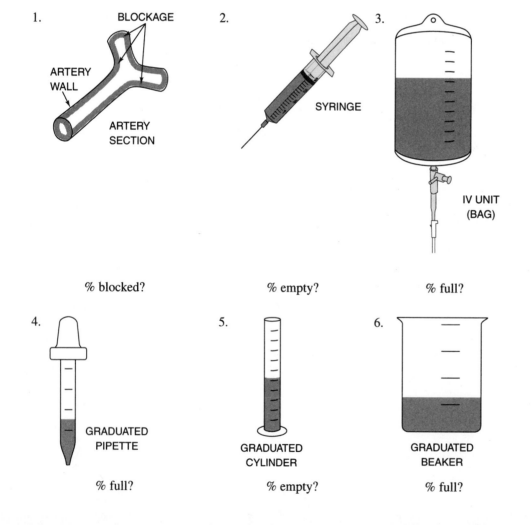

1.   BLOCKAGE

ARTERY
WALL

ARTERY
SECTION

% blocked?

2.   SYRINGE

% empty?

3.   IV UNIT
(BAG)

% full?

4.   GRADUATED
PIPETTE

% full?

5.   GRADUATED
CYLINDER

% empty?

6.   GRADUATED
BEAKER

% full?

In 7–12, circle the best estimate.

7. 21% of 72
    a. 14
    b. 15
    c. 16

8. 71% of 98
    a. 60
    b. 65
    c. 70

9. 15% of 702
    a. 105
    b. 110
    c. 115

10. 5.2% of 88
    a. 4.0
    b. 4.5
    c. 5.0

11. $\frac{1}{10}$% of 5217
    a. 5
    b. 6
    c. 7

12. 1.5% of 1007
    a. 13
    b. 14
    c. 15

In 13–24, estimate each percent:

13. $\frac{13}{25}$

14. $\frac{11}{50}$

15. $\frac{99}{198}$

16. 3 out of 7

17. 8 out of 11

18. 2 out of 9

19. 14.2% of 39.6

20. $1\frac{1}{10}$% of 278.89

21. 0.03% of 7239.93

22. 98% of 135.79

23. 132.1% of 203.14

24. 417% of 101.5

In 25–30, use estimation and mental calculations to insert <, =, or > between each relationship.

25. 3% of 600 _____ 5% of 300

26. $\frac{1}{4}$% of 400 _____ $\frac{1}{2}$% of 600

27. 0.05% of 2500 _____ 0.1% of 1500

28. 0.25% of 12,000 _____ 0.75% of 4 000

29. 150% of 800 _____ 200% of 600

30. 250% of 100 _____ 5% of 5 000

APPLICATIONS

In determining the amount of an available drug that is to be administered, the strength of the drug does not change. For this reason, a direct ratio and proportion can be used. When preparing many other solutions, especially ones desired for topical use, the strength of the solution may be changed to a weaker strength or to the strength ordered to be used. Determining the new strength and the amount of solute to be used requires a different type of proportion. This proportion is:

smaller % strength:larger % strength = smaller volume:larger volume

or

weaker:stronger = solute:solvent

EXAMPLE 1

A 5% boric acid solution is needed in the laboratory. A total of 750 milliliters is required. Calculate the amount of boric acid crystals required to prepare this solution.

smaller % strength:larger % strength = smaller volume:larger volume

5%:100% = x milliliters:750 milliliters

$$5:100 = x:750$$

$$100 (x) = 750 (5)$$

$$100x = 3\ 750$$

$$\frac{100x}{100} = \frac{3\ 750}{100}$$

$$x = 37.5 \text{ grams of boric acid}$$

Note: Boric acid is in crystal form and is measured in grams. For computational purposes, 1 milliliter = 1 gram.

When preparing solutions from a stock solution, the amount of solvent must be considered. The amount of solvent is the difference between the amount of solute and the amount of solution. This is written:

solvent = solution − solute

Notice that this is <u>not</u> simply the solvent. A stock solution (solute) is a liquid and displaces approximately the same volume. This displacement is reflected in the formula:

solvent = solution − solute.

EXAMPLE 2

A laboratory technician prepares 100 milliliters of 25% ethyl alcohol from a 95% ethyl alcohol stock solution.

a. Determine, to the nearer milliliter, the amount of 95% ethyl alcohol that is needed.

b. Determine, to the nearer milliliter, the amount of distilled water that is added.

a. smaller % strength:larger % strength = smaller volume:larger volume

25%:95% = x milliliters:100 milliliters

$$25:95 = x:100$$

$$95 (x) = 25 (100)$$

$$95x = 2\ 500$$

$$\frac{95x}{95} = \frac{2\ 500}{95}$$

$$x = 26 \text{ milliliters of 95\% ethyl alcohol}$$

b. solvent = solution − solute

solvent = 100 milliliters − 26 milliliters

solvent = 74 milliliters of distilled water

22.3 Exercises

1. Calculate the number of grams of iodine crystals that are required to prepare 250 milliliters of a 4% iodine solution.

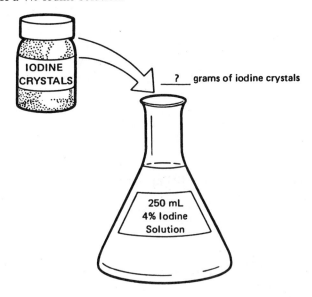

__?__ grams of iodine crystals

250 mL
4% Iodine
Solution

In 2 and 3, a 0.5% boric acid solution is to be prepared from a 5% boric acid stock solution.

2. How much of the 5% boric acid solution is needed to prepare 750 milliliters of the 0.5% boric acid solution?

3. How much distilled water must be added to obtain 750 milliliters of the 0.5% boric acid solution?

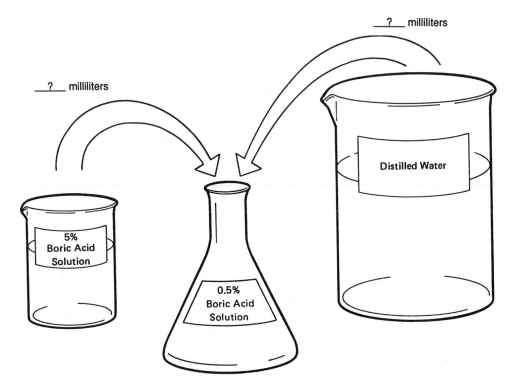

__?__ milliliters

__?__ milliliters

Distilled Water

5%
Boric Acid
Solution

0.5%
Boric Acid
Solution

4. How many milliliters of 40% ethyl alcohol can be prepared from 135 milliliters of 95% ethyl alcohol?

In 5–8, a 40% ethyl alcohol solution is used to make 250 milliliters of a 1:10 ethyl alcohol solution.

5. By expressing the 1:10 ratio as a percent, determine the amount of 40% ethyl alcohol to be used.

6. Determine the amount of distilled water to be used.

7. A stock solution of boric acid contains 1 part boric acid in 200 parts solution. A 1:500 solution is to be prepared. How much of the stock solution must be used to prepare 600 milliliters of the 1:500 solution?

8. A stock solution of 25% ethyl alcohol is available. A total of 250 milliliters of this solution is present in the storage flask. How many 100-milliliter containers of 5% ethyl alcohol can be made from the stock solution?

   Note: Find the amount of 25% ethyl alcohol needed to prepare 100 milliliters of the solution.

Unit **23**

Section Four
Applications to Health Work

Objectives

After studying this unit the student should be able to:
- **Use the basic principles of ratios, proportions and percents to solve health work problems.**

Solutions are prepared for many uses. A solution consists of a solute (the drug) and the solvent (the liquid used to dissolve the drug). Solutions are prepared from pure drugs and from stock solutions.

DEFINITIONS

- Pure drugs are in either a solid form, such as tablets, powders, or crystals, or a liquid form. Pure drugs are 100% strength.
- Stock solutions are solutions that are "on hand" and usually are of a strong strength. The stock solutions are used to make weaker solutions. Stock solution, as the words imply, are always in a liquid form.

Preparing solutions from crystals or powders:
- The drug is usually measured in grams. For purposes of calculation, 1 gram is considered equal to 1 milliliter.
- Weigh the amount of drug desired. Place the drug in a graduated container and add the solvent to obtain the desired amount of solution.
- Since a solid drug does not displace an appreciable volume of liquid, the amount of liquid that is added is considered to be the same as the amount of solution when making large volumes of solutions.

Preparing solutions from liquids or stock solutions:
- The drug is usually measured in milliliters.
- Measure the amount of drug desired. Place the drug in a graduated container and add a sufficient amount of liquid to obtain the desired amount of solution.
- Since a liquid drug displaces approximately the same volume as the liquid, the amount of liquid that is added is the difference between the solution and the solute (drug). That is, for liquids:

  solution = solvent + solute

Preparing solutions from tablets:
- Tablets come in a pre-measured form.
- Determine the number of tablets to be used. Dissolve the desired number of tablets in the amount of solution that is needed.
- The drug (solute) is dissolved in the solvent rather than the solvent being added to the solute. The volume of the solution is not increased appreciably by the solute.

Any solution which is to be sterilized is poured or filtered into a glass flask and sterilized according to prescribed or accepted techniques. Solutions that are used for intravenous or injections, such as glucose or saline solutions, are made from distilled water, are filtered, and then are sterilized.

23.1 Exercises

In 1–6, determine the number of grams of solute that are required to prepare each solution.

1. 500 milliliters of a 0.9% sodium chloride solution

2. 750 milliliters of a 0.2% bichloride of mercury solution

3. 600 milliliters of a 4.5% magnesium sulfate solution

4. 250 milliliters of a 0.75% sodium chloride solution

5. 1 liter of a 0.5% potassium permanganate solution

6. 8 liters of a 35% glucose solution

7. When using 20-milligram (0.020-gram) tablets to prepare 500 milliliters of a 0.75% solution, how many tablets would be required?

   Note: number of tablets $= \dfrac{\text{amont (mass) of solute}}{\text{mass of one tablet}}$

8. Drugs in solution are expressed as weight (mass/volume). Vistaril 100 milligrams/2 milliliters means 100 milligrams of the drug in 2 milliliters of solution. The doctor orders the medication in units of weight (mass) measure. To administer the drug, the prescription must be expressed in units of volume measure. The doctor orders Vistaril 37.5 milligrams. How many milliliters of the Vistaril 100 milligrams/2 milliliters must be administered?

9. The doctor orders scopalomine 0.2 milligram. The ampul contains 0.6 milligram of scopalomine in 1 milliliter of solution. How many milliliters are required to supply this dosage?

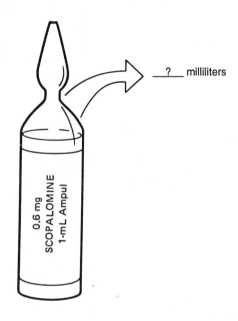

_?_ milliliters

0.6 mg
SCOPALOMINE
1-mL Ampul

In 10–14, tincture of iodine is prepared by dissolving 70 grams of iodine and 50 grams of potassium iodide in 50 milliliters of solution. This is then diluted with 95% ethyl alcohol to make 1 000 milliliters (1 liter) of solution.

10.   What percent of the final solution does the iodine constitute?

11.   What percent of the final solution does the potassium iodide constitute?

12.   What percent of the final solution does the alcohol constitute?

13.   How many grams of potassium iodide are present in 625 milliliters of tincture of iodine?

14.   How many milliliters of water are present in 1 000 milliliters of solution?

> *Note:* It can be assumed that in the 50-milliliter solution of iodine, potassium iodide, and solvent, the solvent is 50 milliliters of water. In 95% ethyl alcohol there is 5% water.

15.   A sodium chloride solution is prepared using 1.5 drams of sodium chloride in 750 milliliters of solution. What is the percent strength?

> *Note*: 1 dram = 4 grams

In 16–19, give the percent strength for each chemical in the solutions.

16.   1.5 grams of merthiolate crystals in 135 milliliters of solution

17.   50 grams of glucose in 625 milliliters of solution

18.   0.9 grams of cocaine in 175 milliliters of solution

19.   3 grams of sodium bromide in 4 fluid ounces of solution

> *Note:* 1 fluid ounce = 31.10 grams

20.   A 0.9% physiological saline solution means that there are 9 grams of sodium chloride in 1 000 milliliters of solution. How many grams are present in 15 milliliters of solution?

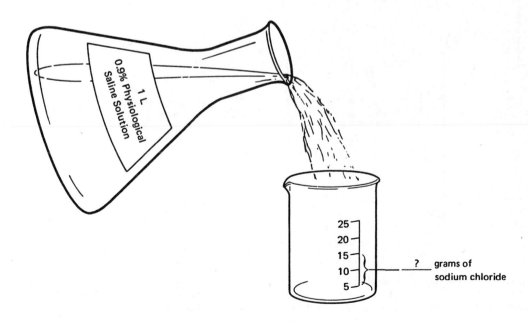

In 21–23, Hayem's solution contains 0.25 grams of mercuric chloride, 2.5 grams of sodium sulfate, and 0.5 grams of sodium chloride in 100 milliliters of solution. Find the amount of each solute in 375 milliliters of an identical solution.

21. mercuric chloride

22. sodium sulfate

23. sodium chloride

In 24 and 25, Gram's iodine solution is prepared by using iodine crystals and potassium iodide. A 300-milliliter solution contains 1 gram of iodine crystals and 2 grams of potassium iodide.

24. How many grams of iodine crystals are in 1 000 milliliters of this solution?

25. How many grams of potassium iodide are in 1 000 milliliters of this solution?

26. It takes 1 gram of iodine crystals and 2 grams of potassium iodide to make 300 milliliters of Gram's iodine solution. An inventory of the laboratory indicates that there are 135 grams of potassium iodide and 450 grams of iodine crystals. What is the maximum amount of Gram's iodine solution that can be prepared from the chemicals not in inventory?

In 27 and 28, a 0.3% zephiran chloride solution is made from a 15% zephiran chloride stock solution.

27. How many milliliters of solute are required to make 750 milliliters of the 0.3% zephiran chloride solution?

28. How many milliliters of solvent are required?

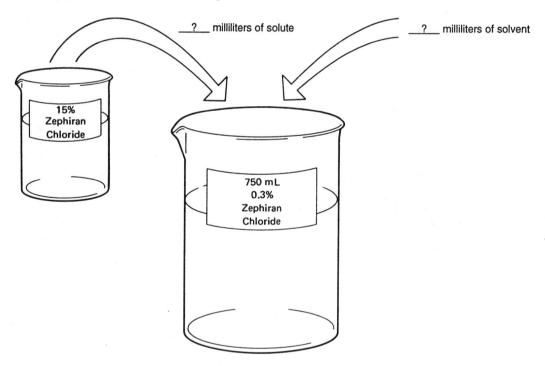

In 29 and 30, a 40% ethyl alcohol preparation is available in the laboratory.

29. Calculate the amount of stock solution needed to prepare 100 milliliters of a 15% ethyl alcohol solution.

30. Calculate the amount of distilled water needed.

In 31–33, a 30% ethyl alcohol solution is available in the laboratory. How much stock solution is required to prepare 50 milliliters of each dilution?

31. 25% ethyl alcohol

32. 10% ethyl alcohol

33. 5% ethyl alcohol

In 34–36, absolute (100%) ethyl alcohol is available as a stock solution. How much stock solution is required to prepare 125 milliliters of each dilution?

34. 25% ethyl alcohol solution

35. 65% ethyl alcohol solution

36. 95% ethyl alcohol solution

37. A 40% magnesium sulfate solution is available. How much water must be added to 60 milliliters of this solution to prepare a 2% solution?

38. A flask contains 57 milliliters of 6% potassium permanganate. How many milliliters of 4% solution can be prepared from 57 milliliters of 6% potassium permanganate?

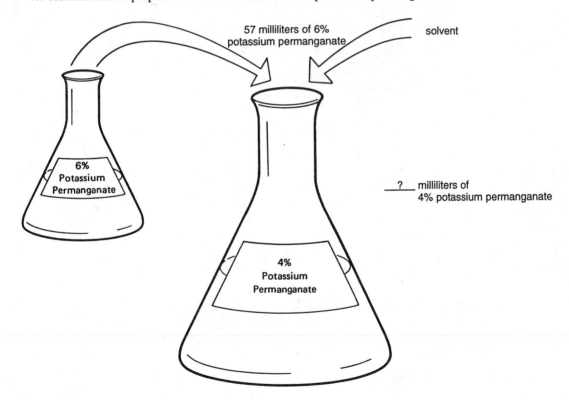

39. How many milliliters of 35% ethyl alcohol can be prepared from 250 milliliters of 40% ethyl alcohol?

A total of 4.750 liters of 95% ethyl alcohol is available in the laboratory. For a series of tests, several dilutions are required. The allocation of the alcohol is: 1.500 liters for 50% ethyl alcohol; 1.250 liters for 25% ethyl alcohol; 0.75 liters for 20% ethyl alcohol; and 1.250 liters for 12% ethyl alcohol. How many milliliters, or liters, will each of the designated amounts prepare when added to distilled water?

40. 50% ethyl alcohol

41. 25% ethyl alcohol

42. 20% ethyl alcohol

43. 12% ethyl alcohol

44. There are 62 grams of sodium chloride in 950 milliliters of a solution. How many grams are in 125 milliliters of the same solution?

In 45–46, using distilled water, boric acid may be dissolved in various dilutions.

45. A boric acid solution is prepared by using 57 grams of boric acid in 10 liters of solution. What is the ratio strength of the solution?

46. A bottle contains a boric acid in a dilution of 1:400 (1 part boric acid to 400 parts solution). How much boric acid is present in 150 milliliters of the solution?

In 47–49, ethyl alcohol may be used in various dilutions for a variety of purposes. The stock solution of 95% ethyl alcohol and 5% water is a 95:100 or 19:20 ethyl alcohol solution.

47. A laboratory technician is requested to prepare 100 milliliters of a 1:4 ethyl alcohol solution (25 parts ethyl alcohol to 100 parts water). How much 95% ethyl alcohol is needed?

48. How much 95% ethyl alcohol is needed to prepare 125 milliliters of a 1:10 ethyl alcohol solution?

49. How much distilled water is needed to prepare 125 milliliters of a 1:10 ethyl alcohol solution?

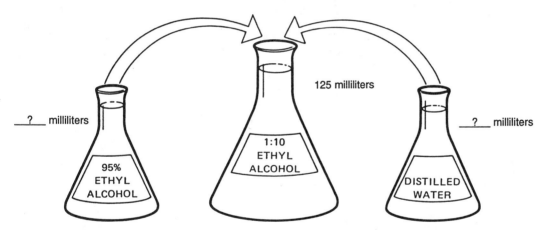

In 50–53, methylene blue contains 0.3 grams of methylene blue in 30.0 milliliters of 95% ethyl alcohol.

50. What percent of the solution is methylene blue?

51. What percent of the solution is ethyl alcohol?

52. How much methylene blue is present in 125 milliliters of solution?

53. How much ethyl alcohol is present in 125 milliliters of solution?

In 54–59, one liter of Ringer's solution contains 0.42 gram of potassium chloride, 9 grams of sodium chloride, 0.24 gram of calcium chloride, 0.20 gram of sodium carbonate, and distilled water.

54. What percent of the solution is composed of potassium chloride?

55. What percent of the solution is composed of sodium chloride?

56. What percent of the solution is composed of calcium chloride?

57. Determine the amount of potassium chloride needed for 750 milliliters of solution.

58. Determine the amount of sodium chloride needed for 750 milliliters of solution.

59. Determine the amount of calcium chloride needed for 750 milliliters of solution.

In 60 and 61, a physiological saline solution is composed of 0.9% sodium chloride and 99.1% distilled water.

60. How much sodium chloride is present in 125 milliliters of solution?

61. How much distilled water is present in 750 milliliters of solution?

In 62 and 63, a clinic consists of 47 full-time and 6 part-time employees.

62. What percent of the individuals are part-time employees?

63. Laboratory technicians make up 16.98% of the staff. How many staff members are laboratory technicians?

In 64–69, a clinic treats 1,262 patients in one month. The cost of operating the clinic (expenditures) is $102,467. The clinic sends out bills (revenue) totalling $127,472. A study reveals that 67% of these bills are paid within 30 days. Insurance payments account for 27% of the money received within the 30 days for the 1,262 patients. Assume all patients were equally insured.

64. How much money is received within the 30 days following the billing?

65. How much of the money received is paid by insurance?

66. Salaries account for 78% of the operating cost. How much is spent on salaries?

67. For the month, insurance premiums are $3,894. What percent of the operating cost is represented by this figure?

68. By what percent do the operating costs exceed the amount of money received within the 30 days from the billing?

69. If all the revenue (total billing) is received, what is the percent of the excess of revenue over expenditure?

70. If the insurance payments increased to cover 32% of the cost, what would this increase represent in total dollars to the clinic?

71. If the clinic gave all patients a 3% discount to patients who paid their bill within 30 days, what dollar amount would this represent for the 67% that did pay within this time?

72. For patients not paying within the 30-day period, a 15% late charge is added. How many additional dollars would the clinic receive if all patients in the 33% late-payment category eventually paid their bill?

SECTION 4: Summary, Review and Study Guide

VOCABULARY

component	percent
cross products	percent of decrease
discount	percent of increase
extremes	proportion
implied component	rate
markup	rate pairs
means	ratio
	terms

CONCEPTS, SKILLS AND APPLICATIONS

Objectives With Study Guide

Upon completion of Section 4 you should be able to:

- **Compare quantities using ratios.**

  Are 10:12 and 15:18 equivalent ratios?

  $$\frac{10}{12} = 0.8\overline{3} \qquad \frac{15}{18} = 0.8\overline{3}$$

  Since their decimal equivalents are the same, the ratios 10:12 and 15:18 are equivalent.

- **Identify mean terms and extreme terms in a proportion.**

  List the means and extremes in the proportion $3{:}5 = x{:}9$
  5 and x are the means.
  3 and 9 are the extremes.

- **Determine if a proportion is an equality using the means and extremes.**

  Is 8 to 5 equal to 10 to 8?

  $$\frac{8}{5} \overset{?}{=} \frac{10}{8}$$

  $5(10) \neq 8(8)$

  Thus, the proportion is not an equality.

- **Determine the unknown in a simple proportion.**

  Solve for a.

  $$\frac{2}{8} = \frac{a}{20}$$

  $$8(a) = 2(20)$$

  $$a = \frac{2(20)}{8}$$

  $$a = 5$$

- **Determine the unknown in a rate pair.**

  If 8 g of sucrose will make 4 L of solution, how many g will be needed for $\frac{1}{2}$ L of solution?

  $$\frac{8}{4} = \frac{n}{\frac{1}{2}}$$

  $$4(n) = 8(\tfrac{1}{2})$$

$$n = \frac{8(\frac{1}{2})}{4}$$

$n = 1$

1 g of sucrose will be needed to make $\frac{1}{2}$ L of solution.

- **Determine the unknown in a proportion which contains an implied component.**

  If Tylenol costs $0.0325 per tablet, how much will 10,000 tablets cost?

$$\frac{0.0325}{1} = \frac{n}{10,000}$$

$$1(n) = 10,000\,(0.0325)$$

$$n = 325$$

10,000 tablets would cost $325.00.

- **Determine the unknown in a proportion which contains fractional components.**

  One-half teaspoon of fructose is used per 2 liters of water. How much fructose would be in 10 liters of water?

$$\frac{\frac{1}{2}}{2} = \frac{n}{10}$$

$$2(n) = \frac{1}{2}(10)$$

$$n = \frac{\frac{1}{2}(10)}{2}$$

$$n = 2\frac{1}{2} \text{ or } 2.5$$

$2\frac{1}{2}$ teaspoons of fructose would be in 10 L of water.

- **Express ratios, fractions, and decimals as percents.**

  Use percent notation to express each of the following: 3:25, $\frac{5}{2}$, and 0.015.

$$3{:}25 = \frac{3}{25} = \left(\frac{3}{25}\right)\left(\frac{4}{4}\right) = \frac{12}{100} = 12\%$$

$$\frac{5}{2} = \left(\frac{5}{2}\right)\left(\frac{50}{50}\right)\frac{250}{100} = 250\%$$

$$0.015 = 0.015\left(\frac{100}{100}\right) = \frac{1.5}{100} = 1.5\%$$

- **Express percents as equivalent fractions or decimals.**

  Express 14.5% and 230% as equivalent fractions, then decimals.

$$14.5\% = \frac{14.5}{100} = \frac{145}{100}$$

$$14.5\% = 0.145$$

$$230\% = \frac{230}{100} = 2.30$$

$$230\% = 2.30$$

- **Express decimal, fractional, and percent equivalents.**

  Express 0.55 and $\frac{1}{50}$ as equivalent percents.

  $$0.55 = 0.55\left(\frac{100}{100}\right) = \frac{55}{100} = 55\%$$

  0.55 equals 55%

  $$\frac{1}{50} = \frac{n}{100}$$

  $$50(n) = 100(1) \text{ or } \frac{1}{50} = \left(\frac{1}{50}\right)\left(\frac{2}{2}\right)$$

  $$n = \frac{100(1)}{50} \qquad \left(\frac{1}{50}\right)\left(\frac{2}{2}\right) = \frac{2}{100}$$

  $$n = 2 \qquad \frac{2}{100} = 2\%$$

  $\frac{1}{50}$ equals 2%

- **Determine unknown quantities involving percents by using proportions.**

  40% of 20 is what number?

  $$\frac{40}{100} = \frac{n}{20}$$

  $$100(n) = 40(20)$$

  $$n = \frac{40(20)}{100} = 8$$

  40% of 20 is 8.

  33 is what percent of 3?

  $$\frac{33}{3} = \frac{n}{100}$$

  $$3(n) = 33(100)$$

  $$n = \frac{33(100)}{3} = 1100$$

  33 is 1100% of 3.

  4 is 0.1 percent of what number?

  $$\frac{0.1}{100} = \frac{4}{n}$$

  $$0.1(n) = 100(4)$$

  $$n = \frac{100(4)}{0.1} = 40$$

  4 is 0.1 percent of 40.

- **Determine unknown quantities involving percents by using percent equations.**

    12% of 60 is what number?

    $12\% \cdot 60 = x$

    $0.12 \cdot 60 = x$

    $\quad\quad 7.2 = x$

    12% of 60 is 7.2.

    7.5 is what percent of 75?

    $7.5 = n\% \cdot 75$

    $\dfrac{7.5}{75} = \dfrac{n\% \cdot 75}{75}$

    $0.1 = n\%$

    $\dfrac{10}{100} = n\%$

    $10\% = n$

    7.5 is 10% of 75.

    16 is 200% of what number?

    $16 = 200\% \cdot y$

    $16 = 2.00 \cdot y$

    $\dfrac{16}{2.00} = \dfrac{2.00y}{2.00}$

    $8 = y$

    16 is 200% of 8.

- **Find the percent of increase or decrease.**

    Compute the percent of increase from 5 to 20.

    $20 - 5 = 15$ (difference)

    $\dfrac{\text{difference}}{\text{original}} \quad \dfrac{15}{5} = \dfrac{n}{100}$

    $5(n) = 15\,(100)$

    $n = \dfrac{15\,(100)}{5}$

    $n = 300$

    From 5 to 20 is a 300% increase.

Compute the percent of decrease from 20 to 5.

$$20 - 5 = 15 \text{ (difference)}$$

$$\frac{\text{difference}}{\text{original}} \quad \frac{15}{20} = \frac{n}{100}$$

$$20(n) = \frac{15\,(100)}{20}$$

$$n = 75$$

From 20 to 5 is a 75% decrease.

• **Calculate discounts and markups.**

Find the cost of a ventilator connection originally priced at $9.95, discounted 30%.

Percent to be paid equals 100% − 30% = 70%

$$\frac{70}{100} = \frac{n}{9.95}$$

$$100(n) = 70(9.95)$$

$$n = \frac{70(9.95)}{100}$$

$$n = 6.97$$

The ventilator connection now costs $6.97.

Find the charged price of a surgical binding originally costing $18.00 that is marked up 250%.

$$18 + 250\% \text{ of } 18 = \text{charged price}$$
$$18 + 2.50(18) = \text{charged price}$$
$$18 + 45 = \text{charged price}$$
$$63 = \text{charged price}$$

The charged price for the surgical binding is $63.00.

• **Estimate answers using percents.**

Using rounded numbers, estimate 19% of 498.

$$\left.\begin{array}{l} 19\% \approx 20\% \\ 498 \approx 500 \end{array}\right\} 500\ (.20) = 100$$

Thus, 19% of 498 is approximately 100.

Review

Major concepts and skills to be mastered for the Section 4 Test.

In 1–10, complete each sentence. Choose from the vocabulary list.

1. The proportional relationship between ratios of unlike measurements is _____ _____.

2. A _____ is an equation which states that two ratios are equal.

3. Proportions are solved using the idea of _____ _____.

4. _____ is a special ratio whose denominator is 100.

5.   In a proportion, the middle terms are called the _____.

6.   A _____ is a ratio of unlike measurements.

7.   The numbers being compared in a ratio are called the _____ of the ratio.

8.   In a proportion, the product of the _____ equals the product of the means.

9.   A _____ is a comparison between numbers expressed as a:b, a to b, or $\frac{a}{b}$.

10. When finding the _____ _____ _____ or _____ _____ _____ a ratio is formed comparing the amount of increase or decrease and the original quantity.

In 11–14, compare each ratio. Use <, =, or >.

11. 5:8 _____ 10:16   12. 3 to 5 _____ 1 to 2

13. $\frac{2}{6}$ ____ $\frac{5}{9}$ 14. 1:4 _____ 6:18

In 15 and 16, list the means and extremes.

15. $\frac{4}{5} = \frac{10}{n}$ 16. $6{:}x = 4{:}8$

In 17–20, determine if each proportion is an equality (=) or an inequality (≠).

17. 10:4 _____ 5:2   18. 3 to 2 _____ 12 to 9

19. 2 to 10 _____ 3 to 20   20. 1:8 _____ 2:15

In 21–24, solve.

21. $\frac{a}{2} = \frac{2}{4}$ 22. $\frac{6}{b} = \frac{3}{4}$

23. $\frac{9}{2} = \frac{c}{5}$ 24. $\frac{2}{6} = \frac{7}{d}$

In 25–28, set up a rate pair and solve.

25. Suppose ten milliliters of solute are contained in 500 milliliters of solution. How much solute would be required to prepare 750 milliliters of this solution with the same strength?

26. Medical service cost at a clinic averages $45.00 per visit. How many patients must be served each month to cover operational costs of $168,750?

27. To give 450 mg of medication using 300 mg tablets, how many tablets would be needed?

28. A clinic is charged 12 cents for each page run on their rented copier. A total of 53 pages are copied in one day. How much is the total charge?

In 29 and 30, set up a proportion and solve.

29. If housekeeping normally takes $\frac{1}{6}$ hour to prepare a patient's room, how many rooms can one housekeeper handle in 2 hours?

30. On a particular weight-loss program, a person is supposed to lose $\frac{1}{2}$ kg per week. At this rate, how long would it take to lose 15 kg?

In 31–36, express each ratio, fraction, or decimal using percent notation.

31. 4:5

32. 8:2

33. $\dfrac{10}{4}$

34. $\dfrac{1}{3}$

35. 0.0025

36. 1.75

In 37–40, express each percent as an equivalent fraction, then as a decimal.

37. 1.35%

38. 60%

39. 125%

40. 0.15%

In 41–44, express each decimal or fraction as an equivalent percent.

41. 0.03

42. $\dfrac{2}{5}$

43. $\dfrac{1}{10}$

44. 2.4

In 45–47, solve using proportions.

45. 500% of 15.5 is what number?

46. 14 is what percent of 42?

47. 300 is 10% of what number?

In 48–50, solve using percent equations.

48. 2.2% of 400 is what number?

49. 250 is what percent of 25?

50. 24 is 20% of what number?

In 51–54, find each percent of increase or decrease.

51. From $2.25 to $7.50.

52. From 0.1% to 0.3%.

53. From 450 to 300.

54. From $10.50 to $5.25.

In 55 and 56, compute each discount or markup.

55. A medical supply wholesaler was running a promotion advertising 12% off all items purchased by the 15th day of the month. The hospital normally spends $23,750 with this supplier. How much could they save?

56. If a roll of tape costs the clinic $1.25, how much should a patient be charged if the tape is marked up 200%.

In 57 and 58, use rounded numbers to estimate each answer.

57. 0.9% of 201

58. 24.7% of $599.99

SECTION 4 TEST

In 1–4, compare each ratio. Use <, =, >.

1. 4:7 _____ 8:12

2. $\dfrac{3}{10}$ _____ $\dfrac{6}{12}$

3. 5 to 8 _____ 6 to 9

4. 1:9 _____ 2:10

In 5–8, solve each proportion.

5. $\dfrac{a}{5} = \dfrac{10}{4}$

6. $\dfrac{3}{b} = \dfrac{6}{10}$

7. $\dfrac{8}{3} = \dfrac{c}{6}$

8. $\dfrac{2}{5} = \dfrac{10}{d}$

In 9–14, set up a rate pair or a proportion and solve.

9. Solution A of Fehling's solution is prepared from copper sulfate and distilled water. One thousand milliliters of solution contain 69.3 grams of copper sulfate. To make 300 milliliters of solution of the same strength would require the use of how much copper sulfate?

10. Lidocaine is an anesthetic often administered prior to suturing (stitching a wound). It often contains epinephrine in small amounts to contain the medication in the area being sutured. A 1:100 solution contains 1 g per 100 mL. How many grams of epinephrine are in 30 mL of a 1:100 000 dilution?

11. Suppose a patient self-administers 50 units of N PH Insulin each day. A one mL vial contains 100 units of insulin? (*Note:* 1 mL = 1 cc)

    a. How many doses are contained in a 10 cc vial?

    b. How many 10-mL vials are needed for a 30-day supply?

12. A patient is given a prescription for Allopurinol. The required dosage is for 0.2 g daily. The pharmacy presently only has 100-mg tablets on hand. How many tablets must the patient take daily to meet the required dosage?

13. Phenobarbital is prescribed for a patient experiencing epileptic seizures. The patient is to take one 50-mg tablet three times each day.

    a. How many total milligrams will be taken in one day?

    b. How many tablets are required to provide a 30-day supply?

    c. How many total milligrams will be taken in 30 days?

14. A patient receives a prescription for a total of 900 mg of Lithium daily. These are to be divided into three equal doses.

    a. What size tablet should the pharmacy provide?

    b. How many tablets should be provided for a one-week supply?

In 15–23, complete the chart of fractional, decimal, and percent equivalents.

	Mixed Number or Fraction	Decimal	Percent
15.	$\dfrac{1}{8}$		
16.		0.1	
17.			5.3%
18.	$\dfrac{14}{2}$		
19.		0.375	
20.			20%
21.	$\dfrac{3}{10}$		
22.		3.0075	
23.			62%

In 24–29, solve using either proportions or percent equations. Round the answer to the nearer tenth if necessary.

24. 25% of 72 is what number?

25. 20 is what percent of 90?

26. 61 is 40% of what number?

27. 450% of 22 is what number?

28. 200 is what percent of 50?

29. 35 is 400% of what number?

In 30 and 31, find each percent of increase or decrease.

30. From 20 to 50

31. From 50 to 20

32. At the last checkup, a patient's weight was 200 lb. One year later, he weighed 215 lb. What was the percent of increase?

33. Recently, one hundred 400-mg Tagamet tablets cost $131.40. Cimetidine, the same medicine in generic form, cost $111.10. What percent is saved by purchasing the generic Cimetidine rather than the brand name Tagamet?

In 34 and 35, find the sale price or the selling price for each item.

34. Regular Price: $1,800
    Discount: 5%
    Sale Price:

35. Cost: $0.05
    Markup: 200%
    Charged Price:

36. See Better Optical is running a grand opening special of 20% off the regular price of an eye exam and one pair of single-vision glasses, costing $120. How much would a customer be charged for an exam and a pair of single-vision glasses?

37. Suppose a hospital marks up the cost of a $0.028 Ibuprofen anti-inflammatory tablet 1000%. How much would a patient be billed for 20 tablets?

In 38–40, use rounded numbers to estimate each answer.

38. 29.6% of 40.5

39. $\dfrac{1}{10}$% of 6900

40. 345% of 15.9

Section Five

Systems of Measure

The Apothecaries' System of Weight

Objectives

After studying this unit the student should be able to:

• **Express Arabic numerals as Roman numerals.**

• **Express apothecaries' units of weight as equivalent apothecaries' units of weight.**

• **Express weight measurements in the apothecaries' system as equivalent metric measurements.**

• **Express weight (mass) measurements in the metric system as equivalent apothecaries' measurements.**

THE APOTHECARIES' SYSTEM OF WEIGHT

Doctors use the apothecaries' system when ordering medications. It is an old English system of measure with whole numbers expressed as Roman numerals. Fractions are expressed as fractions, except for one-half which is expressed as 'ss'. It is important that a health worker be able to read, to understand, and to work with this system. The apothecaries' system has equivalent values within the system and in other systems of measure.

When using the apothecaries' system of measure it is customary to use Roman numerals and fractions. The chart shows some Roman numerals and their base ten equivalents.

I = 1	V = 5	X = 10	L = 50	C = 100

When representing numbers between these values, these principles are used:

• In a sequence, the letters are never repeated more than three times.

• When a letter representing a smaller number precedes a larger one, the value of the first is subtracted from the second.

Subtraction principle in the apothecaries' system.

IV = 5 − 1 *or* 4   XL = 50 − 10 *or* 40
IX = 10 − 1 *or* 9   XC = 100 − 10 *or* 90

• The numerals V and L are not used in a subtractive way.

• When numerals have the same value or when smaller numerals follow larger numerals, the values are added.

Addition principle in the apothecaries' system.

XVI = 10 + 5 + 1 = 16   XXX = 10 + 10 + 10 = 30

When expressing dosages in the apothecaries' system lower case and upper case Roman numerals may be used. When using lower case, be certain to align the dots above the "i" to avoid confusion. Only lower case Roman numerals will be used in this text. The following chart illustrates numbers being written using Roman numerals.

Number	Roman Numeral (capitals)	Roman Numeral (lower case)	Interpretation
one	I	i	1
two	II	ii	$1 + 1 = 2$
three	III	iii	$1 + 1 + 1 = 3$
four	IV	iv	$5 - 1 = 4$
five	V	v	5
six	VI	vi	$5 + 1 = 6$
seven	VII	vii	$5 + 1 + 1 = 7$
eight	VIII	viii	$5 + 1 + 1 + 1 = 8$
nine	IX	ix	$10 - 1 = 9$
ten	X	x	10
eleven	XI	xi	$10 + 1 = 11$
twelve	XII	xii	$10 + 1 + 1 = 12$
thirteen	XIII	xiii	$10 + 1 + 1 + 1 = 13$
fourteen	XIV	xiv	$10 + (5 - 1) = 14$
fifteen	XV	xv	$10 + 5 = 15$
sixteen	XVI	xvi	$10 + 5 + 1 = 16$
seventeen	XVII	xvii	$10 + 5 + 1 + 1 = 17$
eighteen	XVIII	xviii	$10 + 5 + 1 + 1 + 1 = 18$
nineteen	XIX	xix	$10 + (10 - 1) = 19$
twenty	XX	xx	$10 + 10 = 20$

The apothecaries' system of weight has four basic units. The units of weight, the abbreviations or symbol, and the metric equivalent are summarized in this chart.

Unit	Symbol or Abbreviation	Metric Equivalent
grain	gr.	0.065 g
dram	℥ or dr. ap.	3.888 g
ounce	℥ or oz. ap.	31.104 g
pound	lb. ap.	373.242 g or 0.373 kg

When expressing a weight, abbreviations or symbols for the unit may be used. When an abbreviation is used, it appears first, followed by the magnitude of the measurement. The magnitude of the measurement is expressed with lower case Roman numerals or fractions. If the unit is not abbreviated, the magnitude of the measurement is written before the unit and is expressed using Arabic numerals.

EXAMPLE 1

Express eight grains, one-third dram, four ounces, and one-half pound without and with abbreviations or symbols.

- 8 grains *or* gr. viii
- $\frac{1}{3}$ dram *or* ℥ $\frac{1}{3}$

- 4 ounces *or* ℥ iv
- $\frac{1}{2}$ pound *or* lb. ap. ss

Note: The symbol **ss** is used to express $\frac{1}{2}$.

24.1 Exercises

1. Write the numbers 21 through 30 using lower case Roman numerals.

2. In 3–14, use Roman numerals to write the number corresponding to each of the last five years.

Write the meaning of each measurement.

3. ℥ss   4. gr. xix   5. ℥$\frac{1}{4}$ 6. lb. xv

7. ℥xl   8. gr.$\frac{1}{3}$ 9. ℥ iv   10. lb. ap.$\frac{3}{4}$

11. ℥$\frac{5}{6}$ 12. ℥vi   13. lb.$\frac{1}{4}$ 14. ℥$\frac{1}{2}$

Express each measurement by using abbreviations or symbols.

15. $\frac{1}{4}$ ounce   16. 5 drams   17. 4 pounds   18. $\frac{1}{3}$ grain

19. $4\frac{1}{2}$ ounces   20. 10 grains   21. $\frac{3}{4}$ dram   22. 32 grains

23. $10\frac{1}{2}$ pounds   24. 14 ounces   25. $1\frac{1}{2}$ drams   26. $\frac{1}{2}$ pound

APOTHECARIES' EQUIVALENT MEASUREMENTS OF WEIGHT

In the apothecaries' system these units of weight are equivalent.

60 grains (gr.)	= 1 dram (℥ or dr.)
8 drams (℥ or dr.)	= 1 ounce (℥ or oz. ap.)
12 ounces (℥ or oz. ap.)	= 1 pound (lb. ap.)

Proportions are generally used to find equivalent measures.

EXAMPLE 1

Using proportions, express 4 drams as grains.

Method 1

$\cdot\dfrac{drams}{grains} = \dfrac{drams}{grains}$

$\dfrac{1\ dram}{60\ grains} = \dfrac{4\ drams}{a\ grains}$

$1\,(a) = 4\,(60)$

$a = $ gr. ccxl

4 drams equals gr. ccxl.

Method 2

$\cdot\dfrac{drams}{drams} = \dfrac{grains}{grains}$

$\dfrac{1\ dram}{4\ drams} = \dfrac{60\ grains}{a\ grains}$

$1\,(a) = 4\,(60)$

$a = $ gr. ccxl

EXAMPLE 2

Using proportions express 360 grains as drams.

Method 1

$$\bullet\,\frac{\text{grains}}{\text{drams}} = \frac{\text{grains}}{\text{drams}}$$

$$\frac{360 \text{ grains}}{a \text{ drams}} = \frac{60 \text{ grains}}{1 \text{ dram}}$$

$$1\,(360) = 60\,(a)$$

$$360 = 60a$$

$$\frac{360}{60} = \frac{60a}{60}$$

$$\text{\foreignlanguage{}{℈}} \text{vi} = a$$

360 grains equals ℈vi.

Method 2

$$\bullet\,\frac{\text{grains}}{\text{grains}} = \frac{\text{drams}}{\text{drams}}$$

$$\frac{60 \text{ grains}}{360 \text{ grains}} = \frac{1 \text{ dram}}{a \text{ drams}}$$

$$60\,(a) = 1\,(360)$$

$$60a = 360$$

$$\frac{60a}{60} = \frac{360}{60}$$

$$a = ℈\text{vi}$$

EXAMPLE 3

Using proportions, express 0.6 ounce as drams.

Method 1

$$\bullet\,\frac{\text{ounces}}{\text{drams}} = \frac{\text{ounces}}{\text{drams}}$$

$$\frac{1 \text{ ounce}}{8 \text{ drams}} = \frac{0.6 \text{ ounce}}{b \text{ drams}}$$

$$1\,(b) = 0.6\,(8)$$

$$b = 4.8 \text{ drams}$$

0.6 ounce equals 4.8 drams.

Method 2

$$\bullet\,\frac{\text{ounces}}{\text{ounces}} = \frac{\text{drams}}{\text{drams}}$$

$$\frac{1 \text{ ounce}}{0.6 \text{ ounce}} = \frac{8 \text{ drams}}{b \text{ drams}}$$

$$1\,(b) = 0.6\,(8)$$

$$b = 4.8 \text{ drams}$$

EXAMPLE 4

Using proportions, express 32 drams as ounces.

Method 1

$$\bullet\,\frac{\text{ounces}}{\text{drams}} = \frac{\text{ounces}}{\text{drams}}$$

$$\frac{1 \text{ ounce}}{8 \text{ drams}} = \frac{y \text{ ounces}}{32 \text{ drams}}$$

$$8\,(y) = 1\,(32)$$

$$8y = 32$$

$$\frac{8y}{8} = \frac{32}{8}$$

$$y = ℥\text{iv}$$

Method 2

$$\bullet\,\frac{\text{ounces}}{\text{ounces}} = \frac{\text{drams}}{\text{drams}}$$

$$\frac{1 \text{ ounce}}{y \text{ ounces}} = \frac{8 \text{ drams}}{32 \text{ drams}}$$

$$8\,(y) = 1\,(32)$$

$$8y = 32$$

$$\frac{8y}{8} = \frac{32}{8}$$

$$y = ℥\text{iv}$$

32 drams equals ℥ iv.

EXAMPLE 5

Using proportions, express 13 pounds as ounces.

Method 1

- $\dfrac{\text{ounces}}{\text{pounds}} = \dfrac{\text{ounces}}{\text{pounds}}$

$\dfrac{12 \text{ ounces}}{1 \text{ pound}} = \dfrac{z \text{ ounces}}{13 \text{ pounds}}$

$1\,(z) = 12\,(13)$

$z = ℥\text{clvi}$

13 pounds equals $℥$clvi.

Method 2

- $\dfrac{\text{ounces}}{\text{ounces}} = \dfrac{\text{pounds}}{\text{pounds}}$

$\dfrac{12 \text{ ounces}}{z \text{ ounces}} = \dfrac{1 \text{ pound}}{13 \text{ pounds}}$

$1\,(z) = 12\,(13)$

$z = ℥\text{clvi}$

EXAMPLE 6

Using proportions, express $\dfrac{3}{8}$ ounce as pounds.

Method 1

- $\dfrac{\text{pounds}}{\text{ounces}} = \dfrac{\text{pounds}}{\text{ounces}}$

$\dfrac{1 \text{ pound}}{12 \text{ ounces}} = \dfrac{y\ pounds}{\frac{3}{8}\text{ ounce}}$

$12\,(y) = 1\left(\tfrac{3}{8}\right)$

$12y = \dfrac{3}{8}$

$\dfrac{12y}{12} = \dfrac{\frac{3}{8}}{12}$

$y = \dfrac{3}{8} \div 12 \ or \ \dfrac{3}{8} \times \dfrac{1}{12}$

$y = \text{lb.}\ \dfrac{1}{32}$

Method 2

- $\dfrac{\text{ounces}}{\text{ounces}} = \dfrac{\text{pounds}}{\text{pounds}}$

$\dfrac{12 \text{ ounces}}{\frac{3}{8}\text{ ounce}} = \dfrac{1 \text{ pound}}{y \text{ pounds}}$

$12\,(y) = \dfrac{3}{8}\,(1)$

$12y = \dfrac{3}{8}$

$\dfrac{12y}{12} = \dfrac{\frac{3}{8}}{12}$

$y = \dfrac{3}{8} \div 12 \ or \ \dfrac{3}{8} \times \dfrac{1}{12}$

$y = \text{lb.}\ \dfrac{1}{32}$

$\dfrac{3}{8}$ ounce equals lb. ap. $\dfrac{1}{32}$.

24.2 Exercises

In 1–40, express each measurement as an equivalent measurement as indicated.

Grains

1. 0.8 dram
2. ℥ i
3. 0.5 dram
4. ℥ ss
5. ℥ $\dfrac{1}{4}$
6. ℥ ix
7. ℥ xxi
8. ℥ $5\dfrac{1}{2}$

Drams

9. gr. xxx
10. gr. lx
11. 16 ounces
12. ℥ ivss
13. gr. ccxl
14. 12.6 grains
15. ℥ xx
16. ℥ iv
17. gr. $\dfrac{1}{2}$
18. ℥ $\dfrac{1}{4}$
19. gr. $\dfrac{1}{4}$
20. 32 ounces

Ounces

21. $\text{З} \, xii$ 22. 0.4 dram   23. $\text{З} \, clx$ 24. $\text{З} \, cxx$

25. lb. ap. vi   26. 0.5 pound   27. lb. ap. vss   28. lb. ap. v

29. $\text{З} \, 4\frac{4}{5}$ 30. lb. ap. $\frac{1}{4}$ 31. 0.5 dram   32. $\text{З} \, 2\frac{1}{2}$

Pounds

33. $\text{Ʒ} \, lx$ 34. $\text{Ʒ} \, xviii$ 35. 0.6 ounce   36. 600 ounces

37. $\text{Ʒ} \, \frac{4}{5}$ 38. 1.8 ounce   39. $\text{Ʒ} \, \frac{1}{2}$ 40. $\text{Ʒ} \, xiv$

APOTHECARIES'
—METRIC EQUIVALENT MEASUREMENTS OF WEIGHT

Both the apothecaries' and metric systems are used by doctors and health workers. It is important to be able to rapidly and accurately express equivalencies between the two systems. This table provides equivalences that are approximations within the acceptable limits of error. Using slightly different equivalences would produce slightly different answers. The answers, while different, will be very close to each other and within safe limits.

Apothecaries' System		Metric System
1 grain (gr.)	=	60 milligrams (mg)
15 grains (gr.)	=	0.975 g or 1 gram (g)
1 dram (З)	=	3.888 g or 4 grams (g)
1 ounce (З)	=	31.104 grams (g) or 30 g
1 pound (lb. ap.)	=	373.242 grams (g) or 360 g
32 ounces (З)	=	1.119 kg or 1 kg

Note: In the examples and exercises whole number values will be used for metric system equivalencies.

Expressing equivalencies between the two systems is similar to expressing equivalencies within the apothecaries' system. Equivalencies are found by using proportions.

EXAMPLE 1

Using proportions, express 0.5 dram as grams.

Method 1

$\cdot \dfrac{\text{apothecaries'}}{\text{metric}} = \dfrac{\text{apothecaries'}}{\text{metric}}$

$\dfrac{1 \text{ dram}}{4 \text{ grams}} = \dfrac{0.5 \text{ dram}}{w \text{ grams}}$

$1 \, (w) = 0.5 \, (4)$

$w = 2.0 \text{ g}$

0.5 dram equals 2.0 g.

Method 2

$\cdot \dfrac{\text{apothecaries'}}{\text{apothecaries'}} = \dfrac{\text{metric}}{\text{metric}}$

$\dfrac{1 \text{ dram}}{0.5 \text{ dram}} = \dfrac{4 \text{ grams}}{w \text{ grams}}$

$1 \, (w) = 0.5 \, (4)$

$w = 2.0 \text{ g}$

EXAMPLE 2

Using proportions, express $\dfrac{4}{5}$ ounce as kilograms.

Method 1

$\bullet \dfrac{\text{apothecaries'}}{\text{metric}} = \dfrac{\text{apothecaries'}}{\text{metric}}$

$\dfrac{32 \text{ ounces}}{1 \text{ kilogram}} = \dfrac{\frac{4}{5} \text{ ounce}}{b \text{ kilograms}}$

$32(b) = \dfrac{4}{5}(1)$

$32b = \dfrac{4}{5}$

$\dfrac{32b}{32} = \dfrac{\frac{4}{5}}{32}$

$b = \dfrac{4}{5} \div \dfrac{32}{1} \ or \ \dfrac{4}{5} \times \dfrac{1}{32}$

$b = \dfrac{1}{40} \text{ kg}$

Method 2

$\bullet \dfrac{\text{apothecaries'}}{\text{apothecaries'}} = \dfrac{\text{metric}}{\text{metric}}$

$\dfrac{32 \text{ ounces}}{\frac{4}{5} \text{ ounce}} = \dfrac{1 \text{ kilogram}}{b \text{ kilograms}}$

$32(b) \ 1 \left(\dfrac{4}{5}\right)$

$32b = \dfrac{4}{5}$

$\dfrac{32b}{32} = \dfrac{\frac{4}{5}}{32}$

$b = \dfrac{4}{5} \div \dfrac{32}{1} \ or \ \dfrac{4}{5} \times \dfrac{1}{32}$

$b = \dfrac{1}{40} \text{ kg}$

$\dfrac{4}{5}$ ounce equals $\dfrac{1}{40}$ kg.

EXAMPLE 3

Using proportions, express 120 milligrams as grains.

Method 1

$\bullet \dfrac{\text{apothecaries'}}{\text{metric}} = \dfrac{\text{apothecaries'}}{\text{metric}}$

$\dfrac{1 \text{ grain}}{60 \text{ milligrams}} = \dfrac{z \text{ grains}}{120 \text{ milligrams}}$

$60(z) = 120(1)$

$60z = 120$

$\dfrac{60z}{60} = \dfrac{120}{60}$

$z = \text{gr.ii}$

Method 2

$\bullet \dfrac{\text{apothecaries'}}{\text{apothecaries'}} = \dfrac{\text{metric}}{\text{metric}}$

$\dfrac{1 \text{ grain}}{z \text{ grains}} = \dfrac{60 \text{ milligrams}}{120 \text{ milligrams}}$

$60(z) = 120(1)$

$60z = 120$

$\dfrac{60z}{60} = \dfrac{120}{60}$

$z = \text{gr.ii}$

120 milligrams equals gr.ii.

24.3 Exercises

In 1–76, express each measurement as an equivalent measurement as indicated.

Milligrams

1. gr. x
2. gr. ss
3. gr. ivss
4. 0.25 grain
5. gr. xl
6. gr. xvi
7. gr. iss
8. gr. lxxv

Grams

9. gr. lx	10. gr. ivss	11. gr. xc	12. 0.6 grain
13. gr. lxx	14. ʒ xvi	15. ʒ iiiss	16. ʒ $3\frac{3}{5}$
17. ʒ xiv	18. 8.2 drams	19. ʒ ss	20. ℥ v
21. ℥ iiss	22. ℥ xc	23. 0.1 ounce	24. 10 pounds
25. 0.5 pound	26. 4.3 pounds	27. $\frac{1}{4}$ pound	28. $\frac{1}{9}$ pound

Kilograms

29. ℥ lxiv	30. 9.6 ounces	31. 320 ounces	32. ℥ xlviii
33. 960 ounces	34. 240 ounces	35. ℥ xvi	36. 12.8 ounces

Grains

37. 150 mg	38. 40 mg	39. 0.6 mg	40. 600 mg
41. 3.6 mg	42. 12 g	43. 6.5 g	44. 0.3 g
45. 2.5 g	46. 5g	47. 10 g	48. 24 mg

Drams

49. 8.8 g	50. 36 g	51. 0.6 g	52. $4\frac{4}{5}$ g
53. 96 g	54. 12.4 g	55. 8 g	56. 104 g

Ounces

57. 150 g	58. 4.5 g	59. 6 g	60. 80 g
61. 180 g	62. 0.25 kg	63. 5 kg	64. 3.2 kg
65. $\frac{1}{3}$ kg	66. 15 kg	67. 50 g	68. 8 kg

Pounds

69. 0.360 g	70. 720 g	71. 36 g	72. 540 g
73. 1 800 g	74. 180 g	75. 1 500 g	76. 1 080 g

In 77–86, determine the correct relationship. Use the symbols <, >, =.

77. gr. xxv ___?___ 2 g   78. 45 g ___?___ ℥ ii

79. gr. x ___?___ 0.006 g   80. 2.5 g ___?___ gr. xxx

81. 0.5 kg ___?___ lb. ap. i   82. ℥ iv ___?___ 16 g

83. ℥ iiss ___?___ 80 g   84. 6 g ___?___ ℥ iss

85. 5 mg ___?___ gr   86. ℥ v ___?___ 120 g

APPLICATIONS

The medical field is involved with the prevention, diagnosis, and treatment of diseases and illnesses. A physician is responsible for diagnosing the cause of the illness and prescribing the medication. The diagnosis of the illness may require the assistance of laboratory personnel and other health care workers. The prescribed medication is referred to as a *medication order* or a *medicine order.*

A medication order is usually a written document which describes the medicine to be administered to a specific patient. The medication order includes:

- the drug to be used.
- the dose.
- the form of the drug.
- the time to be administered.
- the total number of times to be administered or the dosage.
- the method by which it is to be given.

It may also include directions for checking with the physician in case of complications and/or adverse reactions.

24.4 Exercises

Prescriptions may be written in either the metric or the apothecaries' system. It is often necessary to express the prescription as an equivalent in the other system. In 1–8, express the prescriptions in the apothecaries' system as prescriptions in the metric system. Express prescriptions in the metric system as prescriptions in the apothecaries' system.

1. Atropine gr. $\dfrac{1}{100}$

2. Phenobarb gr. $\dfrac{1}{2}$

3. Phenobarb gr. ss

4. Seconal gr. $1\dfrac{1}{2}$

5. Potassium Triplex ʒ i

6. Vasodilan 10 mg

7. Methadone 7 mg

8. Atenolol 50 mg

9. Atropine sulfate is available in gr. $\dfrac{1}{100}$ tablets. The prescription is for 1 mg of atropine sulfate. How many tablets are required to meet this request?

10. Phenobarbital is available in 5 mg tablets. A patient is given a prescription for gr. ii. How many tablets are required to provide this dosage?

11. Synthroid is available in 0.1 mg tablets. A patient is required to take gr. $\dfrac{1}{20}$. How many tablets provide this dosage?

Unit 25

The Apothecaries' System of Volume

Objectives

After studying this unit the student should be able to:

- **Express apothecaries' units of volume as equivalent apothecaries' units of volume.**
- **Express volume measurements in the apothecaries' system as equivalent metric measurements.**
- **Express volume measurements in the metric system as equivalent apothecaries' measurements.**

THE APOTHECARIES' SYSTEM OF VOLUME

As was stated earlier, when expressing doses in the apothecaries system, lower case Roman numerals rather than capital letters are used. The apothecaries' system of volume has six basic units. The units of volume, the abbreviations or symbols, and an example of the approximate volume of one unit are summarized in this chart.

Unit	Symbol or Abbreviation	Example
minim	m	one drop of water
fluidram	f ʒ	one teaspoon of water
fluidounce	f ʒ	two tablespoons of water
pint	pt.	two glassfuls of water
quart	qt.	four glassfuls of water
gallon	gal.	sixteen glassfuls of water

The use of abbreviations or symbols in expressing volume measurements follows the same guidelines as for weight measurements.

EXAMPLE 1

Express these volume measurements without and with abbreviations or symbols: eleven fluidounces; one-fourth fluidram; four minims; three and one-half pints; five quarts; one-half gallon.

- 11 fluidounces *or* f xi

- fluidram *or* f ʒ $\frac{1}{4}$

- 4 minims *or* m iv

- $3\frac{1}{2}$ pints *or* pt. iiiss

- 5 quarts *or* qt. v

- gallon *or* gal. ss

25.1 Exercises

In 1–20, write the meaning of each measurement.

1. qt. xxv
2. f ℥ viss
3. pt. xl
4. f ℥ xix
5. ɱ x
6. gal. xx
7. f ℥ lxv
8. ɱ l
9. pt. c
10. f ℥ i
11. gal. lxx
12. f ℥ ixss
13. pt. xiv
14. ɱ ivss
15. f ℥ v
16. ɱ xxvi
17. gal. xlv
18. f ℥ xxii
19. f ℥ xc
20. qt. xiii

APOTHECARY EQUIVALENT MEASUREMENTS OF VOLUME

60 minims (ɱ)	=	1 fluidram (f ℥)
8 fluidrams (f ℥)	=	1 fluidounce (f ℥)
16 fluidounces (f ℥)	=	1 pint (pt.)
2 pints (pt.)	=	1 quart (qt.)
4 quarts (qt.)	=	1 gallon (gal.)

In the apothecaries' system, these units of volume are equivalent.
Expressing equivalence of measurement is accomplished by using proportions.

EXAMPLE 1

Express 4 fluidrams as minims.

$$\frac{1 \text{ fluidram}}{60 \text{ minims}} = \frac{4 \text{ fluidrams}}{q \text{ minims}}$$

$$1\,(q) = 4\,(60)$$

$$q = 240 \text{ minims}$$

4 fluidrams equal 240 minims.

EXAMPLE 2

Express 30 minims as fluidrams.

$$\frac{1 \text{ fluidram}}{60 \text{ minims}} = \frac{y \text{ fluidrams}}{30 \text{ minims}}$$

$$60\,(y) = 1\,(30)$$

$$y = \text{f ℥ } ss$$

30 minims equals f ℥ ss.

EXAMPLE 3

Express 40 fluidrams as fluidounces.

$$\frac{1 \text{ fluidounce}}{8 \text{ fluidrams}} = \frac{a \text{ fluidounces}}{40 \text{ fluidrams}}$$

$$8\,(a) = 1\,(40)$$

$$a = \text{f} \overline{\text{3}} v$$

40 fluidrams equal f $\overline{\text{3}}$ v.

EXAMPLE 4

Express 8 fluidounces as fluidrams.

$$\frac{1 \text{ fluidounce}}{8 \text{ fluidrams}} = \frac{8 \text{ fluidounces}}{b \text{ fluidrams}}$$

$$1\,(b) = 8\,(8)$$

$$b = \text{f} \overline{\text{3}} lxiv$$

8 fluidounces equal f $\overline{\text{3}}$ lxiv.

EXAMPLE 5

Express 10 pints as fluidounces.

$$\frac{1 \text{ pint}}{16 \text{ fluidounces}} = \frac{10 \text{ pints}}{r \text{ fluidounces}}$$

$$1\,(r) = 16\,(10)$$

$$r = 160 \text{ fluidounces}$$

10 pints equal 160 fluidounces.

EXAMPLE 6

Express 64 fluidounces as pints.

$$\frac{1 \text{ pint}}{16 \text{ fluidounces}} = \frac{z \text{ pints}}{64 \text{ fluidounces}}$$

$$16\,(z) = 1\,(64)$$

$$z = \text{pt. iv}$$

64 fluidounces equal pt. iv.

EXAMPLE 7

Express 16 pints as quarts.

$$\frac{1 \text{ quart}}{2 \text{ pints}} = \frac{y \text{ quarts}}{16 \text{ pints}}$$

$$2\,(y) = 1\,(16)$$

$$y = \text{qt. viii}$$

16 pints equal qt. viii.

EXAMPLE 8

Express $\frac{1}{4}$ quart as pints.

$$\frac{1 \text{ quart}}{2 \text{ pints}} = \frac{\frac{1}{4} \text{ quart}}{p \text{ pints}}$$

$$1\,(p) = \frac{1}{4}\,(2)$$

$$p = \text{pt. ss}$$

$\frac{1}{4}$ quart equals pt. ss.

EXAMPLE 9

Express 10 quarts as gallons.

$$\frac{1 \text{ gallon}}{4 \text{ quarts}} = \frac{d \text{ gallons}}{10 \text{ quarts}}$$

$$4\,(d) = 1\,(10)$$

$$d = \text{gal. iiss}$$

10 quarts equal gal. iiss.

EXAMPLE 10

Express 20 gallons as quarts.

$$\frac{1 \text{ gallon}}{4 \text{ quarts}} = \frac{20 \text{ gallons}}{y \text{ quarts}}$$

$$1\,(y) = 20\,(4)$$

$$y = \text{qt. lxxx}$$

20 gallons equal qt. lxxx.

25.2 Exercises

In 1–60, express each measurement as an equivalent measurement as indicated.

Minims

1. f℥ x
2. f℥ ss
3. f℥ $\frac{1}{3}$
4. f℥ vi
5. f℥ xvss
6. f℥ xv

Fluidrams

7. ℳ lxxx
8. f℥ ss
9. 3.1 fluidounces
10. 0.6 minim
11. f℥ v
12. f℥ xx
13. 720 minims
14. 1,200 minims
15. 50 minims
16. ℳ $\frac{3}{5}$
17. f℥ $\frac{1}{8}$
18. f℥ ix

Fluidounces

19. f℥ xxiv

20. 480 fluidrams

21. 1.6 fluidrams

22. 320 fluidrams

23. pt. iss

24. pt. x

25. 0.3 pint

26. pt. xl

27. pt. xiv

28. f℥ $\frac{4}{5}$

29. pt. $\frac{1}{4}$

30. pt. $2\frac{1}{2}$

Pints

31. f℥ xxxii

32. 6.4 fluidounces

33. 480 fluidounces

34. f℥ xcvi

35. f℥ xxviii

36. qt. lvss

37. 1.8 quarts

38. 120 quarts

39. qt. xlv

40. qt. $\frac{3}{4}$

41. qt. $1\frac{1}{2}$

42. qt. viii

Quarts

43. pt. xxvi

44. pt. c

45. 14.5 pints

46. pt. lx

47. gal. iiss

48. gal. xxxv

49. gal. xxi

50. 4.2 gallons

51. gal. xxii

52. pt. $\frac{1}{4}$

53. gal. $\frac{1}{4}$

54. pt. $12\frac{1}{2}$

Gallons

55. qt. xvi

56. 3.2 quarts

57. 400 quarts

58. qt. ss

59. qt. lxxxii

60. qt. xxii

In 61–66, determine each correct relationship (<, >, =).

61. f℥ x __?__ f℥ ii

62. pt. ss __?__ f℥ xiii

63. qt. iii __?__ pt. vi

64. f℥ xx __?__ pt. iss

65. f℥ lxx __?__ gal. i

66. f℥ ss __?__ f℥ iv

APOTHECARIES'
—METRIC EQUIVALENT MEASUREMENTS OF VOLUME

Expressing equivalencies between apothecaries' and metric units of volume follow the same procedure as for units of weight. The equivalencies are approximations within the acceptable limits of error.

Apothecaries' System	=	Metric System
15 minims (℩)	=	1 mL (or 1 cm3)
1 fluidram (f℥)	=	4 mL (or 4 cm3)
1 fluidounce (f℥)	=	30 mL (or 30 cm3)
1 pint (pt.)	=	1 L (or 1 000 cm3)
1 quart (qt.)	=	1 L (or 1 000 cm3)
1 gallon (gal.)	=	4 L (or 4 000 cm3)

Expressing equivalencies between the two systems utilizes proportions and the table of equivalencies.

EXAMPLE 1

Express 45 minims as milliliters.

$$\frac{15 \text{ minims}}{1 \text{ mL}} = \frac{45 \text{ minims}}{a \text{ mL}}$$

$$15 (a) = 45$$

$$a = 3 \text{ mL}$$

45 minims equal 3 mL.

EXAMPLE 2

Express 10 fluidrams as milliliters.

$$\frac{1 \text{ fluidram}}{4 \text{ mL}} = \frac{10 \text{ fluidrams}}{b \text{ mL}}$$

$$1 (b) = 4 (10)$$

$$b = 40 \text{ mL}$$

10 fluidrams equals 40 mL.

EXAMPLE 3

Express 6 fluidounces as cubic centimeters.

$$\frac{1 \text{ fluidounce}}{30 \text{ cm}^3} = \frac{6 \text{ fluidounces}}{j \text{ cm}^3}$$

$$1 (j) = 30 (6)$$

$$j = 180 \text{ cm}^3$$

6 fluidounces equal 180 cm^3.

EXAMPLE 4

Express 250 cm^3 as pints.

$$\frac{500 \text{ cm}^3}{1 \text{ pint}} = \frac{250 \text{ cm}^3}{y \text{ pint}}$$

$$500 (y) = 250 (1)$$

$$y = \text{pt. ss}$$

250 cm^3 equal pt. ss.

EXAMPLE 5

Express 10 quarts as liters.

$$\frac{1 \text{ quart}}{1 \text{ L}} = \frac{10 \text{ quarts}}{z \text{ L}}$$

$$1 \, (z) = 1 \, (10)$$
$$z = 10 \text{ L}$$

10 quarts equal 10 L.

EXAMPLE 6

Express 2 liters as gallons.

$$\frac{1 \text{ gallon}}{4 \text{ L}} = \frac{r \text{ fluidrams}}{2 \text{ L}}$$

$$4 \, (r) = 1 \, (2)$$
$$r = \text{gal. ss}$$

2 liters equal gal. ss.

25.3 Exercises

In 1–84, express each measurement as an equivalent measurement as indicated.

Milliliters

1. m lx	2. m viiss	3. 0.6 minim	4. m lxxv
5. m xx	6. f ℥ $\frac{1}{3}$	7. 0.4 fluidounce	8. f ℥ vi
9. f ℥ lx	10. f ℥ xss	11. mv	12. f ℥ xl

Cubic Centimeters

13. f ℥ iv	14. 0.3 fluidram	15. f ℥ vss	16. f ℥ $\frac{1}{4}$
17. f ℥ l	18. 0.25 pint	19. pt. l	20. pt. ss
21. 0.6 pint	22. 1.1 pints	23. 0.6 fluidram	24. pt. viss

Liters

25. qt. xxi	26. 5.6 quarts	27. qt. iiss	28. $7\frac{1}{8}$ quarts
29. qt. xlv	30. gal. ss	31. gal. x	32. gal. $\frac{1}{5}$
33. gal. vi	34. 0.6 gallon	35. qt. viii	36. gal. xxv

Minims

37. 0.5 cm^3	38. 4 cm^3	39. $\frac{1}{3}$ cm^3	40. 2.3 cm^3
41. 15 cm^3	42. 6 cm^3	43. $\frac{1}{4}$ cm^3	44. 2.5 cm^3

Fluidrams

45. 8 cm^3	46. 3.2 cm^3	47. 0.48 cm^3	48. 80 cm^3
49. $\frac{1}{2} \text{ cm}^3$	50. 6 cm^3	51. 4.8 cm^3	52. 22 cm^3

Fluidounces

53. 300 cm^3	54. 0.6 cm^3	55. 1.8 cm^3	56. 45 cm^3
57. 50 cm^3	58. 90 cm^3	59. 3 cm^3	60. 330 cm^3

Pints

61. 1 000 mL	62. 3.5 mL	63. 1 200 mL	64. 500 mL
65. 2 500 mL	66. 250 mL	67. 2 000 mL	68. 100 mL

Quarts

69. 100 L	70. 18.5 L	71. 27 L	72. $8\frac{1}{3} \text{ L}$
73. 0.1 L	74. 15 L	75. 0.3 L	76. 36.5 L

Gallons

77. 8 L	78. 0.25 L	79. 3 L	80. 20 L
81. $\frac{1}{8} \text{ L}$	82. 4 L	83. 16 L	84. 14 L

APPLICATIONS

When a physician prescribes medication, a prescription is written. The prescription is a means of controlling the sale and use of drugs which can be safely and effectively used only under the supervision of a physician. A prescription consists of the superscription, the inscription, the subscription, and the signature.

DEFINITIONS

- The *superscription* includes the patient's name and address, the date, and the symbol Rx which means "I prescribe" or "take thou."
- The *inscription* includes the names and amounts of the drugs that the medication is composed of. In writing the prescription, the most important drug is listed first.
- The *subscription* includes the directions to the pharmacist for preparing the prescription.
- The *signature* includes the directions to the patient for taking the medicine. It may also include the physician's registration number, if required by state law.

When completing the prescription, the physician uses symbols for the inscription, subscription, and signature. Symbols for the inscription may be in metric units or apothecaries' units. Some common symbols used in the subscription and signature are listed on the following page.

Symbol	Meaning	Symbol	Meaning
$\overline{aa}$	(equal parts) of each	q.h.	every hour
a.c.	before meals	q.2.h.	every two hours
ad lib	if desired, freely	q.3.h.	every three hours
b.i.d.	two times a day	q.i.d.	four times a day
$\overline{c}$	with	q.n.	every night
comp.	compound	q.n.s.	quantity not sufficient
dil.	dilute	®	trade name
elix.	elixir	$\overline{s}$	without
ext.	extract	s.c.	subcutaneous (injection)
h.s.	bedtime	Sig. or S.	write on label
I.M.	intramuscular	Sol.	solution
I.V.	intravenous	s.o.s.	if necessary
M	mix	sp.	spirits
p.c.	after meals	s.s.	soap suds
p.o.	by mouth	stat.	immediately
per	through or by	syr.	syrup
p.r.	by rectum	t.i.d.	three times a day
p.r.n.	when required	tr. or tinct.	tincture
q.d.	every day	ung.	ointment

25.4 Exercises

A series of medications are prescribed. In 1–4, find the metric equivalent for each of these amounts.

1. ℳ lv   2. f ℥ ivss   3. f ℥ iii   4. ℳ xv iii

5. How many cubic centimeters of boric acid are required to prepare f ℥ xvi of 4% boric acid.

6. How many cubic centimeters of boric acid are required to prepare f ℥ xxx of 1% boric acid.

In 7–9, each milliliter of Decadron contains 0.004 cm^3 of dexamethasome phosphate.

7. How many minims of dexamethasome phosphate are present in 100 milliliters of Decadron?

8. How many minims of dexamethasome phosphate are present in 1 liter of Decadron?

9. How many minims of dexamethasome phosphate are present in 1.5 liters of Decadron?

In 10–15, convert the following to milliliter equivalents.

10. 23 minims   11. 6 fluidrams   12. 14 fluidounces

13. 1.25 pints   14. 0.75 quart   15. 0.75 gallon

In 16–21, convert the following to fluidounce equivalents.

16. 328 milliliters   17. 125 milliliters   18. 15 milliliters

19. 46 milliliters   20. 750 milliliters   21. 23 milliliters

Household Apothecaries' System of Measure

Objectives

After studying this unit the student should be able to:

- **Express household liquid measurements as approximate equivalent measurements.**
- **Express household measurements as approximate apothecaries' measurements.**
- **Express apothecaries' measurements as approximate equivalent household measurements.**

THE HOUSEHOLD SYSTEM OF LIQUID MEASURE

The household system can be used, within limits, to safely measure amounts of medicine administered at home. To effectively communicate with a patient in terms the patient can understand, the health worker should be familiar with the household system. The approximate equivalent measures which are given are for water. Other substances may have slightly different equivalencies.

Common household measuring articles include droppers, teaspoons, tablespoons, cups, and glasses. In addition, pints and quarts, which have been defined in the apothecaries' system of measure, are usually available. This table summarizes some liquid measure equivalencies. These equivalents should be used only in those instances when the proper measuring instruments are not available.

Approximate Liquid Measure Equivalents		
60 drops	=	1 teaspoonful (t)
4 teaspoonfuls*	=	1 tablespoonful (T)
2 tablespoonfuls	=	1 fluidounce
6 fluidounces	=	1 teacupful
8 fluidounces	=	1 glassful

•In kitchen usage, 3 cooking teaspoonfuls equal one tablespoonful.

Usually equivalent measures can be obtained directly from a table. Using proportions when calculating an equivalent measure can also be helpful.

EXAMPLE 1

What part of a teaspoonful is 20 drops?

$$\frac{60 \text{ drops}}{1 \text{ teaspoonful}} = \frac{20 \text{ drops}}{x \text{ teaspoonful}}$$

$$60\,(x) = 1\,(20)$$

$$x = \frac{20}{60} \text{ or } \frac{1}{3}$$

20 drops $\approx \dfrac{1}{3}$ teaspoonful.

EXAMPLE 2

Nine fluidounces are how many teacupfuls?

$$\frac{6 \text{ fluidounces}}{1 \text{ teacupful}} = \frac{9 \text{ fluidounces}}{y \text{ teacupful}}$$

$$6\,(y) = 1\,(9)$$

$$y = \frac{9}{6} \text{ or } 1\frac{1}{2}$$

9 fluidounces $\approx 1\dfrac{1}{2}$ teacupfuls.

26.1 Exercises

In 1–16, complete using the table of approximate liquid measure equivalents.

1. __?__ drops = 1 teaspoonful

2. 30 drops = __?__ teaspoonfuls

3. __?__ teaspoonfuls = 15 drops

4. 2 tablespoonfuls = __?__ teaspoonfuls

5. __?__ tablespoonfuls = 1 teaspoonful

6. 2 teaspoonfuls = __?__ tablespoonfuls

7. __?__ tablespoonfuls = 1 fluidounce

8. 4 fluidounces = __?__ tablespoonfuls

9. __?__ fluidounces = 1 tablespoonful

10. 1 teacupful = __?__ fluidounces

11. __?__ teacupfuls = 3 fluidounces

12. 12 fluidounces = __?__ teacupfuls

13. __?__ glassfuls = 8 fluidounces

14. 4 fluidounces = __?__ glassfuls

15. __?__ fluidounces = $1\dfrac{1}{2}$ glassfuls

16. 18 fluidounces = __?__ teacupfuls

In 17–28, calculate each approximate equivalent measure. Use the table only if necessary.

17. 4 tablespoonfuls = __?__ fluidounces

18. __?__ teaspoonfuls = 40 drops

19. 15 fluidounces = __?__ teacupfuls

20. __?__ fluidounces = 2 glassfuls

21. 9 teaspoonfuls = __?__ tablespoonfuls

22. __?__ drops = $\dfrac{1}{3}$ tablespoonful

23. 18 fluidounces = __?__ teaspoonfuls

24. __?__ glassfuls = 2 fluidounces

25. 6 teaspoonfuls = __?__ fluidounces

26. __?__ fluidounces = 24 tablespoonfuls

27. 10 tablespoonfuls = __?__ fluidounces

28. __?__ drops = $\dfrac{1}{2}$ teaspoonful

HOUSEHOLD
—APOTHECARIES' EQUIVALENT MEASUREMENTS

The household system of measurement, although not as accurate as the apothecaries' system is frequently used at home by patients. Using familiar measuring instruments, patients can measure drops, teaspoonfuls, tablespoonfuls, ounces, cupfuls, and glassfuls, whereas most patients would not know how to measure a quantity given in apothecaries' units. Since measuring instruments such as droppers, cups, teaspoons, and tablespoons are not the same size, the household system and its use should be avoided by health care professionals.

These tables summarize the approximate weight and liquid measure equivalencies for the apothecaries' and household systems.

Weight		
Apothecaries' System		**Household System**
1 grain (gr.)	=	1 drop
1 dram (℥)	=	1 teaspoonful (t)
4 drams	=	1 tablespoonful (T)
1 ounce (℥)	=	2 tablespoonfuls
6 ounces	=	1 teacupful
8 ounces	=	1 glassful

Liquid Measure		
Apothecaries' System		**Household System**
1 minim (m)	=	1 drop
1 fluidram (℥)	=	1 teaspoonful (t)
4 fluidrams	=	1 tablespoonful (T)
1 fluidounce (℥)	=	2 tablespoonfuls
6 fluidounces	=	1 teacupful
8 fluidounces	=	1 glassful

Proportions using the approximate equivalencies are used to express equivalencies between the two systems. The equivalencies are approximate but are within the limits of error. Keep in mind that drops of different substances vary in size. When minims are ordered, a minim glass or a minim pipette should be used to obtain an accurate measure. When drops are ordered, a medicine dropper may be used.

EXAMPLE 1

Two tablespoonfuls will hold approximately how many drams?

$$\frac{1 \text{ tablespoonful}}{4 \text{ drams}} = \frac{2 \text{ tablespoonfuls}}{x \text{ drams}}$$

$$1\,(x) = 2\,(4)$$

$$x \approx 8$$

2 tablespoonfuls hold approximately eight drams.

EXAMPLE 2

Approximately how many tablespoonfuls are 5 fluidounces?

$$\frac{1 \text{ fluidounce}}{2 \text{ tablespoonfuls}} = \frac{5 \text{ fluidounces}}{x \text{ tablespoonfuls}}$$

$$1\,(x) = 2\,(5)$$

$$x \approx 10$$

Five fluidounces are approximately ten tablespoonfuls.

26.2 Exercises

In 1–36, use the tables of approximate weight and liquid measure equivalencies to find each equivalence.

1. 5 drops = __?__ grains

2. 2 gr. __?__ drops

3. 4 drams = __?__ teaspoonfuls

4. 2 tablespoonfuls = __?__ drams

5. ℥ ii = __?__ T

6. $\frac{1}{2}$ ounce = __?__ tablespoonfuls

7. $\frac{1}{2}$ teacupful = __?__ ounces

8. ℥ vi __?__ teacupfuls

9. 6 ounces = __?__ glassfuls

10. __?__ drops = 3 grains

11. __?__ teaspoonfuls = 2 drams

12. __?__ t = ℥ vi

13. __?__ drams = 1.5 tablespoonfuls

14. __?__ tablespoonfuls = 2 ounces

15. ℥ __?__ = 3 T

16. __?__ ounces = 2 teacupfuls

17. __?__ glassfuls = 4 ounces

18. ℥ __?__ = $1\frac{1}{2}$ glassfuls

19. __?__ minims = 5 drops

20. __?__ teaspoonfuls = 2 fluidrams

21. f℥ __?__ = $\frac{1}{2}$ t

22. __?__ fluidrams = $1\frac{1}{2}$ tablespoonfuls

23. __?__ tablespoonfuls = $\frac{1}{2}$ fluidounce

24. f℥ __?__ 3 T

25. __?__ fluidounces = $1\frac{1}{2}$ teacupfuls

26. __?__ glassfuls = 4 fluidounces

27. f℥ __?__ = $1\frac{1}{2}$ glassfuls

28. 3 drops = __?__ minims

29. ♏ ii __?__ drops

30. 1 fluidram = __?__ teaspoonfuls

31. 2 tablespoonfuls = __?__ fluidrams

32. f℥ ss = __?__ t

33. 2 fluidounces = __?__ tablespoonfuls

34. $\frac{1}{2}$ teacupful = __?__ fluidounces

35. f℥ xii = __?__ teacupfuls

36. 2 fluidounces = __?__ glassfuls

In 37–42, without referring to the tables, complete this chart with the approximate equivalencies. Use symbols or abbreviations when possible.

	Household System	Apothecaries' System	
		Weight	Volume (liquid measure)
37.	1 drop	?	1 minim (℩)
38.	?	1 dram (ʒ)	1 fluidram (fʒ)
39.	1 tablespoonful (T)	?	4 fluidrams
40.	2 tablespoonfuls	1 ounce (℥)	?
41.	?	6 ounces	6 fluidounces
42.	1 glassful	?	8 fluidounces

HOUSEHOLD
—APOTHECARIES' SYMBOLS TRANSLATION

It is important for a health worker to quickly and accurately translate household and apothecaries' symbols. Further understanding of the relationship between the apothecaries' and household systems and additional practice in interpreting and using apothecaries' and household abbreviations and symbols will prove beneficial.

The approximate weight and liquid measure equivalents for the apothecaries' and household systems are summarized in this chart.

Household Units	Apothecaries' Units	
	Weight	Volume (liquid measure)
1 drop	gr. i	℩ i
1 t	ʒ i	fʒ i
1 T	ʒ iv	fʒ iv
2 T	℥ i	f℥ i
1 teacupful	℥ vi	f℥ vi
1 glassful	℥ viii	f℥ viii

Nine ounces are how many teacupfuls?

$$\frac{1 \text{ teacupful}}{℥ \text{ vi}} = \frac{x \text{ teacupfuls}}{℥ \ y \text{ ix}}$$

$$6\,(x) = 9$$

$$x \approx \frac{9}{6} \ or \ 1\frac{1}{2}$$

Nine ounces $\approx 1\frac{1}{2}$ teacupfuls.

EXAMPLE 2

How many fluidounces are contained in $2\frac{1}{2}$ glassfuls?

$$\frac{1 \text{ glassful}}{\text{viii}} = \frac{2\frac{1}{2} \text{ glassfuls}}{y}$$

$$1\,(y) = 2\frac{1}{2}\,(8)$$

$$y = \frac{5}{\overset{1}{\cancel{2}}}\left(\frac{\overset{4}{\cancel{8}}}{1}\right) \text{ or } 20$$

Two and one-half glassfuls contain f ʒ xx.

26.3 Exercises

In 1–45, express each measurement as an equivalent measurement as indicated. Use abbreviations or symbols when possible.

Apothecaries' Weight

1. 4 T
2. 2 teacupfuls
3. 2 glassfuls

4. 6 drops
5. 3 drops
6. 2 t

7. $\frac{1}{2}$ T
8. $\frac{1}{3}$ teacupful
9. $\frac{1}{2}$ glassful

10. 8 t
11. 10 drops
12. 4 teacupfuls

Apothecaries' Liquid Measure

13. 4 t
14. 5 drops
15. 1 glassful

16. 3 T
17. 2 drops
18. 1 t

19. $\frac{1}{2}$ teacupful
20. $\frac{1}{4}$ T
21. $1\frac{1}{2}$ teacupfuls

22. $\frac{1}{4}$ glassful
23. $1\frac{1}{2}$ glassful
24. $\frac{3}{4}$ teacupful

Household

25. gr. v
26. f ʒ vi
27. ɱ ix

28. lb. ap. i
29. qt. ss
30. f ʒ xii

31. ʒ viii
32. gr. xiv
33. ʒ iv

34. f ʒ xxiv
35. ʒ iv
36. qt. i

37. gr. iiss
38. f ʒ xii
39. ɱ vi

40. f ʒ ii
41. qt. iss
42. f ʒ ix

43. lb. ap. $\frac{1}{4}$
44. ɱ xx
45. ɱ xxix

In 46–59, determine the correct relationship (<, >, =).

46. 3 T __?__ ʒ ii
47. 1 t __?__ ʒ iiss

48. f ʒ v __?__ 6 T
49. 1.5 glassfuls __?__ f ʒ x

50. ♏ iv ___?___ 6 drops

51. gr. vii ___?___ 7 drops

52. ʒ vi ___?___ 1 teacupful

53. ʒ iss ___?___ 6 t

54. f ʒ vi ___?___ 12 T

55. f ʒ vi ___?___ $1\frac{1}{3}$ T

56. f ʒ xx ___?___ $2\frac{1}{2}$ glassfuls

57. $1\frac{1}{3}$ teacupfuls ___?___ ʒ vi

58. 2 T ___?___ ʒ viii

59. 17 drops ___?___ ♏ xvii

In 60–71, each dosage is given in apothecaries' units. Determine the most reasonable household equivalent.

60. f ʒ v
   a. 7 t
   b. 1 teacupful
   c. $2\frac{1}{2}$ T

61. qt. i
   a. 5 teacupfuls
   b. 4 glassfuls
   c. 6 teacupfuls

62. ʒ ii
   a. 2 T
   b. 2 t
   c. 1 t

63. ʒ x
   a. 10 T
   b. $1\frac{2}{3}$ teacupfuls
   c. $1\frac{1}{2}$ teacupfuls

64. ʒ vi
   a. 5 t
   b. $1\frac{1}{2}$ T
   c. 7 t

65. ʒ xviii
   a. $2\frac{1}{2}$ glassfuls
   b. 2 glassfuls
   c. 3 teacupfuls

66. f ʒ xxxii
   a. 5 glassfuls
   b. 1 quart
   c. 3 glassfuls

67. ♏ vi
   a. $\frac{1}{8}$ t
   b. 6 drops
   c. $\frac{1}{4}$ t

68. lb. ap. i
   a. 10 T
   b. 1 glassful
   c. 2 teacupfuls

69. ʒ xii
   a. 2 T
   b. 2 glassfuls
   c. 12 t

70. gr. xv
   a. $\frac{1}{2}$ t
   b. $\frac{1}{3}$ t
   c. 15 drops

71. ʒ xx
   a. 10 drops
   b. 40 T
   c. 20 teacupfuls

APPLICATIONS

Household units of measure are found in every home and are used during everyday activities such as cooking, cleaning, or shopping. When purchasing nonprescription drugs such as aspirin, cough medicine, or liquid vitamins, the directions are given in household units of measure. For example, the directions for taking cough medicine may read: "two tablespoons, four times a day." Unfortunately, every person's tablespoon is not the same size. This means that the amount of medication that is taken may vary from person to person.

The health care worker must be aware of the inaccuracy of the measuring instruments in the household system. This awareness should lead the health care worker away from using the household system unless in an emergency or in situations where the system can be used safely; preparing a salt solution for gargle may be done safely with household units. If the household system is to be used, the conscientious health care worker uses caution and forethought.

26.4 Exercises

1. Two cough medicines are compared. Brand **A** contains a 5% concentration of cough inhibitor. Brand **B** contains an 8% concentration of cough inhibitor. The recommended dosage for both brands is one tablespoon every four hours.

   a. After three doses of Brand **A,** how much cough inhibitor will be consumed?

   b. After three doses of Brand **B,** how much cough inhibitor will be consumed?

2. A patient is required to drink 4 glassfuls of water in two hours. How many fluidounces does this represent?

3. A total of 4 drams of medication are added to a glassful of water. What is the percent of medication in the glassful of water? Round the answer to the nearer tenth percent.

4. A patient on a soft diet is allowed to consume:

   1 teacupful of tea

   1 teacupful of soup (broth)

   10 teaspoonfuls of pudding

   Approximately how many fluidounces does this represent?

5. Cough medicine is sold in a variety of bottle sizes. Brand **A** sells for $1.45 in the 8-fluidounce size and for $1.95 in the 12-fluidounce size. Brand **B** sells for $1.55 in the 10-fluidounce size and for $2.25 in the 16-fluidounce size. To provide the same dose as Brand **A,** $\frac{1}{8}$ as much as Brand **B** must be given.

   a. Which brand and size is the best purchase?

   b. How many teaspoons of medication can be obtained from the size and brand representing the best purchase?

Unit 27

Household-Metric System of Measure

Objectives

After studying this unit the student should be able to:

- **Express household measurements as approximate equivalent metric measurements.**
- **Express metric measurements as approximate equivalent household measurements.**

HOUSEHOLD-METRIC EQUIVALENT MEASUREMENTS

As indicated earlier, the household system is not an accurate system of measurement. At present, it is still the more familiar system for patients. It is the responsibility of the health worker to make the approximate equivalents between the household and metric systems and to communicate these equivalencies to the patient.

These tables summarize the approximate weight and liquid measure equivalencies for the metric and household systems.

Weight (Mass)		
Metric System		**Household System**
0.06 gram (g) or 60 milligrams (mg)	=	1 drop
5 grams	=	1 teaspoonful (t)
15 grams	=	1 tablespoonful (T)
180 grams	=	1 teacupful
240 grams	=	1 glassful

Liquid Measure		
Metric System		**Household System**
0.06 milliliters (mL)	=	1 drop
5 milliliters	=	1 teaspoonful (t)
15 milliliters	=	1 tablespoonful (T)
180 milliliters	=	1 teacupful
240 milliliters	=	1 glassful

Note: 1 scant teaspoon = 4 milliliters.

Proportions using the approximate equivalencies are used to express equivalencies between the two systems. The equivalencies are within the limits of error.

EXAMPLE 1

Approximately how many milliliters are 2 teaspoonfuls?

$$\frac{5 \text{ milliliters}}{1 \text{ teaspoonful}} = \frac{x \text{ milliliters}}{2 \text{ teaspoonfuls}}$$

$$1\,(x) = 5\,(2)$$

$$x = 10$$

Two teaspoonfuls are approximately ten milliliters.

EXAMPLE 2

One and one-half teacupfuls contain approximately how many grams?

$$\frac{1 \text{ teacupful}}{180 \text{ g}} = \frac{1\frac{1}{2} \text{ teacupfuls}}{y \text{ g}}$$

$$1\,(y) = 1\frac{1}{2}\,(180)$$

$$y = \frac{3}{2}\left(\frac{180}{1}\right)$$

$$y = 270$$

One and one-half teacupfuls contain approximately 270 grams.

27.1 Exercises

In 1–30, use the tables of approximate weight and liquid measure equivalencies to find each equivalence.

1. 2 drops = __?__ milligrams

2. 5 drops = __?__ mg

3. 2.5 grams = __?__ teaspoonfuls

4. $\frac{1}{3}$ tablespoonful = __?__ grams

5. 37.5 g = __?__ T

6. 120 grams = __?__ teacupfuls

7. $1\frac{1}{3}$ glassfuls = __?__ grams

8. 360 g = __?__ glassfuls

9. __?__ drops = 1.2 grams

10. __?__ teaspoonfuls = 20 grams

11. __?__ t = 12.5 g

12. __?__ grams = 4 tablespoonfuls

13. __?__ teacupfuls = 180 grams

14. __?__ g = $\frac{1}{2}$ teacupful

15. __?__ grams = $\frac{5}{6}$ glassfuls

16. 10 drops = __?__ milliliters

17. 1.8 mL = __?__ drops

18. 7.5 milliliters = __?__ teaspoonfuls

19. $2\frac{1}{2}$ tablespoonfuls = __?__ milliliters

20. 75 mL = __?__ T

21. $\frac{1}{5}$ teacupful = __?__ milliliters

22. 300 milliliters = __?__ glassfuls

23. $\frac{3}{4}$ glassful = __?__ mL

24. __?__ milliliters = 45 drops

25. __?__ teaspoonfuls = 20 milliliters

26. __?__ mL = 3.5 t

27. __?__ milliliters $= 1\dfrac{1}{3}$ tablespoonfuls   28. __?__ teacupfuls = 90 milliliters

29. __?__ mL $= 1\dfrac{1}{2}$ teacupfuls   30. __?__ glassfuls = 80 milliliters

In 31–35, without referring to the tables, complete this chart with the approximate equivalences. Use abbreviations or symbols when possible.

	Household System	Metric System	
		Weight (mass)	Volume (liquid measure)
31.	1 drop	0.06 grams (g)	?
32.	?	5 grams	5 milliliters (mL)
33.	1 tablespoonful (T)	?	15 millimeters
34.	1 teacupful	180 grams	?
35.	?	240 grams	240 milliliters

HOUSEHOLD–METRIC SYMBOLS TRANSLATION

It is important for a health worker to be able to quickly and accurately translate household and metric symbols. Further understanding of the relationship between the metric and household systems, and additional practice in interpreting and using the metric and household abbreviations and symbols, will prove beneficial.

The approximate weight and liquid measure equivalences for the metric and household systems are summarized in this chart.

Household Units	Metric Units	
	Weight (mass)	Volume (liquid measure)
1 drop	0.06 g	0.06 mL
1 t	5 g	5 mL
1 T	15 g	15 mL
1 teacupful	180 g	180 mL
1 glassful	240 g	240 mL

EXAMPLE 1

Approximately how many tablespoonfuls will hold 40 grams?

$$\frac{1\text{ T}}{15\text{ g}} = \frac{x\text{ T}}{40\text{ g}}$$

$$15\,(x) = 40$$

$$x = \frac{40}{15} \text{ or } 2\frac{2}{3}$$

Two and two-thirds tablespoonfuls will hold approximately 40 grams.

EXAMPLE 2

Two-thirds teacupful contains approximately how many milliliters?

$$\frac{1 \text{ teacupful}}{180 \text{ mL}} = \frac{\frac{2}{3} \text{ teacupful}}{x \text{ mL}}$$

$$1(x) = \frac{2}{3}(180)$$

$$x = \frac{2}{\overset{\cancel{3}}{1}}\left(\frac{\overset{60}{\cancel{180}}}{1}\right) \text{ or } 120$$

Two-thirds teacupful contains approximately 120 milliliters.

27.2 Exercises

In 1–40, express each measurement as an equivalent measurement as indicated. Use abbreviations or symbols when possible.

Metric Weight (Mass)

1. $\frac{1}{2}$ glassful

2. $\frac{1}{3}$ teacupful

3. $\frac{1}{2}$ t

4. $\frac{1}{2}$ T

5. 6 drops

6. 2 t

7. 4 drops

8. 2 glassfuls

9. 2 teacupfuls

10. 4 T

Metric Liquid Measure

11. $\frac{1}{2}$ teacupful

12. $\frac{1}{4}$ glassful

13. 1 t

14. 3 T

15. 5 drops

16. 4 t

17. 2 drops

18. 1 glassful

19. $1\frac{1}{2}$ teacupfuls

20. $\frac{1}{5}$ T

Household

21. 30 g

22. 0.007 5 L

23. $2\frac{1}{2}$ mL

24. 0.48 kg

25. 0.015 L

26. 60 mg

27. 60 mL

28. 0.960 L

29. 0.6 mL

30. 30 mL

31. 90 mL

32. 1 g

33. 0.06 g

34. 45 g

35. 360 mL

36. 0.72 kg

37. 0.011 25 L

38. 1.5 g

39. 120 mg

40. 0.18 kg

In 41–52, determine the correct relationship ($<$, $>$, $=$).

41. 12 g __?__ 1 T

42. 360 g __?__ 2 teacupfuls

43. 1.5 t __?__ 0.008 L

44. 4 500 mg __?__ 2 t

45. 5 glassfuls __?__ 1 kg

46. 1.5 T __?__ 0.025 kg

47. $2\frac{1}{2}$ teacupfuls __?__ $\frac{1}{2}$ L

48. $1\frac{1}{2}$ mL __?__ $\frac{2}{3}$ t

49. 0.080 L __?__ $\frac{1}{4}$ glassful

50. $\frac{1}{3}$ T __?__ 1 000 mg

51. 150 mL __?__ $\frac{2}{3}$ teacupful

52. $\frac{1}{2}$ t __?__ 2 mL

In 53–64, each dosage is given in metric units. Determine the most reasonable household equivalent.

53. 6 mL
 a. 1 T
 b. 5 T
 c. $1\frac{1}{5}$ t

54. 120 mL
 a. 5 T
 b. $\frac{2}{3}$ teacupful
 c. $\frac{3}{4}$ teacupful

55. 10 g
 a. 2 t
 b. 1 T
 c. $\frac{1}{2}$ T

56. 3 g
 a. $\frac{1}{8}$ T
 b. $\frac{3}{5}$ t
 c. $\frac{1}{4}$ t

57. 5 mL
 a. $\frac{1}{4}$ t
 b. $\frac{1}{2}$ t
 c. 1 t

58. 30 mL
 a. $\frac{1}{4}$ teacupful
 b. 5 t
 c. 2 T

59. 2 500 mg
 a. $\frac{1}{2}$ t
 b. $\frac{1}{2}$ T
 c. $\frac{1}{4}$ T

60. 120 g
 a. 1 teacupful
 b. $\frac{1}{2}$ glassful
 c. $\frac{1}{3}$ glassful

61. $\frac{1}{2}$ L

    a. 3 teacupfuls

    b. $1\frac{1}{2}$ glassfuls

    c. 2 glassfuls

62. 0.007 5 kg

    a. 1.5 t

    b. $1\frac{1}{2}$ T

    c. $1\frac{3}{4}$ t

63. 0.240 kg

    a. 12 tablespoonfuls

    b. 72 teaspoonfuls

    c. 1 glassful

64. 0.750 kg

    a. $4\frac{1}{2}$ glassfuls

    b. $4\frac{1}{5}$ teacupfuls

    c. 12 teaspoonfuls

APPLICATIONS

Household measures are not accurate and should not be used when administering medication in medical treatment facilities. The lack of accuracy is partly due to the many designs and capacities for teaspoons, tablespoons, teacups and glasses. Even the size of a drop can vary according to the size of the opening in the pipette. Gradually, with the passage of time, household measures will be replaced with more refined measures. In the interim, the health care worker should understand and be able to administer medication in the household system.

Nonprescription liquid medication is usually prescribed by the teaspoon or the tablespoon. Drops for eyes and ear treatments are not uncommon. Dietary requirements have been indicated in teacup or glass sizes. Today, more pharmaceutical companies are supplying graduated plastic cups and pipettes for more accurate measurement of medication. Most often, these instruments are graduated in metric units.

27.3 Exercises

Several brands of eye drops are on the market. Three brands are compared. It is determined that 15 ml of each contains similar medication but in different amounts.

Brand **A** contains 5 mg of medication per 15 mL.

Brand **B** contains 7 mg of medication per 15 mL.

Brand **C** contains 10 mg of medication per 15 mL.

1. How much medication will a patient receive if one drop of Brand **A** is placed in each eye?

2. How much medication will a patient receive if one drop of Brand **B** is placed in each eye?

3. How much medication will a patient receive if one drop of Brand **C** is placed in each eye?

4. The recommended dose for each treatment is 0.04 mg per eye. Brand **A** is dispensed in a 15-mL container and costs $1.59. Brand **C** is dispensed in a 10-mL container and costs $1.79. To achieve the recommended dose, which brand is the better buy?

In 5 and 6, a child receives a prescription for medication to relieve an upset stomach. The recommended dose is one teaspoonful every three hours.

5. Express, in metric units, the total amount of medication that is consumed within a 24-hour period.

6. Express, in apothecaries' units, the total amount of medication that is consumed within a 24-hour period.

In 7–9, a pharmacist dispenses a prescription for 50 mL of medication.

7. How many drops can be obtained from this prescription?

8. How many scant teaspoonfuls can be obtained from this prescription?

9. How many tablespoonfuls can be obtained from this prescription?

In 10–12, once a day, vitamins are added to a baby's formula. If ten drops are added to each amount of formula, what percent vitamins is present in a formula containing:

10. 250 milliliters?   11.   30 milliliters?   12.   500 milliliters?

In 13 and 14, a prescription for 20 mL of medication is requested. Each time the medication is administered, two drops of medication are added to one glassful of milk.

13. If all the medication is to be used, how many fluidounces of milk are needed?

14. How many milliliters of milk are needed if all the medication is to be used?

In 15 and 16, a patient is to receive 16 mL of cough medicine in a 24-hour period.

15. How many scant teaspoonfuls of cough medicine does this represent?

16. How many fluidrams of cough medicine does this represent?

In 17 and 18, a patient is given a prescription and directed to take one tablespoonful each hour for six hours.

17. What are the total milliliters taken?

18. How many drops are taken?

In 19–21, eight glassfuls of water are to be taken during an 8-hour period.

19. How many kilograms are taken?

20. How many grams are taken?

21. A patient is directed to put four drops of medication into each ear every three hours. How many milliliters must the doctor prescribe to provide the patient enough for five treatments over a 12-hour period?

Section 5: Summary, Review, and Study Guide

VOCABULARY

Apothecary System	grain (gr.)	pint (pt.)
dram (ℨ)	quart (qt.)	pound (lb. ap.)
drop	Household System	Roman numerals
fluidounce (fℨ)	one-half (ss)	tablespoonful (T)
fluidram (fℨ)	Metric System	teacupful
gallon	minim (m)	teaspoonful (t)
glassful	ounce (ℨ)	

CONCEPTS, SKILLS, AND APPLICATIONS

Objectives With Study Guide

Upon completion of Section 5 you should be able to:

- **Express measurements using apothecaries' abbreviations and symbols.**

  Write the meaning of each measurement.

  lb. ap. xx   means 20 pounds

  gr. $\dfrac{3}{4}$ means $\dfrac{3}{4}$ grain

  ℨ vi   means 6 drams

  ℨ $\dfrac{1}{3}$ means $\dfrac{1}{3}$ ounce

Apothecaries' System Weight Equivalents				Metric System
		1 grain (gr.)	=	60 milligrams (mg)
		15 grains (gr.)	=	1 gram (g)
60 grains (gr.)	=	1 dram (ℨ)	=	4 grams (g)
8 drams (ℨ)	=	1 ounce (ℨ)	=	30 grams (g)
12 ounces (ℨ)	=	1 pound (lb.)	=	360 grams (g)
		32 ounces (ℨ)	=	1 kilogram (kg)

- **Express Apothecaries' units of weight as equivalent apothecaries units of weight.**

  Using accepted abbreviations express each measurement as an equivalent measurement.

  2 drams as grains

  $$\frac{1\ dram}{60\ grains} = \frac{2\ drams}{x\ grains}$$

  $$x = 120$$

  2 drams equal gr. cxx.

  gr. xl as drams

  $$\frac{40\ grains}{b\ drams} = \frac{60\ grains}{1\ dram}$$

  $$b = \frac{2}{3}\ dram$$

  gr. xl equal ℨ$\frac{2}{3}$.

  4.5 ounces as drams

  $$\frac{1\ ounce}{8\ drams} = \frac{4.5\ ounces}{a\ drams}$$

  $$a = 36$$

  4.5 ounces equal ℨ xxxvi.

  lb. ap. ss as ounces

  $$\frac{12\ ounces}{1\ pound} = \frac{c\ ounces}{\frac{1}{2}\ pound}$$

  $$c = 6$$

  lb. ap. ss equals ℨ vi.

- **Express weight measurements in the Apothecaries' System as equivalent metric measurements.**

  Using accepted abbreviations, express each measurement as an equivalent measurement.

  0.4 grain to milligrams

  $$\frac{1 \text{ grain}}{60 \text{ milligrams}} = \frac{0.4 \text{ grain}}{y \text{ milligrams}}$$

  $$y = 24$$

  0.4 gr equals 24 mg.

  ℥ xv to grams

  $$\frac{1 \text{ dram}}{4 \text{ grams}} = \frac{15 \text{ drams}}{t \text{ grams}}$$

  $$t = 60$$

  ℥ xv equal 70 g.

  ℥ iv to grams

  $$\frac{1 \text{ ounce}}{30 \text{ grams}} = \frac{4 \text{ ounces}}{b \text{ grams}}$$

  $$b = 120$$

  ℥ iv equal 120 g.

  2.5 pounds to kilograms

  $$\frac{1 \text{ pound}}{0.360 \text{ kilograms}} = \frac{2.5 \text{ pounds}}{x \text{ kilograms}}$$

  $$x = 0.9$$

  2.5 pounds equal 0.9 kg.

- **Express weight (mass) measurements in the Metric System as equivalent Apothecaries' measurements.**

  Using accepted abbreviations, express each measurement as an equivalent measurement.

  180 mg to grains

  $$\frac{1 \text{ grain}}{60 \text{ milligrams}} = \frac{d \text{ grains}}{180 \text{ milligrams}}$$

  $$d = 3$$

  180 mg equal gr. iii.

  20 g to drams

  $$\frac{1 \text{ dram}}{4 \text{ grams}} = \frac{a \text{ drams}}{20 \text{ grams}}$$

  $$a = 5$$

  20 g equal ℥ v.

  0.75 kg to ounces

  $$\frac{1 \text{ kilogram}}{32 \text{ ounces}} = \frac{0.75 \text{ kilograms}}{c \text{ ounces}}$$

  $$c = 24$$

  0.75 kg equals 0.9 ℥ xxiv.

  2160 g to pounds

  $$\frac{360 \text{ grams}}{1 \text{ pound}} = \frac{2160 \text{ grams}}{t \text{ pounds}}$$

  $$t = 6$$

  2 160 g equal lb. ap. vi.

Apothecaries' System Volume Equivalents				Metric System
		15 minims (m)	=	1 mL or (1 cm³)
60 minims (m)	=	1 fluidram (f ℥	=	4 mL or (4 cm³)
8 fluidrams (f ℥	=	1 fluidounce (f ℥	=	30 mL or (30 cm³)
16 fluidounces (f ℥	=	1 pint (pt.)	=	500 mL or (500 cm³)
2 pints (pt.)	=	1 quart (qt.)	=	1 L or (1000 cm³)
4 quarts (qt.)	=	1 gallon (gal.)	=	4 L or (4000 cm³)

- **Express Apothecaries' units of volume as equivalent Apothecaries' units of volume.**

  Using accepted abbreviations, express each measurement as an equivalent measurement.

  4 fluidrams to minims

  $$\frac{1 \text{ fluidram}}{60 \text{ minims}} = \frac{4 \text{ fluidrams}}{r \text{ minims}}$$

  $$r = 240$$

  f ʒ iv equal m ccxl.

  5 fluidounces to fluidrams

  $$\frac{1 \text{ fluidounce}}{8 \text{ fluidrams}} = \frac{5 \text{ fluidounces}}{a \text{ fluidrams}}$$

  $$a = 40 \text{ fluidrams}$$

  f ʒ v equal f ʒ xl.

  2 pints to fluidounces

  $$\frac{1 \text{ pint}}{16 \text{ fluidounces}} = \frac{2 \text{ pints}}{c \text{ fluidounces}}$$

  $$c = 32$$

  pt. ii equal f ʒ xxxii.

  5 quarts to pints

  $$\frac{1 \text{ quart}}{2 \text{ pints}} = \frac{5 \text{ quarts}}{x \text{ pints}}$$

  $$x = 10$$

  qt. v equal pt. x.

  3 gallons to quarts

  $$\frac{1 \text{ gallon}}{4 \text{ quarts}} = \frac{3 \text{ gallons}}{y \text{ quarts}}$$

  $$y = 12$$

  gal. iii equal qt. xii.

- **Express volume measurements in the Apothecaries System as equivalent Metric measurements.**

  Using accepted abbreviations, express each measurement as an equivalent measurement.

  30 minims to milliliters

  $$\frac{15 \text{ minims}}{1 \text{ mL}} = \frac{30 \text{ minims}}{q \text{ mL}}$$

  $$q = 2$$

  m xxx equal 2 ml.

  12 fluidrams to cubic centimeters

  $$\frac{8 \text{ fluidrams}}{30 \text{ cm}^3} = \frac{12 \text{ fluidrams}}{p \text{ cm}^3}$$

  $$p = 45$$

  f ʒ xxii equal 45 cm^3.

  0.5 pints to cubic centimeters

  $$\frac{2 \text{ pints}}{1000 \text{ cm}^3} = \frac{0.5 \text{ pint}}{b \text{ cm}^3}$$

  $$b = 250$$

  pt. ss equals 250 cm^3.

  1.5 gallons to liters

  $$\frac{4 \text{ liters}}{1 \text{ gallon}} = \frac{x \text{ liters}}{1.5 \text{ gallons}}$$

  $$x = 16$$

  1.5 gallons equal 6 liters.

Apothecaries' System (Weight)	Household System Weight		Apothecaries' System (Liquid)	Household System Liquid	
4 drams	= 1 tablespoonful (T)		4 fluidrams	= 1 tablespoonful (T)	
1 grain (gr)	= 1 drop		1 minim (℩)	= 1 drop	
1 dram (ℨ)	= 1 teaspoonful (t)		1 fluidram (fℨ)	= 1 teaspoonful (t)	
1 ounce (ℨ)	= 2 tablespoonfuls		1 fluidounce (fℨ)	= 2 tablespoonfuls	
6 ounces	= 1 teacupful		6 fluidounces	= 1 teacupful	
8 ounces	= 1 glassful		8 fluidounces	= 1 glassful	

Approximate Liquid Measure Equivalents		
60 drops	=	1 teaspoonful (t)
4 teaspoonfuls	=	1 tablespoonful (T)
2 tablespoonfuls	=	1 fluidounce
6 fluidounces	=	1 teacupful
8 fluidounces	=	1 glassful

- **Express volume measurements in the Metric System as equivalent Apothecaries' measurements.**

Using accepted abbreviations, express each measurement as an equivalent measurement.

10 cm^3 to minims

$$\frac{4 \text{ cm}^3}{60 \text{ minims}} = \frac{10 \text{ cm}^3}{d \text{ minims}}$$

$$d = 150$$

10 cm^3 equal mcl.

1500 cm^3 to fluidounces

$$\frac{16 \text{ fluidounces}}{500 \text{ cm}^3} = \frac{y \text{ fluidounces}}{1\,500 \text{ cm}^3}$$

$$y = 48$$

1 500 cm^3 equal 48 fluidounces.

1 250 ml to pints

$$\frac{1 \text{ pint}}{500 \text{ ml}} = \frac{x \text{ pint}}{1\,250 \text{ ml}}$$

$$x = 2.5$$

1 250 ml equal 2.5 pints.

5 L to quarts

$$\frac{1 \text{ quart}}{1 \text{ L}} = \frac{p \text{ quarts}}{5 \text{ L}}$$

$$p = 5$$

5 L equal qt. v.

- **Express Household liquid measurements as approximate equivalent measurements.**

Using accepted abbreviations, express each measurement as an equivalent measurement.

4 tablespoonfuls to fluidounces

$$\frac{2 \text{ tablespoonfuls}}{1 \text{ fluidounce}} = \frac{4 \text{ tablespoonfuls}}{x \text{ fluidounces}}$$

$$x = 2$$

2 tablespoonfuls are approximately equal to 2 fluidounces.

12 fluidounces to glassfuls

$$\frac{8 \text{ fluidounces}}{1 \text{ glassful}} = \frac{12 \text{ fluidounces}}{a \text{ glassfuls}}$$

$$a = 1\frac{1}{2}$$

8 fluidounces are approximately equal to $1\frac{1}{2}$ glassfuls.

- **Express Household weight measurements as approximate equivalent Apothecaries' measurements.**

  Using accepted abbreviations, express each measurement as an approximate equivalent measurement.

  3 teacupfuls to ounces

  $$\frac{1 \text{ teacupful}}{6 \text{ ounces}} = \frac{3 \text{ teacupfuls}}{x \text{ ounces}}$$

  $$x = 18$$

  3 teacupfuls are approximately 18 ounces.

  6 drops to grains

  $$\frac{1 \text{ drop}}{1 \text{ grain}} = \frac{6 \text{ drops}}{y \text{ grains}}$$

  $$y = 6$$

  6 drops are approximately 6 grains.

- **Express Apothecaries' liquid measurements as approximate Household measurements.**

  Using accepted abbreviations, express each measurement as an approximate equivalent measurement.

  5 fluidrams to tablespoonfuls

  $$\frac{\text{f } \mathfrak{Z} \text{i}}{1 \text{ T}} = \frac{\text{f } \mathfrak{Z} \text{v}}{b \text{ T}}$$

  $$b = 5$$

  f $\mathfrak{Z}$v are approximately equal to 5 tablespoonfuls.

  5 fluidounces to tablespoonfuls

  $$\frac{\text{f } \mathfrak{Z} \text{i}}{2 \text{ T}} = \frac{\text{f } \mathfrak{Z} \text{v}}{x \text{ T}}$$

  $$x = 10$$

  f $\mathfrak{Z}$ v are approximately equal to 10 tablespoonfuls.

Weight (Mass)		Liquid Measure	
Metric System	**Household System**	**Metric System**	**Household System**
0.06 gram (g) or 60 milligrams (mg) = 1 drop		0.06 milliliters (mL) = 1 drop	
5 grams	= 1 teaspoonful (t)	5 milliliters	= 1 teaspoonful (t)
15 grams	= 1 tablespoonful (T)	15 milliliters	= 1 tablespoonful (T)
180 grams	= 1 teacupful	18 milliliters	= 1 teacupful
240 grams	= 1 glassful	240 milliliters	= 1 glassful

- **Express Household measurements as approximate equivalent Metric measurements.**

  Using accepted abbreviations, express each measurement as an approximate equivalent measurement.

  10 drops to milliliters

  $$\frac{1 \text{ drop}}{0.06 \text{ mL}} = \frac{10 \text{ drops}}{p \text{ mL}}$$

  $$p = 0.6$$

  10 drops are approximately 0.6 mL.

  4 teaspoonfuls to grams

  $$\frac{1 \text{ t}}{5 \text{ g}} = \frac{4 \text{ t}}{a \text{ g}}$$

  $$a = 20$$

  4 t are approximately 20 g.

- **Express Metric measurements as approximate equivalent Household measurements.**

Using accepted abbreviations, express each measurement as an approximate equivalent measurement.

360 grams to glassfuls

$$\frac{240 \text{ g}}{1 \text{ glassful}} = \frac{360 \text{ g}}{b \text{ glassfuls}}$$

$$b = 1\frac{1}{2}$$

360 g are approximately $1\frac{1}{2}$ glassfuls.

30 milliliters to tablespoonfuls

$$\frac{15 \text{ mL}}{1 \text{ T}} = \frac{30 \text{ mL}}{x \text{ T}}$$

$$x = 2$$

30 mL are approximately 2 tablespoonfuls.

Review

Major concepts and skills to be mastered for the Section 5 test.

In 1–10, complete each sentence. Choose from the vocabulary list.

1. ℨ is the symbol for _____.

2. Fifteen _____ are equivalent to 1 mL or 1 cm³ in the Metric System.

3. Gal. _____ stands for one-half gallon.

4. In the _____ 60 grains (gr) are equivalent to 1 dram (ℨ).

5. Thirty grams in the _____ are equivalent to one ounce in the Apothecary System.

6. The abbreviation for fluidounce is _____.

7. _____ is the abbreviation for tablespoonful.

8. One _____ is considered to contain 8 fluidounces.

9. One teacupful is equivalent to f ℥ vi in the _____.

10. Thirty-two written using _____ is xxxii.

11. Using Roman numerals, write the first 7 multiples of seven.

In 12–15, write each measurement using apothecaries' abbreviations or symbols.

12. 12 ounces   13. 4 drams   14. $\frac{1}{5}$ pound   15. 6 grains

In 16–23, compare. Use the symbols <, >, or =.

16. gr. lxx __?__ ℨ ii

17. ℨ v __?__ gr. ccc

18. ℨ x __?__ ℥ ii

19. lb. ap. $\frac{1}{5}$ __?__ ℨ xxv

20. lb. ap. ss __?__ ℥ viii

21. ℨ xxiv __?__ ℥ iii

22. gr. lx __?__ ℨ i

23. ℥ xx __?__ lb. ap. $\frac{1}{4}$

In 24–31, express each measurement as an equivalent measurement. Where appropriate, use accepted abbreviations.

24. 540 grams to pounds

25. 80 grams to drams

26. 20 ounces to kilograms

27. 90 milligrams to grains

28. 1.5 kilograms to pounds   29.   9 ounces to grams

30. 3 grams to grains   31.   540 grains to grams

In 32–38, compare. Use the symbols, >, <, or =.

32. lb. ap. ss __?__ 600 g   33.   gr. xx __?__ 2 g

34. $\frac{1}{2}$ kg __?__ ʒ cxx   35.   10 mg __?__ gr. v

36. 4 g __?__ gr. lx   37.   ʒ viii __?__ 0.4 kg   38.   90 g __?__ ʒ iv

In 39–46, express each measurement as an equivalent measurement. Where appropriate, use accepted abbreviations.

39. 6 quarts to gallons   40.   4 fluidounces to fluidrams

41. 90 minims to fluidrams   42.   3 pints to quarts

43. 2 fluidrams to fluidounces   44.   24 fluidounces to pints

45. 6 pints to gallons   46.   1.5 pints to fluidounces

In 47–54, compare. Use the symbols <, >, or =.

47. gal. ii __?__ 8 L   48.   pt. ss __?__ 250 cm^3

49. ♏ xxv __?__ f ʒ ii   50.   qt. iv __?__ 3 L

51. 1 L __?__ gal.$\frac{1}{2}$ 52.   0.5 fluidram __?__ 8 mL

53. 750 mL __?__ pt. ii   54.   2 mL __?__ ♏ xxx

In 55–62, express each measurement as an equivalent measurement. Where appropriate, use accepted abbreviations.

55. 0.6 cm^3 to minims   56.   5 L to gallons

57. 60 mL to fluidounces   58.   1 000 mL to pints

59. 0.5 gallon to cubic centimeters 60.   12 milliliters to fluidrams

61. 2.6 pints to cubic centimeters 62.   45 minims to milliliters

In 63–68, compare. Use the symbols <, >, or =.

63. f ʒ xxii __?__ 1 glassful   64.   t __?__ 1 T

65. 1$\frac{1}{2}$ teacupful __?__ f ʒ xv   66.   120 drops __?__ f ʒ i

67. 3 T __?__ f ʒ iii   68.   1 glassful __?__ 2 teacupfuls

In 69–74, express each measurement as an equivalent measurement. Where appropriate, use accepted abbreviations.

69. 6 drams as tablespoonfuls   70.   10 ounces as tablespoonfuls

71. 2 glassfuls as ounces   72.   3 drams as teaspoonfuls

73. 4 teacupfuls as glassfuls   74.   4 teaspoonfuls as tablespoonfuls

In 75–80, compare. Use the symbols <, >, or =.

75.   f ℥ vi __?__ 1 T

76.   1 teacupful __?__ f ℥ viii

77.   mv __?__ 7 drops

78.   T __?__ f ℥ xx

79.   1 teacupful __?__ $\frac{1}{2}$ glassful

80.   f ℥ xxii __?__ $2\frac{1}{2}$ T

Household Units	Metric Units	
	Weight (mass)	**Volume (liquid measure)**
1 drop	0.06 g	0.06 mL
1 t	5 g	5 mL
1 T	15 g	15 mL
1 teacupful	180 g	180 mL
1 glassful	240 g	240 mL

In 81–86, express measurement as an equivalent measurement. Where appropriate, use accepted abbreviations.

81.   2 teaspoonfuls to milliliters

82. $1\frac{1}{2}$ teacupfuls to grams

83.   3 glassfuls to milliliters

84.   10 drops to milliliters

85. $\frac{1}{2}$ teacupful to milliliters

86.   3 teaspoonfuls to grams

In 87–92, compare. Use the symbols <, >, or =.

87.   2 teacupfuls __?__ 450 g

88. $\frac{1}{2}$ T __?__ 10 mL

89.   10 g __?__ 2 t

90.   3 glassfuls __?__ 600 mL

91.   120 mg __?__ 2 drops

92.   200 mL __?__ $1\frac{1}{2}$ teacupfuls

SECTION 5 TEST

In 1–10, indicate whether each statement is True or False.

1. The use of fractions and Roman numerals are customary in the Apothecaries' System of measure.

2. The symbol for dram is ʒ.

3. ʒ ss means one-half ounce.

4. Sixteen written using Roman Numerals is XIV.

5. CL equals 50

6. XCIV equals 94.

7. 180 grains (gr) = 3 drams (ʒ).

8. A patient should be able to get approximately 50 drops of ear medication from a 50 mL prescription.

9. One-half liter of liquid is approximately $1\frac{1}{2}$ glassfuls.

10. 20 drams (ʒ) = 2.5 ounces (ʒ).

In 11–20, choose the appropriate approximate equivalent measure(s) from Column II and write the letter(s) in the blank under Column I. Answers may be used more than once.

Column I		Column II	
11. 1 fluidram	_____	a.	1 gram
12. 2 liters	_____	b.	1 glassful
13. 4 drams	_____	c.	3 quarts
14. 15 grains	_____	d.	1 pint
15. 60 drops	_____	e.	6 fluidounces
16. 240 grams	_____	f.	1 teaspoonful
17. 16 fluidounces	_____	g.	1 gallon
18. 4 tablespoonfuls	_____	h.	4 pints
19. 15 milliliters	_____	i.	1 tablespoonful
20. 1 teacupful	_____	j.	15 drops
		k.	none of the above

In 24–35, complete this table of equivalent measures.

Apothecaries'	Household	Metric (Liquid)	Metric (Mass)
℥ i	21.	22.	23.
24.	1 glassful	25.	26.
27.	28.	180 mL	29.
30.	31.	32.	240 g
4 fluidrams	33.	34.	35.

In 36–45, calculate each equivalence.

36. 5 drops = __?__ minims

37. __?__ teacupfuls = 12 fluidounces

38. pt. ss = __?__ cm³

39. ℥ viii = __?__ t

40. gal. ii = __?__ L

41. __?__ g = $\frac{1}{2}$ teacupful

42. 4 glassfuls = __?__ kg

43. __?__ mL = 40 drops

44. 2 T = __?__ ℥

45. f ℥ i = __?__ T

In 46–55, compare. Use <, >, or =.

46. $\frac{2}{3}$ t __?__ 150 mL

47. f ʒ v __?__ f ℥ ii

48. pt. vi __?__ qt. iv

49. lb. ap. i __?__ 1 glassful

50. 5 tablespoonfuls __?__ 75 mL

51. 3 t __?__ 20 g

52. ℥ i __?__ gr. c

53. m xvi __?__ 15 drops

54. ℥ ix __?__ $1\frac{1}{2}$ teacupfuls

55. 20 g __?__ 25 mL

Section Six

Organizing and Reporting Data

Unit 28

Interpreting Charts and Graphs

Objectives

After studying this unit the student should be able to:

- Collect and arrange data in a table.
- Analyze data displayed on bar graphs, broken-line graphs, circle graphs, and pictographs.
- Determine the range mean, median and mode of a set of data.
- Calculate the variance and standard deviation of a set of numbers.

COLLECTING AND ARRANGING DATA

Every day, the health worker handles numerical information. It is important that the health worker is able to understand, analyze, and interpret this information. At times it is necessary to use a variety of sources when collecting information about a particular topic. The information that is collected is referred to as data. *Data* represents a group of facts about a specific topic. The science of analyzing and interpreting data is *statistics*. Analyzing and interpreting data depend on the data collected. If different sets of data are collected, the statistics may vary.

EXAMPLE 1

Janet gathers data about the hospitals in a state. Based on the number of beds available, she determines the five largest hospitals.

St. Anne 325   Proctor 200   Mennonite 450
General 1 500   Brokaw 775

This data is arranged for easier reading and interpretation. It is arranged from the largest hospital to the fifth largest hospital.

Hospital	Number of Beds Available
General	1,500
Brokaw	775
Mennonite	450
St. Anne	325
Proctor	200

EXAMPLE 2

In the community of Williamsville, find the number of deaths in each age group during a three-week period: 0–10 yr.; 11–25.; 26–35 yr.; 36–50 yr.; 51–64 yr.; 65+ yr.

Jeff researches this question, gathers data, and records it in the table.

Deaths in Williamsville Between August 14 and September 5	
Age Group	Number
0–10 yr.	1
11–25 yr.	0
26–35 yr.	2
36–50 yr.	10
51–64 yr.	3
65+ yr.	6

28.1 Exercises

In 1–5, the boiling point, in degrees Celsius for some substances are: acetone 57 °C; camphor 205 °C; ethyl alcohol 78.3 °C, methyl alcohol 64.7 °C; chloroform 61.2 °C; ether 34.6 °C; glycerin 291 °C.

1. Arrange the data in a table.

2. Which substance has the highest boiling point?

3. Which substance has the lowest boiling point?

4. Which substance has a boiling point about five times higher than acetone?

5. Which two substances have the closest boiling point temperatures?

In 6–10, find the population for the five largest cities in the country.

6. Arrange the data in a table.

7. What is the difference in population between the largest and smallest cities?

8. The largest city is __?__ % larger than the smallest city.

9. The largest city is __?__ % larger than the second smallest city.

10. The largest city is __?__ % larger than the second largest city.

In 11–15, collect and arrange data about these topics.

11. The six most common communicable diseases in the Western Hemisphere.

12. The six most common causes of death in a country, state, or province.

13. The doctor-patient ratio in six cities or regions in a country, state, or province.

14. Give another way that the data in Example 2 could be displayed.

INTERPRETING CHARTS AND GRAPHS

Statistical data can be displayed in many ways. The most common method is with a chart or graph. Charts and graphs are visual representations of data. Among the many types of graphs are the bar graph, the broken-line graph, the circle graph, and the pictograph.

EXAMPLE 1

The occurrence of influenza in Williamsville is documented for 1990 through 1995. The number of reported cases per thousand population is recorded as:

1990 – 10	1992 – 18	1994 – 6
1991 – 27	1993 – 9	1995 – 47

This data is represented with different types of graphs.

Bar Graph

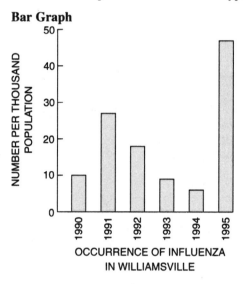

Broken-line Graph

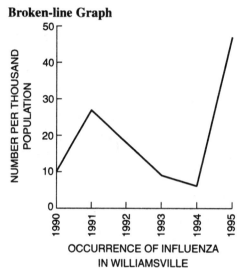

Circle Graph

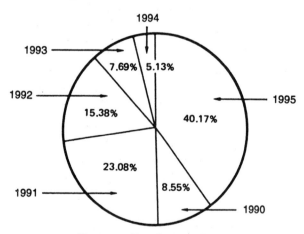

Percent distribution of 117 influenza cases in Williamsville during a five-year period.

Pictograph

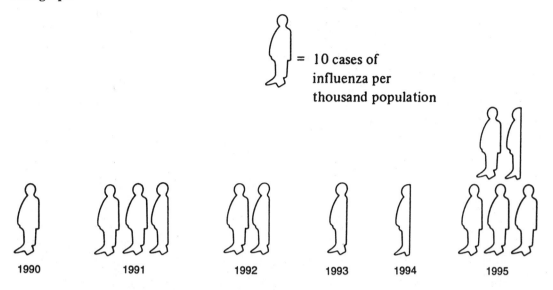

When interpreting charts and graphs, it is important to analyze the method used. Data that is displayed incorrectly can lead to errors in interpretation of the chart or graph. It is also important to understand the relationships or comparisons that a graph is displaying. Analyzing the subject and graphic scales should precede analyzing the data.

In analyzing a graph:

- Read the title of a graph.
- Determine what information is given.
- Determine the value of each major unit on the vertical and horizontal scales.

INTERPRETING BAR GRAPHS

Bar graphs visually represent quantities by comparing bars of varying lengths and uniform widths. Bar graphs are best used in showing the size or the amount of different items at the same time, or the size or the amount of the same item at different times.

EXAMPLE 2

This bar graph shows the occurrences of influenza in Williamsville during the years 1990 through 1995.

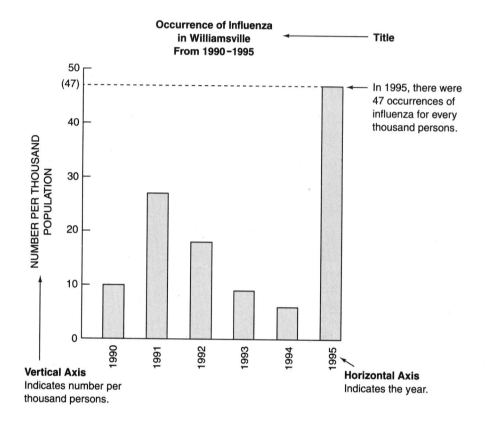

In this bar graph, the number of occurrences of influenza (same item) is shown for different times. This bar graph can compare the number of occurrences for each year. For example, there were three times more occurrences of influenza in 1992 than in 1994. Notice that additional computation is needed to find this comparison. It could not be read directly from the graph.

INTERPRETING LINE GRAPHS

There are three basic types of line graphs. The *straight-line graph* and the *curved-line graph* are used for related facts where there is regularity in the changes. The *broken-line graph* is used for related data where the changes are irregular. The advantage of using a line graph is that other values may be determined without additional computations.

EXAMPLE 1

Temperature is measured every hour and displayed in a broken-line graph.

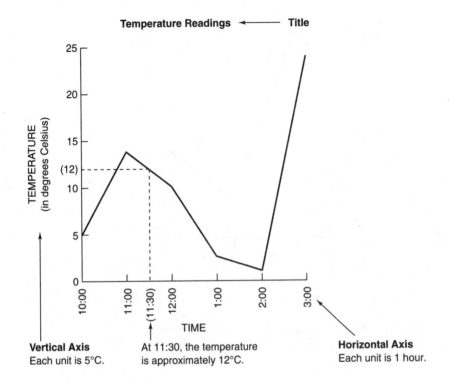

This broken-line graph shows the relationship of the temperature and the time of day. Readings for values not on the axis can be found without additional computations; for example, the temperature reading at 11:30 is approximately 12 °C. Reading values not on the axis is possible because the temperature constantly changed.

When data is accumulated over a period of time and then recorded, as in the occurrences of influenza, the broken-line graph only shows a trend or pattern. In such cases, the broken-line graph is not useful for computations.

EXAMPLE 2

This broken-line graph shows the number of automobile accidents in a community during selected months of a year.

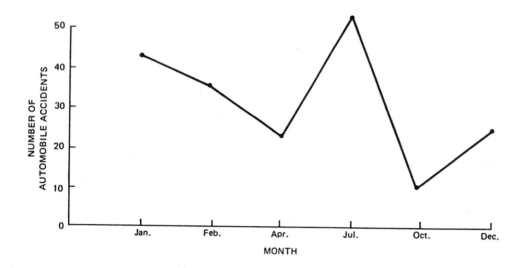

This graph only shows a trend or pattern for the given months. It does not represent the months in sequential order or all the months of the year. No information or trend can be determined for the months of March, May, June, August, September, or November.

The same data can be represented differently.

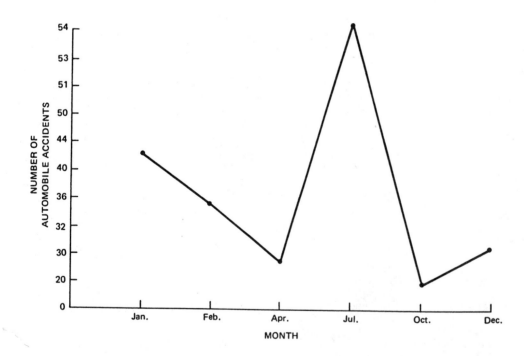

Notice the lack of uniform value intervals along the vertical axis. The same data gives a totally different visual presentation.

28.2 Exercises

Analyze each graph and supply the requested information.

A survey is conducted in a community of 5,000. The occurrence of common communicable diseases from January 1 through June 30 is determined. This bar graph shows the accumulated data.

Occurrence of Common Communicable Diseases

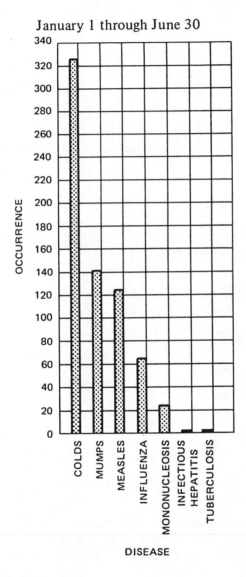

January 1 through June 30

DISEASE

1. What is the subject of the graph?
2. What information is given on the vertical scale?
3. Which disease occurs most often?
4. How many cases of measles are reported?
5. What disease occurs about five times as often as mononucleosis?
6. How effective would a broken-line graph be in representing this data?

A community study is conducted to discover the relationship between persons 35 years or older with smoking and/or drinking habits and common forms of cancer. The survey reveals that approximately 60% of the community in this age group has smoking and/or drinking habits. The total population of the community is 37,495 with 15,962 males and 21,533 females. In the age group 35 and over, 9,483 are males and 10,986 are females.

Relationship of Cancer and Smoking/Drinking Habits

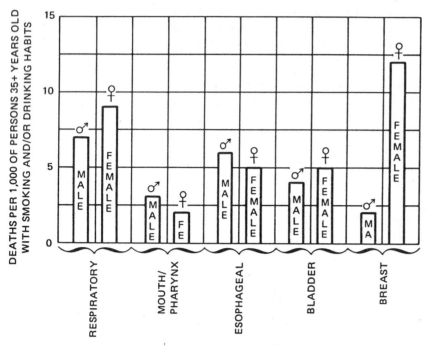

7. How many males involved in this study have smoking and/or drinking habits?

8. How many females involved in this study have smoking and/or drinking habits?

How many persons in this community, age 35+, with smoking and/or drinking habits actually died from:

		Males	Females
9.	respiratory cancer?	?	?
10.	mouth/pharynx cancer?	?	?
11.	esophageal cancer?	?	?
12.	bladder cancer?	?	?
13.	breast cancer?	?	?

14. What can be concluded from this study about smoking and/or drinking habits and their relationship with deaths caused by cancer?

Liquids, when heated above room temperature, tend to cool over a period of time. A 125-mL beaker is filled with 100 mL of water and heated to 60 °C. It is then allowed to cool and the temperature is recorded in ten-minute intervals. The temperatures are recorded on a broken-line graph.

Cooling Pattern of Water

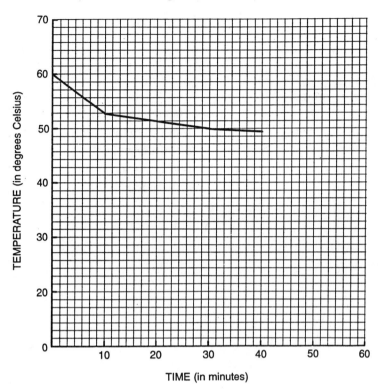

15. What is the temperature at 20 minutes?

16. In which time period is the decrease in temperature the greatest?

17. How many degrees does the temperature decline in 30 minutes?

18. Assuming that the room temperature is 23 °C, is it possible to project the pattern of cooling over a longer period of time without actually making the measurements?

19. What variables must be considered when projecting the cooling pattern of a liquid?

INTERPRETING CIRCLE GRAPHS

Circle graphs illustrate how one part is related to another part or to the whole. The circle graph is used primarily for comparison purposes and is impractical when determining other numerical values.

EXAMPLE 1

A circle graph is prepared to show the percent distribution of influenza cases in a school.

Influenza Cases in a School

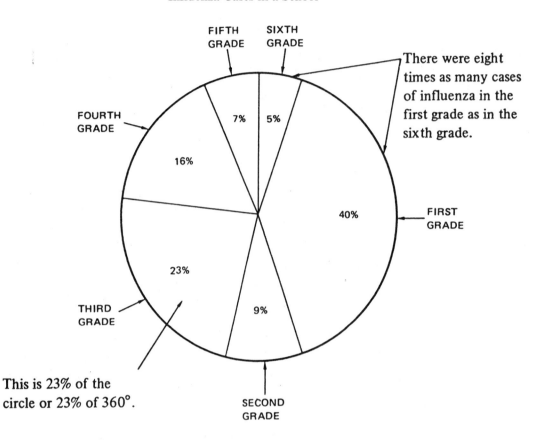

This graph compares the influenza cases in each grade to the total influenza cases. The fourth grade students had 16% of the influenza cases. This does *not* mean that 16% of the fourth grade students had influenza, but rather that 16% of all students having influenza were fourth grade students. Comparisons can also be made between different grades. For example, there were eight times as many cases of influenza in the first grade as in the sixth grade. In order to find the actual number of cases for each grade, the percent is multiplied by the total number of cases.

INTERPRETING PICTOGRAPHS

Pictographs are probably the easiest graphs to read but are the most difficult to draw. The pictograph shows approximations and is used for comparisons. Accurate numerical values usually cannot be obtained from pictographs.

EXAMPLE 1

A pictograph is designed to display the influenza cases in Williamsville during a 5-year period.

Influenza Cases in Williamsville for 1990–1995

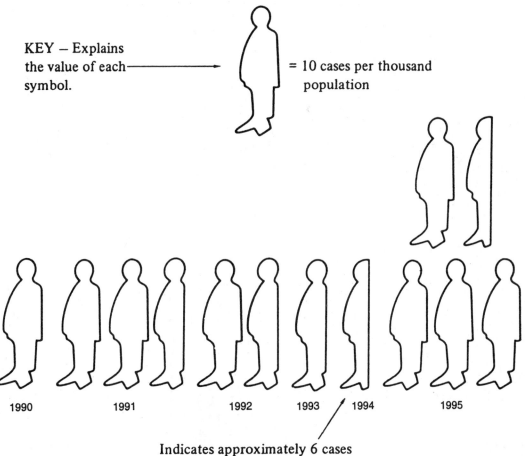

KEY — Explains
the value of each
symbol. ⟶ = 10 cases per thousand
population

1990   1991   1992   1993   1994   1995

Indicates approximately 6 cases
per thousand people.

Notice the difficulty in distinguishing between the actual numbers of cases. Five cases looks like six cases or even seven cases. The graph is best used to determine the years in which the most or least number of cases occurred or years in which the greatest change occurred.

Displaying statistical data with pictographs can be misleading. The comparisons may be misrepresented or the presentation may be inaccurate.

EXAMPLE 2

This pictograph is supposed to show that three times as many people live in City *B* as in City *A*.

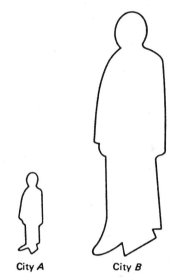

City **A**   City **B**

COMPARATIVE POPULATIONS OF TWO CITIES

The figure representing City *B* is actually nine times larger than the one representing City *A*. As the figure was increased in height it was also increased in width.

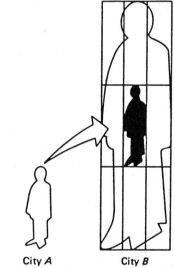

City **A**   City **B**

COMPARATIVE POPULATIONS OF TWO CITIES

A better method of comparison would be to assign each figure a value. If each figure represents a population of 25,000, the pictograph is then a valid comparison.

City **A**   City **B**

COMPARATIVE POPULATIONS OF TWO CITIES

28.3 Exercises

Analyze each graph and supply the requested information.

Within one portion of the current hospital budget, funds are allocated as presented in this circle graph.

Current Expenditures

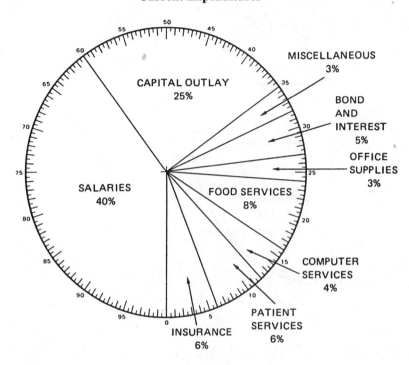

In 1–9, this portion of the budget contains $1,462,983. How many actual dollars exist in each area?

1.	Salaries	2.	Capital Outlay	3.	Miscellaneous
4.	Bond and Interest	5.	Office Supplies	6.	Food Services
7.	Computer Services	8.	Patient Services	9.	Insurance

In 10–12, there are 360° in a circle. How many degrees are allowed for each category in the circle graph? Round the answer to the nearer tenth if necessary.

10. Salaries   11. Capital Outlay   12. Food Services

The emergency room deals with a variety of patients. The data from the monthly report is displayed in a circle graph. The number of patients that are admitted is 130.

Emergency Room Injuries

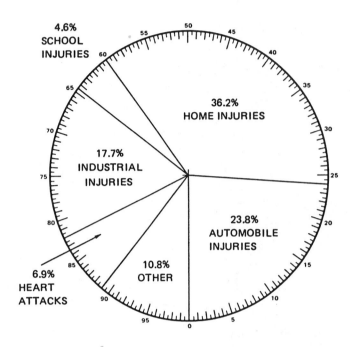

In 13–18, find the number of cases in each category.

13. Automobile Injuries 14. Home Injuries 15. School Injuries
16. Industrial Injuries 17. Heart Attacks 18. Other

In 19–24, commercially prepared circle graph paper is available and is divided into increments of 0.5%. It is sometimes necessary to construct graphs with the aid of a compass and protractor. Assuming this approach must be followed, how many degrees must be allotted for each of the categories? Round each answer to the nearer tenth.

Note: There are 360° in a circle.

19. Automobile Injuries 20. Home Injuries 21. School Injuries
22. Industrial Injuries 23. Heart Attacks 24. Other

In 25–30, the county health department records the number of babies born during a year. This data is used to construct a pictograph.

Births for a Twelve-Month Period

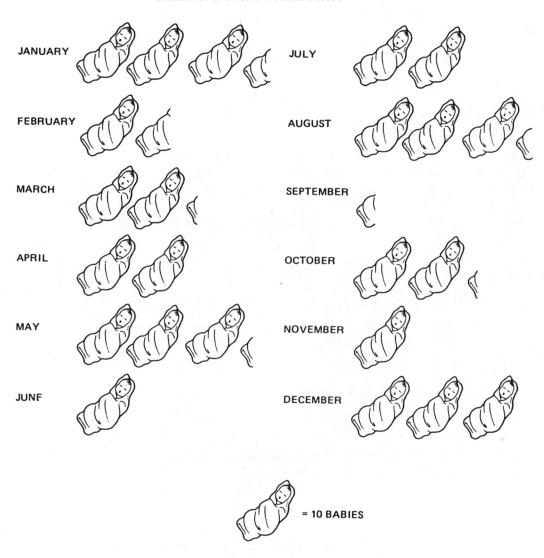

= 10 BABIES

25. What does each symbol represent?
26. Approximately how many babies are born in September?
27. Approximately how many babies are born in the last quarter of the year?
28. Approximately how many more babies are born in May than in November?
29. In which month are the most babies born?
30. Approximately how many times as many babies are born in December as in November?

RANGE, MEAN, MEDIAN, MODE, VARIANCE, AND STANDARD DEVIATION

When data is analyzed and presented, it is often useful to make comparative descriptive comparisons. For the health worker, there are six descriptive measures that are useful. When calculating these descriptive measures, it is necessary to arrange the data in rank order from the smallest value to the largest value.

DEFINITIONS

- The *range* is the difference between the largest and smallest data items.
- The *mean* is the arithmetic average of a set of numbers. It is the sum of the numbers divided by the number of data items in the set. In a set of numbers, the mean need not be one of the numbers in the set.

  In symbols, the mean is denoted as:

  $$\overline{X} = \frac{\sum_{i=1}^{n} x_i}{n}$$

  Where $\overline{X}$ is the mean (read "X bar");
  $\sum_{i=1}^{n} x_i$ is the sum of the data items in the set and n is the number of data items.

- The *median* is a positional average. If there is an odd number of data items in a set, the median is the middle number. If there is an even number of data items in the set, the median is the mean of the two middle numbers. Since the median is a positional average, an extremely large or small data item does not affect it.
- The *mode* is the number (or numbers) that occur most frequently in a set of data items. A set with more than one mode is termed *bimodal*.
- The *variance* of the set of data items measures how variable the data items are within the set. One formula for variance is:

  $$\sigma^2 = \frac{\sum_{i=1}^{n} x_i^2}{n} - \overline{X}^2$$

  Where σ^2 (read "sigma squared") stands for variance;
  $\sum_{i=1}^{n} x_i^2$ is the sum of the data items squared in the set;
  n is the number of data items; and $\overline{X}$ is the mean squared.

- The *standard deviation* of a set of data items is a number that expresses the degree to which the data items in a set are dispersed. The standard deviation is found by taking the positive square root of the variance.

  $$\sigma = \sqrt{\sigma^2}$$

Note: Most scientific calculators have formulas built in to find the mean variance, and standard deviation. Check your owner's manual for specific details.

EXAMPLE 1

Compare the following three sets of numbers by finding their range, mean, median, mode, variance, and standard deviation.

	Set I 2,6,6,6,6,6,10	Set II 2,3,4,6,8,9,10	Set III 2,2,2,6,10,10,10
Range	$10 - 2 = 8$	$10 - 2 = 8$	$10 - 2 = 8$
Mean	$\overline{X} = \frac{2+6+6+6+6+6+10}{7} = 6$	$\overline{X} = \frac{2+3+4+6+8+9+10}{7} = 6$	$\overline{X} = \frac{2+2+2+6+10+10}{7}$
Median	6	6	6
Mode	6	no mode	2 and 10

Variance

x_i	x_i^2
2	4
6	36
6	36
6	36
6	36
6	36
10	100
Sum = 284	

x_i	x_i^2
2	4
3	9
4	16
6	36
8	64
9	81
10	100
Sum = 310	

x_i	x_i^2
2	4
2	4
2	4
6	36
10	100
10	100
10	100
Sum = 348	

$$\sigma^2 = \frac{\sum_{i=1}^{n} x_i^2}{n} - \overline{X}^2$$

$$\sigma^2 = \frac{284}{7} - 6^2$$

$$\sigma^2 = \frac{284}{7} - 36$$

$$\sigma^2 = \frac{284}{7} - \frac{252}{7}$$

$$\sigma^2 = \frac{32}{7} \text{ or } 4.57$$

$$\sigma^2 = \frac{\sum_{i=1}^{n} x_i^2}{n} - \overline{X}^2$$

$$\sigma^2 = \frac{310}{7} - 6^2$$

$$\sigma^2 = \frac{310}{7} - 36$$

$$\sigma^2 = \frac{310}{7} - \frac{252}{7}$$

$$\sigma^2 = \frac{58}{7} \text{ or } 8.29$$

$$\sigma = \sqrt{4.57}$$

$$\sigma^2 = \frac{\sum_{i=1}^{n} x_i^2}{n} - \overline{X}^2$$

$$\sigma^2 = \frac{348}{7} - 6^2$$

$$\sigma^2 = \frac{348}{7} - 36$$

$$\sigma^2 = \frac{348}{7} - \frac{252}{7}$$

$$\sigma^2 = \frac{96}{7} \text{ or } 13.7$$

$$\sigma = \sqrt{8.29}$$

$$\sigma = \sqrt{13.7}$$

$$\sigma \approx 2.14 \qquad\qquad \sigma \approx 2.88 \qquad\qquad \sigma \approx 3.70$$

These graphs illustrate the distribution of data items and standard deviations for the three sets of data.

Set I

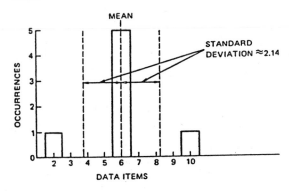

A standard deviation of 2.14 means that most of the data items are within 2.14 units of each side of the mean.

Set II

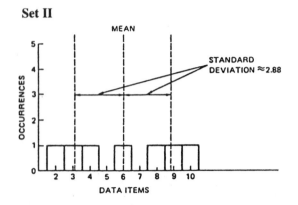

A standard deviation of 2.88 means that most of the data items are within 2.88 units of each side of the mean.

Set III

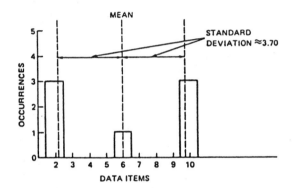

A standard deviation of 3.70 means that most of the data items are within 3.70 units of each side of the mean.

The three sets of numbers appear to be the same when comparing their ranges, means, and medians. Only by further analyzing their modes, variances, and standard deviations can any meaningful comparisons be made. Clearly, the data items in Set III are more widely dispersed than either those in Sets I or II.

28.4 Exercises

1. Arrange the data from smallest to largest and find the range, mean, median, mode or modes, variance and standard deviation.

   a. 8, 1, 2, 4, 5, 3, 9, 4, 5, 4

   b. 13. 10, 4, 5, 1, 5, 10, 5, 10

2. Multiply each number in this set by ten.
   8, 1, 2, 4, 5, 3, 9, 4, 5, 4

   How does the mean of the new set of numbers compare with the original set?

3. Replace one of the numbers in this set so the mean of the numbers in the new set is 10.
   8, 4, 10, 8, 40

4. Replace one of the numbers in this set so the median of the new set is 6.
   5, 6, 8, 12

5. Is the mode always a member of the set of data?

6. Is the median always a member of the set of data?

7. A laboratory coat manufacturer has equipment to make only one size coat—small, medium, or large. The equipment cannot make all three. To sell the most coats, the manufacturer must properly determine the market needs. Which measure (mean, median, mode) should the manufacturer use to determine production?

8. In a community college, the average age of a group of ten people is 20 years. Nine of the people are students and the other person is the teacher. Which measure best typifies the age of the group: mean, median, mode?

9. A study is made to determine the effects that a mother's diet has on the birth weight of her child. To undertake the study, two groups of mothers are established, the control group and the experimental group. Each group of mothers has the same general characteristics in terms of height, weight prior to pregnancy, and physical condition. Each mother in the control group continues with the diet of her choice. Each mother in the experimental group receives a carefully planned diet designed to encourage proper development of the child. At birth, the weights, in kilograms are recorded and arranged from smallest to largest. Find the range, mean, median, mode, variance and standard deviation. Round each answer to the nearer whole number. Can you draw any conclusions concerning the experimental versus the control group?

Control Group				**Experimental Group**			
2.2	2.7	3.1	3.6	2.5	2.9	3.3	3.7
2.3	2.7	3.3	3.7	2.6	2.9	3.3	3.7
2.4	2.8	3.3	3.8	2.6	2.9	3.3	3.9
2.4	2.9	3.3	3.9	2.6	3.0	3.7	4.0
2.5	2.9	3.3	4.2	2.7	3.0	3.7	4.1
2.5	3.0	3.5	4.4	2.9	3.1	3.7	4.1
2.6	3.1			2.9			

10. The Laboratory Electronic Technicians' Union is striking for higher wages. The union announces that the average worker receives a salary of only $9,000 while management claims the average wage is $22,636. Both sides use this data: $9,000; $9,000; $9,000; $10,500; $12,000; $13,500; $18,000; $18,000; $45,000; $45,000; $60,000.

a. Which measure (mean, median, mode) is each side using?

b. Which measure is the more appropriate? Explain.

APPLICATIONS

When performing laboratory tests or experiments, or when taking any measurement, there is some error. The error may be due to many factors including lack of accuracy in the measuring instrument or in the laboratory procedure. The amount of error is therefore taken into account and limits within which an error may be tolerated are established.

The allowable limits of error in a procedure may be determined for the normal range of values. This calculation uses the mean and the range of the normal biological range. The percent of allowable limits of error is:

$$\text{Percent of Allowable Limits of Error} = \frac{\dfrac{\text{range}}{4}}{\text{mean}} \times 100$$

EXAMPLE 1

The normal range for hematocrit for adult males is 42 mL/100 mL to 52 mL/100 mL. Find the percent of allowable limits of error, to the nearer tenth percent.

$$\text{Percent of Allowable Limits of Error} = \frac{\dfrac{\text{range}}{4}}{\text{mean}} \times 100$$

$$\text{Percent of Allowable Limits of Error} = \frac{\dfrac{10}{4}}{47} \times 100$$

$$\text{range} = 52 - 42 \ or \ 10$$

$$\text{mean} = \frac{42 + 52}{2} \ or \ 47$$

$$\text{Percent of Allowable Limits of Error} = \frac{2.5}{47} \times 100$$

$$\text{Percent of Allowable Limits of Error} = 5.3\%$$

The percent of allowable limits of error is used to determine how a derived measurement compares with an accepted value. It may also be used to determine if a second, or follow-up, test or procedure shows a significant difference or is an allowable difference. The accepted limit is found by multiplying the accepted value by the percent of allowable limits of error. If the derived measurement falls within these limits, the measurement is acceptable.

EXAMPLE 2

A sample of blood has a hematocrit value (derived measurement of 45 mL/100 mL). The known concentration (accepted measurement) of the original sample is 47 mL/100 mL. Is the derived measurement within the allowable limits of error? (The percent of allowable limits of error is 5.3%.)

$47 \times 5.3\% = 47 \times 0.053 \ or \ 2.4$
Upper Limit $= 47 + 2.4 \ or \ 49.4$ mL/100 mL
Lower Limit $= 47 - 2.4 \ or \ 44.6$ mL/100 mL

Since 45 mL/100 mL falls within the limits, it is an allowable value.

Note: This information gains importance when interpreting laboratory data. For example, in monitoring the progress of a patient, a change from 47 mL/100 mL to 45 mL/100 mL may not be a significant difference since the 45 mL/100 mL is within the allowable limits. The percent of allowable limits of error is used to determine values that fall within the normal range and also to determine the significance of differences that falls in the abnormal range.

28.5 Exercises

In order to determine the normal range for calcium content in the blood, 235 specimens of serum are tested. The results are illustrated in this bar graph.

Calcium Content of the Blood

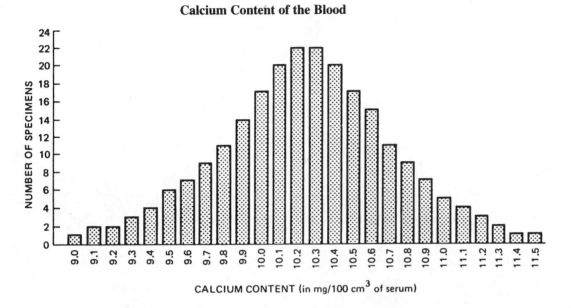

CALCIUM CONTENT (in mg/100 cm^3 of serum)

1. What is the normal range of values as indicated by this graph?

2. Find the range of the data items.

3. Find the mode(s).

4. Find the median.

5. The mean of the data items is calculated to be 10.25 mg/100 cm of serum. Using this value, find the percent of allowable limits of error to the nearer tenth percent.

6. Using the mean as the accepted measurement, determine the allowable limits of error.

7. Is a derived reading of 9.8 within the allowable limits of error?

8. Is a derived reading of 11.1 within the allowable limits of error?

In 9–11, a serum specimen indicates that the chloride in the body is 98 milliequivalents per liter of serum (derived measurement). The known concentration is 103 milliequivalents per liter and the normal range is 100 to 106 milliequivalents per liter.

9. Find the percent of allowable limits of error to the nearer tenth percent.

10. Find the allowable limits of error to the nearer tenth.

11. Is the derived measurement within the allowable limits of error?

In 12–17, the degree of acidity or alkalinity of the urine is expressed by pH values. The pH values for ten specimens are:

4.8	6.4
5.2	6.4
5.3	7.8
5.9	7.8
6.4	8.0

12. Find the range of the pH values.

13. Find the mode(s).

14. Find the median value.

15. Determine the mean.

16. Calculate the standard deviation.

17. Using the mean and the range, calculate the percent of allowable limits of error to the nearer tenth.

Unit 29

Computers— Tools of the 21st Century

Objectives

After studying this unit the student should be able to:
- **Describe the essential value of the computer to the field of health services.**
- **Apply mathematic skills to create interactive tables for data management.**
- **Develop the background essential for confident use of computers as a tool in providing health care services.**

THE COMPUTER AS A TOOL

Computers are the basic tools of the 21st century. They assist the user in accessing data stored in the computer or data available thousands of miles away by way of telephone lines and/or satellites. Computers make it possible for patients to remain at home and basic medical information can be electronically monitored and transmitted to computers at the hospital or clinic. New uses for computers in providing health care services are rapidly increasing. Computers, somewhere, now have stored data on virtually every citizen in the U.S. This may be a scary thought, but when used properly it can help us in many ways.

Even if you have never used a computer, it is possible to understand its usefulness to your work. A computer system consists of hardware and software. The hardware may include the computer (processor), monitor (visual screen), printer, as well as a modem (to send and receive information), scanner (to enter data into the computer) and a wide variety of other input devices. Software is a set of electronic instructions that directs the operation of the computer and how it processes the data entered into the computer.

WORD PROCESSING, DATA MANAGEMENT, SPREADSHEETING

Word processing, data management, and spreadsheeting are three common applications of the computer. Students in middle and high school commonly use computers to do some word-processing. This is a valuable experience when preparing to enter advanced education and the work place. In many schools, computers have replaced the typewriter and students are learning to type and use computer software at the same time. Because of the need to know how to use a computer beyond high school, more students are now enrolling in basic keyboarding classes. Some schools even require the course for graduation.

Data management allows the storage, processing and retrieval of information. Numerical data can be processed to produce a variety of graphs. Many types of data may be stored for each patient in the clinic. Selected pieces of information from each patient's record can be used to develop reports essential to providing better services in the clinic.

Spreadsheeting makes full use of the math skills you have developed in Units 1 through 28. Its effective use will also depend upon your knowledge of the tasks to be met by the clinic. Since this type of software is directly related to the math you have been studying, the remainder of this unit will deal with its use.

SPREADSHEETING—A POWERFUL TOOL

FOR MATHEMATICAL CALCULATIONS

A spreadsheet is really quite simple. It consists of a grid on a screen as shown below. Letters are used across the top of the grid. Each of these letters represents a column that extends below it. Along the left-hand side of the grid is a series of numbers in ascending order. Each number represents a row of information extending to the right. If you follow column **C** down to row 10 you will find a box. This box is known as the coordinate of column **C,** row 10 and is termed a cell. Information can be entered into each cell. This information may be either a label or a value.

This grid of cells allows you to set up a table for the purpose of doing a wide variety of

	A	B	C	D	E
1					
2					
3					
4					
5					
6					
7					
8					
9					
10					

Figure 29-1

mathematical calculations. The remainder of Unit 29 will introduce sample applications of the spreadsheet and ask you to try your skill at setting up a table. The single mathematics function of multiplication will be used in both the examples and problems. It is possible to use multiplication (∗), division(/), addition (+), and subtraction (−) all within the same formula, if the problem requires this type of calculation. The formula is not limited to the size of the cell. If you are actually working with a computer, consult the manual relative to the spreadsheet program you are using. All spreadsheet programs are similar in how they operate but may have some unique procedures for entering the data.

EXAMPLE 1

Patients entering the clinic are provided a medical service. It has been determined that there are two levels of services that may be provided with the staff and equipment available. Additionally, the cost of the service is also determined by the length of time the patient must remain in the treatment process. Periodically, due to increased staff, materials, and equipment costs, it is necessary to adjust the cost of the treatment. To determine exact cost quickly and accurately, a chart must be established for staff use during billing.

The first step is to create the basic table layout. We know, through study, that the cost must fall into two levels. We also know the cost should increase for each 15 minutes of service. Further, it has been determined that the basic cost for the service is $45.00. From this, it is possible to determine the percent of increase for each 15 minutes of service at each level of service. The following figure shows the creation of the basic table including the labels and values.

	A	B	C	D	E
1		**Physical Medical Services — Out Patient**				
2					
3		**Level I**		**Level II**	
4	**Time**	**%**	**Cost**	**%**	**Cost**
5					
6	90 Min	1.25		2.00	
7	75 Min	1.20		1.90	
8	60 Min	1.15		1.80	
9	45 Min	1.10		1.70	
10	30 Min	1.05		1.60	
11	15 Min	1.00		1.50	
12					
13					
14		Base Cost:	45.00		
15					

Figure 29-2A

Once the basic table has been created, it is possible to enter the formulae necessary to make the table do the automatic calculations. Figure 29-2B shows the formulae entered under each cost column. Notice that, in this case, the calculations are simply the coordinate found in either the **B** or **C** columns times the basic cost found in coordinate C14. The method used to enter the formula may vary slightly from one spreadsheet program to another. In this case, the = sign tells the computer a value is being entered rather than a label. The full formula, such as =B6*C14, is really telling the computer to take the percent shown in B6 (1.25) times the base cost shown in C14 (45.00).

	A	B	C	D	E
1		**Physical Medical Services — Out Patient**				
2					
3		**Level I**		**Level II**	
4	**Time**	**%**	**Cost**	**%**	**Cost**
5					
6	90 Min	1.25	=B6*C14	2.00	=D6*C14
7	75 Min	1.20	=B7*C14	1.90	=D7*C14
8	60 Min	1.15	=B8*C14	1.80	=D8*C14
9	45 Min	1.10	=B9*C14	1.70	=D9*C14
10	30 Min	1.05	=B10*C14	1.60	=D10*C14
11	15 Min	1.00	=B11*C14	1.50	=D11*C14
12					
13					
14		Base Cost:	45.00		
15					

Figure 29-2B

Figure 29-2C indicates the actual costs after entry of these formulae.

	A	B	C	D	E
1		Physical Medical Services — Out Patient			
2					
3		Level I		Level II	
4	Time	%	Cost	%	Cost
5					
6	90 Min	1.25	56.25	2.00	90.00
7	75 Min	1.20	54.00	1.90	85.50
8	60 Min	1.15	51.75	1.80	81.00
9	45 Min	1.10	49.50	1.70	76.50
10	30 Min	1.05	47.25	1.60	72.00
11	15 Min	1.00	45.00	1.50	67.50
12					
13					
14		Base Cost:	45.00		
15					

Figure 29-2C

Now, let's assume the cost of the service must be increased. The cost of staff and materials indicates that the new base rate must be $60.00. The new base cost would be entered into coordinate C14 and instantly the entire chart will change to reveal what is indicated in Figure 29–2D.

	A	B	C	D	E
1		Physical Medical Services — Out Patient			
2					
3		Level I		Level II	
4	Time	%	Cost	%	Cost
5					
6	90 Min	1.25	75.00	2.00	120.00
7	75 Min	1.20	72.00	1.90	114.00
8	60 Min	1.15	69.00	1.80	108.00
9	45 Min	1.10	66.00	1.70	102.00
10	30 Min	1.05	63.00	1.60	96.00
11	15 Min	1.00	60.00	1.50	90.00
12					
13					
14		Base Cost:	45.00		
15					

Figure 29-2D

After a period of several months, it is noted that increased costs require a change in the level of charges to the patient. This time, it is not necessary to change the base cost, but the percent of charge beyond 15 minutes at **Level I.** Figure 29-2E reveals the new charges after the percents have been changed in columns **B** and **D.** Note that it is not necessary to change any of the original formulae and that the calculations on a computer are automatic.

	A	B	C	D	E
1		Physical Medical Services — Out Patient			
2					
3		Level I		Level II	
4	Time	%	Cost	%	Cost
5					
6	90 Min	1.75	105.00	3.15	189.00
7	75 Min	1.60	96.00	2.95	177.00
8	60 Min	1.45	87.00	2.75	165.00
9	45 Min	1.30	78.00	2.55	153.00
10	30 Min	1.15	69.00	2.35	141.00
11	15 Min	1.00	60.00	2.15	129.00
12					
13					
14		Base Cost:	45.00		
15					

Figure 29-2E

29.1 Exercises

1. It is found that one of the medical supplies is sometimes distributed in different amounts to each patient, depending on the nature of his/her treatment. The most frequent pattern of distribution is packages of 10, 50, and 100. Furthermore, there is a base cost of $16.00 for the first 10, but after that the cost per unit is $1.45 for 11 through 50, and then $1.30 for units 51 through 100. Create a spreadsheet to show the formulae for automatically calculating the cost.

	A	B	C	D	E
1					
2					
3					
4					
5					
6					
7					
8					
9					
10					

Figure 29-1

2. A salary schedule is established for a classification of clinical workers based on their years of experience and their level of training. It is determined that the base salary for beginners with no experience and **Level I** training will be $15,000 for 12 months of work. The workers would receive a salary increase of 2.00% for each additional year of experience, up to five years, at which time they would not receive additional increases without further training. **Level II** workers, with more training, would receive 5.00% more than a **Level I** worker at the beginning step and a 3.00% increase each year up to the tenth year. They, too, would need additional training to move to the next level. **Level III** workers would receive 5.00% more than a **Level II** worker at the beginning step and a 5.00% increase for each additional year of experience up to the fifteenth year of service. At this point, a **Level III** worker, at the fifteenth year, would represent the top of the salary schedule. The only way that the pay at any step could be increased would be to increase the base step for the **Level I** worker. This would, therefore, affect all other salaries since they are a function of the first step for a **Level I** worker. It would also be possible to increase salaries by changing the percent differences between the various levels and years of experience. Create your table to allow automatic calculations of the entire schedule simply by changing the amount paid to a **Level I** worker with no experience.

Section 6: Summary, Review, and Study Guide

VOCABULARY

bar graph	mean	spreadsheets
broken-line graph	median	standard deviation
circle graph	mode	statistics
data	pictograph	variance
data management	range	word processing

CONCEPTS, SKILLS, AND APPLICATIONS

Objectives With Study Guide

Upon completion of Section 6 you should be able to:

• **Collect and arrange data in a table.**

A control group of 32 people had their systolic blood pressure recorded by a health care worker studying the effects of smoking on blood pressure. The following readings were obtained:

136, 165, 184, 128, 129, 189, 175, 139, 153,

150, 132, 202, 190, 147, 127, 163, 183, 167,

153, 125, 143, 189, 162, 151, 165, 188, 171,

165, 187, 133, 201, 149

Arrange the data in a table and tally.

Range	Tally	Frequency
120-129	IIII	4
130-139	IIII	4
140-149	III	3
150-159	IIII	4
160-169	IIII I	5
170-179	III	3
180-189	IIII I	6
190-199	I	1
200-209	II	2

• **Analyze data displayed in bar graphs, broken-line graphs, circle graphs and pictographs.**

Bar graphs are used to make comparisons.

The multiple-bar graph below compares the cost of a semi-private room versus a private room at St. Joseph hospital over a 15-year period.

Hospital Room Cost 1980–1995

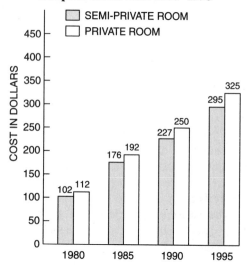

In which 5-year time period did cost increase the most?
Costs increased the most from 1980-1985.

How much more did a private room cost in 1995 than it did in 1980?
A private room cost $213 more in 1995 than in 1980.

What was the percent of increase in the cost of a semi-private room between 1985 and 1995?
The cost of a semi-private room increased 68% between 1985 and 1995.

Suppose between 1995 and the year 2000 the cost of a semi-private room increases 35% while the cost of a private room increases 40%. How much will each cost, to the nearer dollar, in the year 2000?
A semi-private room will cost $398 while a private room will cost $455.

Broken-line graphs show change over time.

Seat belts, when used properly save lives.

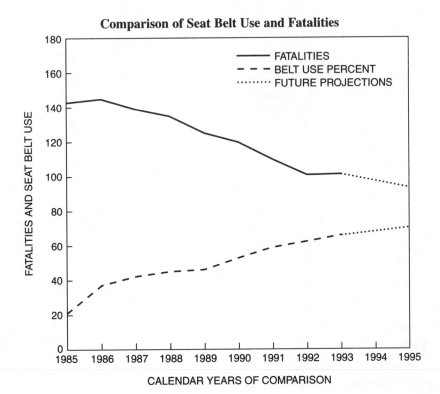

Comparison of Seat Belt Use and Fatalities

Per Ten Billion Vehicle Miles

Calendar Year	Seat Belt Use %	Fatalities Per 10 Billion Vehicle Miles
1985	21	142.7
1986	37	144.8
1987	42	139.4
1988	45	134.6
1989	46	125.2
1990	53	119.8
1991	59	110.0
1992	62	100.6
1993	66	101.1

Explain what the numbers along the left column on the preceding page represent.
They represent both percent and the number of driver fatalities per ten billion miles traveled.

Give two reasons why the use of seat belts is gradually increasing.
Car makers are installing more automotive restraint systems.
More states are enacting laws concerning seat belt usage.

According to the graph, driver fatalities decreased sharply from 1986 to 1993. Why did this happen?
Sealt belt usage rose sharply

Do you think there is likely to be a sharp decline in driver fatalities due to seat belt usage in the future?
Highly unlikely.

Circle graphs compare parts to parts or parts to the whole.

Recent spending on health care.

Distribution of Health Care Spending Per Dollar

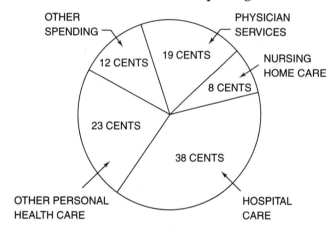

The parts of the graph total what percent?
The parts of the graph total 100%.

For which area is the least money spent?
Nursing home care spent the least money.

Which two areas added together consume 50% of each dollar?
Hospital care along with other spending consume 50% of each dollar.

In which area was approximately five times as much money spent as was spent on nursing home care?
Approximately five times as much is spent on hospital care in relation to nursing home care.

As the population ages, which area will likely increase the most?
Nursing home care will likely increase the most.

Pictographs are used to compare quantities.

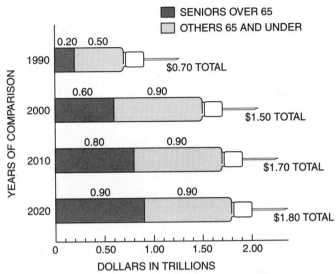

Health Care Costs
Seniors Over 65 Compared to All Others
(in Dollars Trillions)

In 1990 what percent of health care expenditures were by senior citizens over age 65?
28.57% of health care expenditures were by citizens over age 65.

In the year 2020 what percent of health care expenditures will be by people over age 65?
In the year 2020 approximately 50% of all health care dollars will be spent by people over age 65.

- **Determine the range, mean, median and mode of a set of data.**

  The ages of 40 patients treated in one day at the See Better Eye Clinic in Whereisit, IL, are listed below.

72	50	16	37	21
87	41	22	4	8
63	70	83	55	67
53	72	43	17	25
59	7	19	81	68
52	10	11	72	80
67	72	80	21	47
80	43	62	5	49

Find: the range 77
         the mean 47.3
         the median 51
         the mode 72

Which measure best represents the age of the patients? mode

Classify the ages of the patients into the following categories:

children . . . . less than 20   9
young adults 20-39   5
middle age 49-45   9
seniors 56-75   11
elderly . . . . greater than 75   6

On this particular day which category had the largest number of patients? seniors

- **Calculate the variance and standard deviation of a set of numbers.**

Using your scientific calculator, find the standard deviation and variance for the patients who were treated at the See Better Eye Clinic.

$$\sigma = 26.32 \qquad \sigma^2 = 692.87$$

Review

Major concepts and skills to be mastered for the Section 6 Test.

In 1–10, complete each sentence. Choose from the vocabulary list.

1. When placed in order the _____ is the middle most number in the set.

2. A _____ illustrates how one part is related to another part of the whole.

3. The science of analyzing and interpreting data is called _____ .

4. The three most common applications of computers are _____ , _____ _____ , and _____ .

5. The _____ is the arithmetic average of a set of numbers.

6. A _____ typically shows approximations and is used for comparisons.

7. A number that expresses the degree to which the data items in a set are dispersed is called the _____ .

8. The _____ is the data item that occurs most frequently in a set of data items.

9. A group of facts about a specific topic is called _____ .

10. When arranged in rank order the _____ is the difference between the largest and smallest data items.

Fifty adults were studied concerning the effect of diet on cholesterol. After six months, their cholesterol levels were recorded.

183	241	240	225	171
227	220	229	260	250
286	268	295	208	175
205	243	168	236	211
247	275	273	296	279
235	170	215	217	253
258	197	254	253	307
307	290	303	283	236
166	238	190	311	327
270	281	316	178	218

Cholesterol Level	Tally	Frequency (# of patients)
165-179	ⅢⅢ I	6
180-194		
195-209		
210-		
-239		
240-254		
-269		
270-		
-299		
300-314		
-329		

1. Using the data above, complete the table to the right.

2. How many patients had readings above 239?

3. If a reading of below 210 is considered acceptable, what percent of the group had levels that were not acceptable?

4. Which category had the largest number of patients?

5. How many patients had a cholesterol level between 225 and 299?

In 1993 about 3400 accidental deaths were recorded nationwide for children under age five (5). The six leading causes are summarized in the following bar graph.

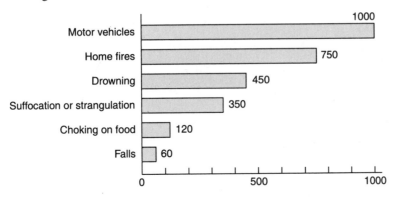

6. What percent of the 3400 accidental deaths were caused by drowning?

7. Pre-school age children die disproportionately to their numbers in house fires. Give two reasons why this is true.

8. How many more deaths were due to home fires than falls?

9. What could be done to lower the number of deaths of children under five (5) in motor vehicle accidents?

The following broken-line graph shows the percentage of adults who said they were regular smokers from 1976 to 1993.

Percentage of Adults Who Said They Were Regular Smokers

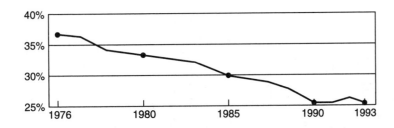

10. In 1985 the adult population in the U.S. was 238,736,000. Approximately how many of those people smoked regularly?

11. Between 1980 and 1990 what percent of adults who smoked regularly quit smoking?

12. Suppose the adult population in 1980 was 226,546,000 and in 1990 it was 248,709,000. Based upon your answer to question 11, approximately how many adults quit smoking in that 10-year time period?

13. What do you believe has contributed most to the decline of adult smokers who said they were regular smokers?

During a one-month period, a walk-in clinic treats a total of 1,262 patients. The illnesses and conditions are grouped into categories and are placed on a circle graph.

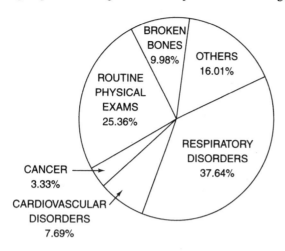

14. How many people had routine physical exams?

15. Find the number of people treated for respiratory disorders or cancer.

16. How many people are treated for broken bones or cardiovascular disorders?

17. Find the number of people treated for respiratory disorders or cardiovascular disorders or other conditions.

The following graph shows the life expectancy at birth for all races in the U.S. for selected years 1950, 1960, 1970, 1980 and 1990.

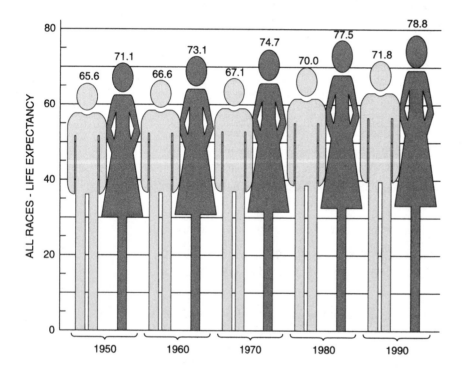

18. Was the life expectancy of women less than that of men for any year shown?

19. In what year did women's life expectancy first go over 70 years?

20. In what year do men's life expectancy equal or exceed 70 years?

21. In the year 2000 would you expect to find more men or more women over age 75?

22. Give at least two reasons why life expectancy, in general, is increasing.

Researchers undertook a project in a small community to determine the height of individuals born from January 1, 1970 through December 31, 1970. The measurements are:

Males						Females				
6'2"	5'10"	5'6"	6'0"	5'6"		5'1"	5'9"	4'10"	5'4"	5'3"
6'1"	6'3"	6'0"	5'11"	5'4"		5'4"	5'7"	5'0"	5'7"	5'0"
5'10"	6'3"	6'1"	5'10"	5'9"		6'0"	5'4"	5'4"	5'5"	5'6"
5'4"	6'7"	5'1"	5'11"	6'0"		5'9"	6'0"	5'7"	5'4"	5'2"
5'9"	5'3"	5'8"	5'5"	5'11"		5'2"	5'4"	5'5"	4'11"	5'9"
5'11"	5'0"	5'9"	5'9"	5'5"		5'11"	4'11"	5'6"	5'2"	4'11"
6'0"	5'1"	6'0"	6'1"	6'2"		5'10"	5'3"	5'11"	5'0"	5'3"
6'4"	5'10"	5'11"	6'4"	5'8"		4'11"	5'7"	5'6"	4'10"	5'10"
5'11"	5'11"	5'10"	5'11"	6'5"		5'3"	5'5"	5'5"	5'3"	4'11"
5'6"	6'2"	6'0"	5'8"	5'10"		5'6"	5'4"	5'2"	5'4"	5'3"

Determine each measure for males and females. (Express the measurements in inches for easier calculation.)

		Males		Females
23.	Range	a. ___?___	b. ___?___
24.	Mean	a. ___?___	b. ___?___
25.	Median	a. ___?___	b. ___?___
26.	Mode	a. ___?___	b. ___?___

27. Calculate the standard deviation for each set of data to the nearer hundredth. (Express the measurements in inches for easier calculation.)

28. Describe the comparison between the two sets of data.

SECTION 6 TEST

Listed below is the recent average daily charge and length of stay for seven hospitals in a central Illinois tri-county region for two selected illnesses.

Hospital	Average Daily Charge for Heart Attack Patients Over 65	Average Stay (Days)	Average Daily Charge for Psychoses Patients Under 65	Average Stay (Days)
A	$1394	7.9	$815	15.2
B	$ 975	8.1	$652	10.6
D	$1182	5.3	$753	11.5
E	$1160	7.6	$659	8.3
F	$1187	9.2	$717	12.7
G	$1376	7.3	$772	13.8

Plot on the same graph:

1. The average daily charge for heart attack patients over 65.
2. The average daily charge for psychoses patients under 65.

Calculate each of the following:

	Range	Mode	Median	Mean	Standard Deviation
Average daily charge for heart attack patients over 65	3. _____	4. _____	5. _____	6. _____	7. _____
Average stay of heart attack patients over 65	8. _____	9. _____	10. _____	11. _____	12. _____
Average daily charge for patients under 65 treated for psychoses	13. _____	14. _____	15. _____	16. _____	17. _____
Average stay for patients under 65 treated for psychoses	18. _____	19. _____	20. _____	21. _____	22. _____

23. What conclusions can be drawn from this information?

In a large metropolitan area 5,000 households were surveyed to identify those households that had inadequate health insurance coverage. The results follow:

Household Classification	# of Households Surveyed
Employed full time	2,400
Unemployed	300
Going to school	450
Retired	400
Homemakers	800
Employed part time	650

24. Calculate the percent surveyed for each classification.

25. Construct a circle graph that represents this data.

Appendix
Section I

DENOMINATE NUMBERS

Denominate numbers are numbers that include units of measurement. The units of measurement are arranged from the largest units at the left to the smallest unit at the right.

For example: 6 yd. 2 ft. 4 in.

All basic operations of arithmetic can be performed on denominate numbers.

I. EQUIVALENT MEASURES

Measurements that are equal can be expressed in different terms. For example, 12 in. = 1 ft. If these equivalents are divided, the answer is 1.

$$\frac{1 \text{ ft.}}{12 \text{ in.}} = 1 \qquad \frac{12 \text{ in.}}{1 \text{ ft.}} = 1.$$

To express one measurement as another equal measurement, multiply by the equivalent in the form of 1.

To express 6 inches in equivalent foot measurement, multiply 6 inches by 1 in the form of $\frac{1 \text{ ft.}}{12 \text{ in.}}$. In the numerator and denominator, divide by a common factor.

$$6 \text{ in.} = \frac{\overset{1}{\cancel{6 \text{ in.}}}}{1} \times \frac{1 \text{ ft.}}{\underset{1}{\cancel{12 \text{ in.}}}} = \frac{1}{2} \text{ ft. or 0.5 ft.}$$

To express 4 feet in equivalent inch measurement, multiply 4 feet by 1 in the form of $\frac{12 \text{ in.}}{1 \text{ ft.}}$.

$$4 \text{ ft.} = \overset{4}{\cancel{4 \text{ ft.}}} \times \frac{12 \text{ in.}}{\underset{1}{\cancel{1 \text{ ft.}}}} = \frac{48 \text{ in.}}{1} = 48 \text{ in.}$$

Per means division, as with a fraction bar. For example, 50 miles per hour can be written $\frac{50 \text{ miles}}{1 \text{ hour}}$.

II. BASIC OPERATIONS
A. ADDITION

> **SAMPLE: 2 yd. 1 ft. 5 in. + 1 ft. 8 in. + 5 yd. 2 ft.**

1. Write the denominate numbers in a column with like units in the same column.

2. Add the denominate numbers in each column.

3. Express the answer using the largest possible units.

$$
\begin{array}{llll}
 & 2 \text{ yd.} & 1 \text{ ft.} & 5 \text{ in.} \\
 & & 1 \text{ ft.} & 8 \text{ in.} \\
+ & 5 \text{ yd.} & 2 \text{ ft.} & \\
\hline
 & 7 \text{ yd.} & 4 \text{ ft.} & 13 \text{ in.}
\end{array}
$$

$$
\begin{array}{rcll}
7 \text{ yd.} & = & 7 \text{ yd.} & \\
4 \text{ ft.} & = & 1 \text{ yd.} & 1 \text{ ft.} \\
13 \text{ in.} & = + & & 1 \text{ ft.} \quad 1 \text{ in.} \\
\hline
7 \text{ yd.} \quad 4 \text{ ft.} \quad 13 \text{ in.} & = & 8 \text{ yd.} & 2 \text{ ft.} \quad 1 \text{ in.}
\end{array}
$$

B. SUBTRACTION

SAMPLE: 4 yd. 3 ft. 5 in. – 2 yd. 1 ft. 7 in.

1. Write the denominate numbers in columns with like units in the same column.

$$\begin{array}{rrr} 4 \text{ yd.} & 3 \text{ ft.} & 5 \text{ in.} \\ -\ 2 \text{ yd.} & 1 \text{ ft.} & 7 \text{ in.} \\ \hline \end{array}$$

2. Starting at the right, examine each column to compare the numbers. If the bottom number is larger, exchange one unit from the column at the left for its equivalent. Combine like units.

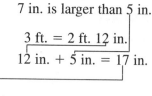

7 in. is larger than 5 in.

3 ft. = 2 ft. 12 in.

12 in. + 5 in. = 17 in.

3. Subtract the denominate numbers.

$$\begin{array}{rrr} 4 \text{ yd.} & 2 \text{ ft.} & 17 \text{ in.} \\ -\ 2 \text{ yd.} & 1 \text{ ft.} & 7 \text{ in.} \\ \hline 2 \text{ yd.} & 1 \text{ ft.} & 10 \text{ in.} \end{array}$$

4. Express the answer using the largest possible units.

2 yd. 1 ft. 10 in.

C. MULTIPLICATION

— *By a constant*

SAMPLE: 1 hr. 24 min. X 3

1. Multiply the denominate number by the constant.

2. Express the answer using the largest possible units.

$$\begin{array}{rr} 1 \text{ hr.} & 24 \text{ min.} \\ \times & 3 \\ \hline 3 \text{ hr.} & 72 \text{ min.} \end{array}$$

$$\begin{array}{rrclrr} 3 \text{ hr.} & & = & 3 \text{ hr.} & \\ & 72 \text{ min.} & = & 1 \text{ hr.} & 12 \text{ min.} \\ \hline 3 \text{ hr.} & 72 \text{ min.} & = & 4 \text{ hr.} & 12 \text{ min.} \end{array}$$

— *By a denominate number expressing linear measurement*

SAMPLE: 9 ft. 6 in. X 10 ft.

1. Express all denominate numbers in the same unit.

$$9 \text{ ft. } 6 \text{ in.} = 9\frac{1}{2} \text{ ft.}$$

2. Multiply the denominate numbers. (This includes the units of measure, such as ft. × ft. = sq. ft.)

$$9\frac{1}{2} \text{ ft.} \times 10 \text{ ft.} =$$

$$\frac{19}{2} \text{ ft.} \times 10 \text{ ft.} =$$

95 sq. ft.

— *By a denominate number expressing square measurement*

SAMPLE: 3 ft. X 6 sq. ft.

1. Multiply the denominate numbers. (This includes the units of measure, such as ft. × ft. = sq. ft. and sq. ft. × ft. = cu. ft.)

3 ft. × 6 sq. ft. = 18 cu. ft.

— By a denominate number expressing rate

SAMPLE: 50 miles per hour X 3 hours

1. Express the rate as a fraction using the fraction bar for *per.*

$$\frac{50 \text{ miles}}{1 \text{ hour}} \times \frac{3 \text{ hours}}{1} =$$

2. Divide the numerator and denominator by any common factors, including units of measure.

$$\frac{50 \text{ miles}}{\cancel{1 \text{ hour}}_{1}} \times \frac{\cancel{3 \text{ hours}}^{3}}{1} =$$

3. Multiply numerators. Multiply denominators.

$$\frac{150 \text{ miles}}{1} =$$

150 miles.

4. Express the answer in the remaining unit.

D. DIVISION

— By a constant

SAMPLE: 8 gal. 3 qt. ÷ 5

1. Express all denominate numbers in the same unit.

8 gal. 3 qt. = 35 qt.

2. Divide the denominate number by the constant.

35 qt. ÷ 5 = 7 qt.

3. Express the answer using the largest possible units.

7 qt. = 1 gal. 3 qt.

— By a denominate number expressing linear measurement

SAMPLE: 11 ft. 4 in. ÷ 8 in.

1. Express all denominate numbers in the same unit.

11 ft. 4 in. = 136 in.

136 in. ÷ 8 in. =

2. Divide the denominate numbers by a common factor. (This includes the units of measure, such as inches ÷ inches = 1.)

$$\frac{\cancel{136 \text{ in.}}^{17}}{\cancel{8 \text{ in.}}_{1}} = \frac{17}{1} = 17.$$

— By a linear measure with a square measurement as the dividend

SAMPLE: 20 sq. ft. ÷ 4 ft.

1. Divide the denominate numbers. (This includes the units of measure, such as sq. ft. ÷ ft. = ft.)

20 sq. ft. ÷ 4 ft.

$$\frac{\cancel{20 \text{ sq. ft.}}^{5 \text{ ft.}}}{\cancel{4 \text{ ft.}}_{1}} = \frac{5 \text{ ft.}}{1}$$

2. Express the answer in the remaining unit.

5 ft.

— By denominate numbers used to find rate

SAMPLE: 200 mi. ÷ 10 gal.

1. Divide the denominate numbers.

$$\frac{\cancel{200 \text{ mi.}}^{\;20 \text{ mi.}}}{\cancel{10 \text{ gal.}}_{\;1 \text{ gal.}}} = \frac{20 \text{ mi.}}{1 \text{ gal.}}$$

2. Express the units with the fraction bar meaning *per.*

$$\frac{20 \text{ mi.}}{1 \text{ gal.}} = 20 \text{ miles per gallon.}$$

Note: Alternate methods of performing the basic operations will produce the same result. The choice of method is determined by the individual.

Section II

TABLE I — METRIC RELATIONSHIPS

The base units in SI metrics include the meter and the gram. Other units of measure are related to these units. The relationship between the units is based on powers of ten and uses these prefixes:

kilo (1 000) deci (0.1) hecto (100) centi (0.01) deka (10) milli (0.001)

These tables show the most frequently used units with an asterisk (*).

METRIC LENGTH MEASURE

10 millimeters	=	1 centimeter (cm)*
10 centimeters (cm)	=	1 decimeter (dm)
10 decimeters (dm)	=	1 meter (m)*
10 meters (m)	=	1 dekameter (dam)
10 dekameters (dam)	=	1 hectometer (hm)
10 hectometers (hm)	=	1 kilometer (km)*

METRIC AREA MEASURE

100 square millimeters (mm^2)	=	1 square centimeter (cm^2)*
100 square centimeters (cm^2)	=	1 square decimeter (dm^2)
100 square decimeters (dm^2)	=	1 square meter (m^2)*
100 square meters (m^2)	=	1 square dekameter (dam^2)
100 square dekameters (dam^2)	=	1 square hectometer (hm^2)*
100 square hectometers (hm^2)	=	1 square kilometer (km^2)

METRIC VOLUME MEASURE FOR SOLIDS

100 cubic millimeters (mm^3)	=	1 cubic centimeter (cm^3)*
100 cubic centimeters (cm^3)	=	1 cubic decimeter (dm^3)*
100 cubic decimeters (dm^3)	=	1 cubic meter (m^3)*
100 cubic meters (m^3)	=	1 cubic dekameter (dam^3)
100 cubic dekameters (dam^3)	=	1 cubic hectometer (hm^3)
100 cubic hectometers (hm^3)	=	1 cubic kilometer (km^3)

METRIC VOLUME MEASURE FOR FLUIDS

10 milliliters (mL)*	=	1 centiliter (cL)
10 centiliters (cL)	=	1 deciliter (dL)
10 deciliters (dL)	=	1 liter (L)*
10 liters (L)	=	1 dekaliter (daL)
10 dekaliters (daL)	=	1 hectoliter (hL)
10 hectoliters (hL)	=	1 kiloliter (kL)

METRIC VOLUME MEASURE EQUIVALENTS

1 cubic decimeter (dm^3)	=	1 liter (L)
1 000 cubic centimeters (cm^3)	=	1 liter (L)
1 cubic centimeter (cm^3)	=	1 milliliter (mL)

METRIC MASS MEASURE

10 milligrams (mg)*	=	1 centigram (cg)
10 centigrams (cg)	=	1 decigram (dg)
10 decigrams (dg)	=	1 gram (g)*
10 grams (g)	=	1 dekagram (dag)
10 dekagrams (dag)	=	1 hectogram (hg)
10 hectograms (hg)	=	1 kilogram (kg)*
1 000 kilograms (kg)	=	1 megagram (Mg)*

- To express a metric length unit as a smaller metric length unit, multiply by a positive power of ten such as 10, 100, 1 000, 10 000, etc.

- To express a metric length unit as a larger metric length unit, multiply by a negative power of ten such as 0.1, 0.01, 0.001, 0.000 1, etc.

- To express a metric area unit as a smaller metric area unit, multiply by 100, 10 000, 1 000 000, etc.

- To express a metric area unit as a larger metric area unit, multiply by 0.01, 0.000 1, 0.000 001, etc.

- To express a metric volume unit for solids as a smaller metric volume unit for solids, multiply by 1 000, 1 000 000, 1 000 000 000, etc.

- To express a metric volume unit for olids as a larger metric volume unit for solids, multiply by 0.001, 0.000 001, 0.000 000 001, etc.

- To express a metric volume unit for fluids as a smaller metric volume unit for fluids, multiply by 10, 100, 1 000, 10 000, etc.

- To express a metric volume unit for fluids as a larger metric volume unit for fluids, multiply by 0.1, 0.01, 0.001, 0.000 1, etc.

- To express a metric mass unit as a smaller metric mass unit, multiply by 10, 100, 1 000, 10 000, etc.

- To express a metric mass unit as a larger metric mass unit, multiply by 0.1, 0.0l, 0.001, 0.000 1, etc.

Metric measurements are expressed in decimal parts of a whole number. For example, one half millimeter is written as 0.5 mm.

In calculating with the metric system, all measurements are expressed using the same prefixes. If answers are needed in millimeters, all parts of the problem should be expressed in millimeters before the final solution is attempted. Diagrams that give dimensions in different prefixes must first be expressed using the same unit.

TABLE II
HOUSEHOLD EQUIVALENTS

Approximate Liquid Measure Equivalents		
60 drops	=	1 teaspoonful (t)
4 teaspoonfuls*	=	1 tablespoonful (T)
2 tablespoonfuls	=	1 fluidounce
6 fluidounces	=	1 teacupful
8 fluidounces	=	1 glassful

TABLE III
APOTHECARIES EQUIVALENTS

Equivalent Measurements of Volume		
60 minims (m)	=	1 fluidram (f ʒ)
8 fluidrams (f ʒ)	=	1 fluidounce (f ʒ)
16 fluidounces (f ʒ)	=	1 pint (pt.)
2 pints (pt.)	=	1 quart (qt.)
4 quarts (qt.)	=	1 gallon (gal.)

Equivalent Measurements of Weight		
60 grains (gr.)	=	1 dram (ʒ)
8 drams (ʒ)	=	1 ounce (ʒ)
12 ounces (ʒ)	=	1 pound (lb.)

TABLE IV

Metric System			Apothecaries' System				Household System
Weight (mass)	Volume		Weight (mass)	Symbol	Volume (liquid)	Symbol	
	Solid	Fluid					
0.06 g or 60 mg	60 mm³	0.06 mL	1 grain (gr.)	gr. i	1 minim	m i	1 drop
1 g	1 cm³	1 mL	15 grains	gr. xv	15 minims	m xv	
4 g	4 cm³	4 mL	1 dram	ʒ i	1 fluidram	f ʒ i	1 scant teaspoonful
5 g	5 cm³	5 mL					1 teaspoonful (t)
15 g	15 cm³	15 mL	4 drams	ʒ iv	4 fluidrams	f ʒ iv	1 tablespoonful (T)
30 g	30 cm³	30 mL	1 ounce	ʒ i	1 fluidounce	f ʒ i	2 tablespoonfuls
180 g	180 cm³	180 mL	6 ounces	ʒ vi	6 fluidounces	f ʒ vi	1 teacupful
240 g	240 cm³	240 mL	8 ounces	ʒ viii	8 fluidounces	f ʒ viii	1 glassful
360 g	360 cm³		12 ounces	ʒ xii			2 teacupfuls
			1 pound	lb. i			
500 g	500 cm³	500 mL			1 pint	pt. i	2 glassfuls
1 000 g or 1 kg	1 dm³	1 L	32 ounces	ʒ xxxii	1 quart	qt. i	4 glassfuls
4 kg	4 dm³	4 L			1 gallon	gal. i	16 glassfuls

Section III

MATHEMATICAL AND HEALTH RELATED FORMULAS

Areas

square: $A = s^2$

rectangle: $A = lw$

triangle: $A = \dfrac{1}{2}bh$

circle: $A = \pi r^2$

Circumference

$r = \dfrac{1}{2}d$

$C = \pi d$

$C = \pi r$

Roman Numerals

I 1	VI 6	XI 11	L 50	CD 400	$\overline{X}$ 10,000
II 2	VII 7	XIX 19	LX 60	D 500	$\overline{L}$ 50,000
III 3	VIII 8	XX 20	XC 90	CM 900	$\overline{C}$ 100,000
IV 4	IX 9	XXX 30	C 100	M 1,000	$\overline{D}$ 500,000
V 5	X 10	XL 40	CC 200	$\overline{V}$ 5,000	$\overline{M}$ 1,000,000

Doses and Dosages

Young's rule: Child's Dose $= \dfrac{\text{Child's Age (in years)}}{\text{Child's Age (in years)} + 12} \times \text{Adult Dose}$

Fried's rule: Infant's Dose $= \dfrac{\text{Age (in months)}}{150 \text{ pounds}} \times \text{Adult Dose}$

Clark's rule: Child's Dose $= \dfrac{\text{Weight of Child (in pounds)}}{150 \text{ pounds}} \times \text{Adult Dose}$

or

Child's Dose $= \dfrac{\text{Weight of Child (in kilograms)}}{68 \text{ kilograms}} \times \text{Adult Dose}$

Solutions

ratio strength of solutions $= \dfrac{\text{amount of drug}}{\text{amount of solution}}$

percent of strength by volume $= \dfrac{\text{volume of solute}}{\text{volume of solution}} \times 100$

percent strength by weight (mass) $= \dfrac{\text{mass of solute}}{\text{volume of solution}} \times 100$

$\dfrac{\text{amount of solute}}{\text{amount of first solution}} = \dfrac{\text{amount of solute}}{\text{amount of second solution}}$

smaller % strength:larger % strength = smaller volume:larger volume

or

weaker:stronger — solute:solvent

desired solution:solution on hand = amount of solute:amount of solution

$$D \; : \; H \; = \; q \; : \; Q$$

or

$$\frac{D}{H} = \frac{q}{Q}$$

Volumes

rectangular solid: $V = lwh$

cylindrical solid: $V = \pi r^2 h$

equivalences: $1 \text{ cm}^3 = 0.002 \text{ L} \textit{ or } 1 \text{ cm}^3 = 1 \text{ mL}$

$1\,000 \text{ cm}^3 = 1 \text{ L}$

At 4 °C and standard pressure (760 millimeters), the volume and mass of water are equivalent relationships.

$1\,000 \text{ cm}^3 = 1\,000 \text{ mL} = 1\,000 \text{ g}$

or

$1 \text{ dm}^3 = 1 \text{ L} = 1 \text{ kg}$

Glossary

Agar — A dried gelatine-like product obtained from certain species of algae, especially Gelideum. Since it is unaffected by bacterial enzymes, it is widely used as a solidifying agent for bacterial culture media.

Agar-agar — A bacterial culture medium made from certain species of seaweed, not specific in nature.

Autotrophic — Self-nourishing or capable of growing in the absence of organic compounds.

Bacteria — Any of the class Schizomycetes of microscopic plants having round, rod-like, spiral, or filamentous single-celled or noncellular bodies. Bacteria often aggregate into colonies or move by means of flagella; live in soil, water, organic matter, or in the bodies of plants and animals; are autotrophic, saprophytic, or parasitic in nutrition; are important to humans because of the chemical effects. Pathogenic bacteria are commonly called germs.

Bacteria culture — A cultivation of bacteria in a prepared nutrient medium.

Bacteriology — The science that deals with bacteria and the relationships to medicine, industry, and agriculture.

Cell — The basic unit in the human body. It consists of a nucleus, a cytoplasm, and a membrane.

Chromatography — The separation of various organic chemical compounds, such as carotene and chlorophyll, by differential or selective absorption.

Compound — A substance that consists of two or more elements chemically united in a definite proportion so that the elements lose individual characteristics.

Cresols — A yellowish-brown liquid obtained from coal tar and not containing more than 5% phenol. It is used as a disinfectant for inanimate articles and areas which do not come into contact with food.

Czapek solution agar — A culture medium for various fungi consisting essentially of a balanced and buffered mixture of inorganic salts, a sugar, and water being used with added agar as a solid.

Dosage — The total quantity of a drug that is to be administered during a given period of time.

Dose — The portion of the drug that is to be administered at one time.

Drug — A substance or a mixture of substances that have been found to have a definite value in the detection, prevention, or treatment of disease.

Enzyme — A substance that initiates and accelerates a chemical reaction.

Erlenmeyer flask — A flask with a conical body, broad base, and narrow neck.

Fehling's solution — A solution used for detecting the presence of sugar in urine.

Flask — A laboratory vessel usually made of glass and having a constricted neck.

Gram's iodine solution — A solution containing iodine, potassium iodine, and water which is used in the straining of bacteria.

Hayem's solution — A solution used to dilute the blood prior to counting the blood cells.

Hemacytometer — A piece of apparatus used in counting blood cells.

Hematocrit — The volume of red blood cells (erythrocytes) packed by centrifugation in a given volume of cells.

Intracellular water — The body water that is within the cells. It comprises about $\frac{2}{3}$ of the total body water.

Lugol's solution — Strong iodine solution used in iodine therapy.

Lysol solution — A proprietory preparation of a mixture of cresols used for disinfection of inanimate objects.

Manometer — A device used to determine the rate at which oxygen is used by a bacteria culture.

Medicine glass — A graduated container. It may be graduated in milliliters, ounces, or drams.

Medium — A substance used for the cultivation of microorganisms or cellular tissue.

Microbiology — The study of simple forms of living matter which cannot be seen by the naked eye.

Microorganism — Minute living body not visible to the naked eye, especially a bacterium or a protozoon.

Minim glass — A graduated container used to measure small portions. It is graduated in minims and drams.

Milliequivalences per liter (mEq/L) — A measure of electrolyte concentration which refers to the amount of electrolytes in each liter of body fluid.

Mycology — The science of fungi.

Nutrient — Food that supplies the body with its necessary elements. Carbohydrates, fats, proteins, and alcohol provide energy; water electrolytes, minerals, and vitamins are essential to the metabolic processes.

Pathogenic — Disease producing.

Petri dish — Shallow dish with cover that is used to hold solid media for culturing bacteria.

Physiological saline solution — An isotonic sterile solution containing sodium chloride in distilled water; 0.85% salt solution. This solution may be used for dehydration, shock or hemorrhaging.

Physiological solution — A solution which matches a person's body chemistry.

Pipette — A narrow glass tube with both ends open. It is used for transferring and measuring liquids.

Protozoa — A phylum of the animal kingdom which includes all of the unicellular forms. Most consist of a single cell or of an aggregation of nondifferentiated cells but not forming a tissue.

Pure drug — A medical substance which is not mixed, combined, or diluted; having a strength of 100%.

Ratio strength of a solution — The ratio of the amount of solute (drug) to the amount of solution. This ratio may also be expressed as a percent or a fraction.

Ringer's solution — A solution resembling the blood serum in its salt constituents. For topical use on burns and wounds.

Saline — Containing or pertaining to salt.

Solute — The substance that is dissolved in the liquid to form a solution.

Solution — A liquid containing one or more dissolved substances.

Solvent — The liquid in which the substance or substances are dissolved to form a solution.

Stock solution — A solution which is kept on hand. It has a strength of less than 100% and is usually diluted further.

Substrate — The substance that an enzyme acts upon.

Viable — Capable of living.

Vial — A small bottle.

Virus — Minute organisms that are not visible with ordinary light microscopy. Parasitic in nature and dependent upon the nutrients inside the cells for metabolic and reproductive needs.

Yeast — Any of the spherical-shaped or oval-shaped fungi which reproduce by budding. Plant cells which are capable of fermenting carbohydrates.

Descriptive Glosssary

Area — Area is the number of square units needed to cover a surface.

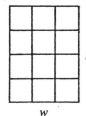

$l = 4$ in

$w = 3$ in

area $= 3 \times 4 = 12$ in^2

Each square equals 1 in^2

Bar graph — A bar graph is used to make comparisons.

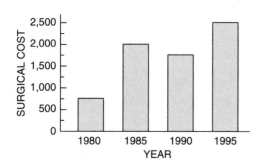

Base — The base is the number that is used as a factor when a number is written in exponential form.

$3^4 = 3 \cdot 3 \cdot 3 \cdot 3$

Base is 3

Base — The base of a triangle is the length of the side on which the triangle sits.

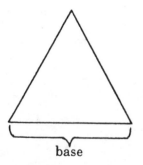

base

342

Broken-line graph — Broken-line graphs show change over time.

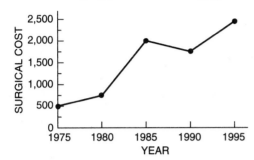

Circle graph — A circle graph illustrates how one part is related to another part of the whole.

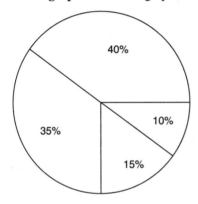

Circumference — The distance around a circle is the circumference.

The circumference is found by multiplying the diameter by π.

$\pi \approx 3.14$

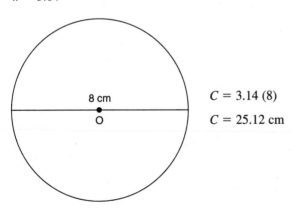

$C = 3.14\,(8)$

$C = 25.12$ cm

Common denominator — A common denominator is a common multiple of the denominators of two or more fractions.

The common denominator (least) of $\frac{2}{3}$ and $\frac{4}{5}$ is 15, since 15 is the least common multiple (LCM) of 3 and 5.

Common factor — A common factor of two or more numbers is a factor that is common to both numbers.

   3 is a common factor of 6 and 9.

Common multiple — A common multiple is a number which is a multiple of each of two or more numbers.

   Common multiples of 2 and 4 are 4, 8, 12, 16 and so on.

Compatible numbers — Compatible numbers are two numbers that are easy to divide mentally.

   10 and 5, as well as 7 and 49, are compatible numbers.

Composite number — A composite number is any number with more than two factors.

   20 is a composite number that has 1, 2, 4, 5, 10 and 20 as factors.

Denominator — In a fraction the number below the fraction bar represents the number of equal parts to make a whole.

   In $\frac{3}{4}$, 4 is the denominator.

Diameter — The diameter is the distance across a circle through its center.

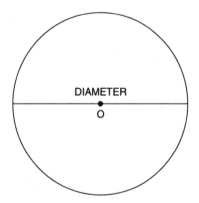

Difference — The result or answer to a subtraction problem is called the difference.

   In 52 − 37 = 15,

   15 is the difference.

Discount — A discount is the dollar amount by which the regular price is reduced.

   Regular price: $28

   Discount: 25%

   Sale price: $21

Divisible — A number is divisible by another number if the quotient is a whole number and the remainder is zero.

20 is divisible by 5

Equivalent fractions — Equivalent fractions are two or more fractions which represent the same part of the whole or the same part of a group.

$\frac{3}{5}$ and $\frac{6}{10}$ are equivalent fractions.

Estimation — Estimation is the process of finding a number close to the exact amount.

A liter is approximately equal to one part.

Exponent — An exponent indicates the number of times the base is used as a factor.

$5^3 = 5 \cdot 5 \cdot 5$

Exponent is 3.

Expression — An expression is a symbol or combination of symbols representing a mathematical relationship.

The formula for the area of a rectangle is length times width.

In symbols $A = lw$

Extremes — In a proportion the outside terms are called the extremes.

In $\frac{2}{3} = \frac{6}{9}$, 2 and 9 are the extremes.

Factor tree — A factor tree is an organizational method used to find a number's prime factors.

Factors — Numbers being multiplied to find a product are called factors.

In 4 times 3 equal 12, 4 and 3 are factors.

Fraction — A fraction is a comparison between a part of a whole and the whole.

$\frac{1}{12}$ of the nurses on 6 West at Deland Hospital staff are male.

Greatest common factor (GCF) — The greatest common factor (GCF) of two or more numbers is the largest factor common to all numbers.

10 and 15 have a greatest common factor (GCF) of 5.

Height —The height of a triangle is the line segment from the opposite vertex intersecting the base at a right angle.

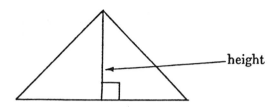

Improper fraction — An improper fraction contains a numerator that is greater than or equal to its denominator.

$\frac{7}{2}$ is an improper fraction.

Least common denominator (LCD) — The least common denominator is the least common multiple of the denominator of two or more fractions.

For $\frac{2}{5}$ and $\frac{3}{4}$ the LCD is 5 • 2 • 2, or 20.

Least common multiple (LCM) — The least common multiple is the least of the common multiples of two or more numbers.

The least common multiple (LCM) of 12 and 18 is 36.

Lowest terms — A fraction is in lowest terms if the greatest common factor (GCF) of both the numerator and denominator is one (1).

$\frac{2}{5}$ is in lowest terms. One (1) is the GCF common to 2 and 5.

Markup — A markup is the dollar amount by which the original cost is increased to cover expenses and profit in arriving at the charged price.

Cost:　　　　$40

Markup:　　　50%

Charged price: $60

Mean — The mean is the arithmetic average that is found by summing the data items divided by the number of data items.

The mean of 2, 3, 4, 5 is 4 because $\dfrac{2 + 3 + 5 + 6}{4 \text{ data items}} = \dfrac{16}{4} = 4$

Means — In a proportion the middle terms are called the means.

In $\dfrac{3}{4} = \dfrac{6}{8}$, 4 and 6 are the means.

Median — The median is the middle most data item in a set of data items arranged in rank order.

In 4, 7, 9, 15 and 18, the median is 9.

Mode — The data item that occurs most often in a set of data items.

In 2, 5, 7, 7, 7, 9 and 11, the mode is 7.

Multiple — A multiple is the product of a given number and another factor.

The multiples of 9 are 9, 18, 27, 36, 45, and so on.

Multiplicative inverse — A number times its multiplicative inverse is equal to one.

$\left(\dfrac{4}{5}\right)\left(\dfrac{5}{4}\right) = 1.$ $\dfrac{5}{4}$ is the multiplicative inverse of $\dfrac{4}{5}$.

Numerator — In a fraction the number above the fraction bar that represents the number of parts being used is the numerator.

In $\dfrac{5}{6}$, 5 is the numerator.

Percent — A percent is a ratio that compares a number to one hundred. The symbol for percent is %.

$\dfrac{25}{100} = 0.25 = 25\%$

Percent of decrease — $\dfrac{\text{amount of decrease}}{\text{original quantity}} \times 100$

Original: 60

Ending: 20

Percent of decrease: $\dfrac{40}{60} \times 100 = 66.\overline{6}\%$

Percent of increase — $\dfrac{\text{amount of increase}}{\text{original quantity}} \times 100$

Original: 20

Ending: 60

Percent of decrease: $\dfrac{40}{20} \times 100 = 200\%$

Period — The period of a repeating decimal is the number of digits in the repeating block.

In $\dfrac{3}{11} = 0.272727\ldots$, the period is 2.

Pi (π) — Pi is the ratio of the circumference of a circle to its diameter.

$$\pi = \frac{\text{circumference}}{\text{diameter}} \approx 3.1415926\ldots$$

Pictograph — A pictograph is used to compare quantities.

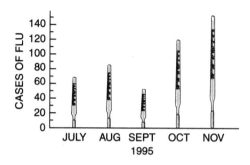

Power — A power is a number expressed using an exponent.

The power 3^5 is read "three to the fifth power" meaning $3 \cdot 3 \cdot 3 \cdot 3 \cdot 3 = 243$.

Prime number — A prime number is a whole number, other than 1, having only 1 and itself as factors.

11 is a prime number because it has exactly two factors, 1 and 11.

Product — The result or answer to a multiplication problem is called the product.

In $14 \times 3 = 52$, 52 is the product.

Proportion — A proportion is an equation which states that two ratios are equal.

$\frac{2}{4} = \frac{8}{16}$ is a proportion.

Quotient — The result or answer to a division problem is called the quotient.

In $6\overline{)30}^{5}$, 5 is the quotient.

Radius — A radius is the distance from the center of a circle to any point on the circle.

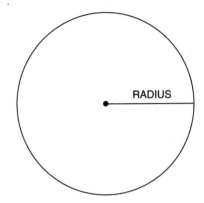

Range — The range is the difference between the largest and smallest numbers in a set of data.

In 2, 5, 6, 9, 13 and 16, the range is 16−2 or 14.

Rate — A rate is the ratio of two measurements with unlike measurements.

10 adhesive surgical dressings for $17.00.

Rate pairs — Rate pairs are the proportional relationship between ratios of unlike measurements.

$$\frac{40 \text{ calories}}{1 \text{ slice of bread}} = \frac{X \text{ calories}}{4 \text{ slices of bread}}$$

Ratio — A ratio is a comparison of two numbers using division.

Three different ways are used to express a ratio:

$\frac{25}{50}$, 25 to 50 and 25:50

Reciprocal — A number times its reciprocal is equal to one.

$\left(\frac{3}{4}\right)\left(\frac{4}{3}\right) = 1.$ $\frac{4}{3}$ is the reciprocal of $\frac{3}{4}$.

Repeating decimal — In a repeating decimal, a digit or a sequence of digits repeats again and again.

$\frac{1}{3} = 0.3333\ldots$ is a repeating decimal.

Repetend — A repetend is a symbol that is used to shorten the way an infinitely repeating decimal is written.

$0.3333\ldots$ can be written as $0.\overline{3}$ and is read as "point 3 repetend."

Rounding — Rounding is an expression used when referring to an approximation for a number.

3.68 rounded to the nearer tenth is 3.7.

Scientific notation — A number is expressed in scientific notation when it is written as a product of a number from 1 to 10 (not including 10) and a power of 10.

0.005128 is written as 5.128×10^{-3} using scientific notation.

Significant digits — Significant digits are digits which indicate the number of units that are reasonably sure of having been counted in making a measurement.

5192 rounded to 3 significant digits is 5190.

Standard deviation — Standard deviation is a number that expresses the degree to which data items in a set are dispersed.

The standard deviation of 2, 3, 5, 6, 9 and 14 is 1.71.

Statistics — Statistics is the study of numerical data.

The number of highway deaths per million miles traveled is statistics.

Sum — The result or answer to an addition problem is called the sum.

In $21 + 13 = 34$, 34 is the sum.

Terminating decimal — A terminating decimal is a decimal that stops or whose remainder is zero.

$\frac{35}{100} = 0.35$ is a terminating decimal.

Truncate — To truncate means to cut a number off at a certain place value position.

$\frac{5}{11} = 5 \div 11 = 0.45454545\underline{5}$ the last digit is cut off and rounded.

Vertex — In a triangle a vertex is the common end-point of 2 line segments.

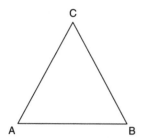

Each point A, B, and C is a vertex.

Volume — Volume is the number of cubic units needed to fill the space inside a figure.

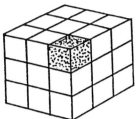 Each cube is 1 cm^3.

The volume of the rectangular solid is 27 cm^3.

Health Occupation Information

Health occupations involve the categories of physicians, nurses, dentists, and therapists as well as the all-important behind-the-scenes categories of technologists, technicians, administrators, and assistants. Each category of health occupations is related to another category and no occupation stands alone.

Health occupations are classified as dental occupations; medical professions; medical technologist, technician, and assistant occupations; nursing occupations; therapy and rehabilitation occupations; and other health occupations. Brief descriptions of job titles in each classification will be presented. For further information, consult governmental publications such as the *Dictionary of Occupational Titles* and the *Occupational Outlook Handbook* or consult private organizations. The *Occupational Outlook Handbook* lists some such organizations and also lists sources of additional information for each particular occupation.

DENTAL OCCUPATIONS

Dental care — preventative and corrective — is an integral part of a person's overall health. There are four key health occupations involving dental care.

- *Dentists* examine teeth and tissues of the mouth to diagnose disease or abnormalities. They also concentrate on preventative medicine through properly educating their patients.

- *Dental hygienists* maintain medical and dental records for each patient. They also scale, clean, and polish the teeth as well as advise the patient on certain prophylaxis procedures. They are often responsible for preparing x-rays of the teeth.

- *Dental assistants* work with the dentists during the treatment of patients. This involvement includes providing the many pieces of equipment and supplies necessary for the treatment. Dental assistants also play a major role in preparing the patient for treatment and providing pre-treatment and post-treatment instructions.

- *Dental laboratory technicians* make dentures, fabricate crowns to restore teeth, construct bridges to replace missing teeth, and make other dental orthodontic appliances. The field of dental technology involves many areas of specialty, but the main concern of all dental laboratory technicians is to provide the patient with a natural appearance.

MEDICAL PROFESSIONS

Medical professionals prevent, cure, and alleviate disease. The group of medical professions is composed of physicians, osteopaths, chiropractors, optometrists, podiatrists, and veterinarians.

Physicians, osteopaths, and chiropractors treat diseases that affect the entire body; chiropractors and osteopaths specialize in the manipulation of muscles and bones. Optometrists specialize in care for the eyes, and podiatrists care for foot disease and deformities. Veterinarians provide care for animals.

All of these medical professionals complete from six to nine years of postsecondary education. Within each category there are assistants and/or technicians and technologists whose occupation usually requires less postsecondary education.

MEDICAL TECHNOLOGIST, TECHNICIAN, AND ASSISTANT OCCUPATIONS

The development of sophisticated diagnostic tools and techniques for treatment of diseases, along with the advances in medical sciences and technology, have produced the need for medical technologists, technicians, and assistants.

- *Electrocardiograph (EKG) technicians* are responsible for correctly setting up the equipment and preparing the patient for an electrocardiogram. They also manipulate the switches of the electrocardiograph and move electrodes across the patient's chest. Electrocardiograph technicians sometimes conduct other tests such as vectorcardiograms and phonocardiograms.

- *Electroencephalographic (EEG) technicians* have the primary responsibility of conducting EEG examinations of a patient. The EEG examination is involved with the gathering of information about the electrical impulses given off by the patient's brain. This information is vital in diagnosing and treating certain types of disorders associated with the nervous system.

- *Medical assistants* aid the doctor in the care and treatment of patients. Their duties may range from making appointments to carrying out various laboratory procedures. They will prepare patients for examination and may gather preliminary data.

- *Medical laboratory workers* are on three different levels: medical laboratory assistants, technicians, and technologists. The duties at each level increase as does the amount of post-secondary education needed. In general, medical laboratory workers perform tests under the direction of pathologists, physicians, or scientists who specialize in clinical chemistry, microbiology, or other biological sciences. They also analyze the blood, tissue, and fluids from the human body.

- *Medical record technicians* are responsible for maintaining up-to-date records for each patient. This requires a system for recording and retrieving data. Much of the recordkeeping is now carried out with the aid of a computer. The records are important for prompt, efficient, and complete treatment of the patient, and also serve as a historical record of the patient.

- *Operating room technicians* may be involved in preoperative, operative, and postoperative activities. This requires them to prepare the operating room and patient for surgery. During the operation, they must supply the necessary equipment for the surgery. Following the surgery, they may be required to assist in patient and operating room cleanup.

- *Optometric assistants* may serve both as secretaries and technicians. They may be responsible for conducting preliminary tests for the optometrist. They may also aid patients in the selection of eyeglasses and provide supervision in certain exercises recommended by the optometrist. In some offices, highly trained optometric technicians may cut and polish eye glass lenses or contact lenses.

- *Radiologic (x-ray) technologists,* working with radiologists, apply x-rays and radioactive substances to patients for diagnostic and therapeutic purposes. The findings of this team may be used by the physician in further treatment of the patients.

- *Respiratory therapy workers,* sometimes called inhalation therapy workers, treat patients who have cardiorespiratory problems. The treatment they provide ranges from giving temporary relief to asthma or emphysema patients to giving emergency care in case of stroke, heart failure, shock, or drowning. This treatment is given by therapists, technicians, or assistants depending on the severity of the problem and the complexity of the treatment. Therapists have the highest level of expertise, followed by technicians, then followed by assistants who are in the process of training.

NURSING OCCUPATIONS

The nursing field consists of registered nurses (RN); licensed practical nurses (LPN); nurse aides or nursing assistants; orderlies; and attendants. The nursing field accounts for one-half of the total employment in the health occupations.

- *Nurses* assume the role of caring for the sick, aiding in the prevention of disease and promoting good health. Nursing assignments may entail a variety of duties and there are many variations in the nursing assignments.

- *Nurse aides or nursing assistants, orderlies and attendants* care for sick and injured people in various ways. They are the all-essential "back-up" team and aid in the efficiency and effectiveness of health care facilities.

THERAPY AND REHABILITATION OCCUPATIONS

Therapy allows handicapped persons to learn to build satisfying and productive lives. People involved in therapy and rehabilitation occupations aid the handicapped in this therapy.

Rehabilitation occupations can be divided into three areas: occupational therapy, physical therapy, and speech therapy and audiology.

- *Occupational therapists* help mentally and physically disabled patients by planning and directing educational, vocational, and recreational activities. They teach manual and creative skills; business and industrial skills, and daily routines such as eating, dressing, and writing. Occupational therapists may design special equipment and adaptive devices.

- *Occupational therapy assistants* may work with both physically and mentally handicapped patients. They assist the therapist in helping the patients adjust to their physical and/or mental limitations. They may serve as a teacher of arts and crafts as well as provide vocational training.

- *Physical therapists* aid people who have muscle, nerve, joint, and bone diseases or injuries in overcoming their disabilities. They perform and interpret tests, prescribe and evaluate treatment, and aid the disabled person in accepting and adjusting to handicaps.

- *Physical therapy assistants and aides* work with patients who are recovering from physical disorders which may be the result of disease, surgery, or accident. Supervised by physical therapists, the assistants help in conducting both physical exercises and educational programs for the patients.

- *Prosthetists and orthotists* work with the physical therapists and the doctors in the development of mechanical aids that allow patients to regain some degree of normal function. This involves the use of many kinds of materials to recreate lost limbs or to prepare braces that strengthen weakened limbs.

- *Speech pathologists* diagnose and treat speech and language problems of children and adults. The speech problems may be the result of hearing loss, brain injury, mental retardation, or emotional problems. Speech pathologists may also conduct research to develop diagnostic and treatment techniques or apparatus. *Speech therapists* work under the supervision of speech pathologists.

- *Audiologists* perform diagnostic evaluations of hearing, prescribe and execute habilitation and rehabilitation treatment, and conduct research. They may design and develop clinical and research procedures and apparatus. Since speech and hearing are so interrelated, competence in one field mandates having a working knowledge in the other field and sometimes performing the same duties.

OTHER HEALTH OCCUPATIONS

The cure and prevention of diseases and illnesses, together with proper nutrition and well-being, unite technology, science, and medicine. There are many occupations in technology and science that contribute to the prevention of disease and the goal to cure and prevent illnesses has encouraged the development of many technological and scientific occupations.

- *Biomedical equipment technicians* serve as a valuable link between the field of medicine and the world of technology. Their assignments include the operation and maintenance of many complex pieces of equipment used in the field of health services. The technicians also play a significant role in the development of new equipment to better serve the patients.

- *Cytotechnologists* assist the pathologist in preparing living or dead cells for detailed examination. Special dyes are used to prepare and stain the cells. The staining allows for detailed identification of specific structures within the cell. A wide variety of medical disorders requires this type of examination.

- *Dietitians* are involved with the careful planning of nutritious and appetizing meals which will aid in maintaining or recovering good health. Dietitians often specialize in a certain area such as administrative, clinical, research, or nutrition. Dietitians are aided by *dietetic technicians* who are responsible for the nutritional care of patients. Dietetic technicians may be involved directly with the patient as they gather information relating to nutritional needs. Some technicians may work in food service administration, research, or with health agencies.

- *Environmental health technicians* work to identify various environmental factors that may influence the continued good health of the community, nation, or world. Water and air pollution are only two of the many areas of concern. These technicians also take an active role in the safety of recreational areas and water-processing systems.

- *Histologic technicians* work closely with the surgeon to determine the cause of numerous disorders. They may prepare frozen sections of tissue removed from the patient. These small sections are then examined with a microscope. The ability to carry out this study while the patient is still on the operating table is vital to the patient's health. This information then becomes useful to the surgeon in determining the extent of surgery to be conducted.

- *Medical or dental secretaries* must have many of the skills common to an executive secretary. Further, they must have a knowledge of medical terminology. The duties may vary but they must be able to type, take shorthand, arrange meetings, and maintain certain financial records. They handle the clerical duties essential to the efficient schedule of a physician.

- *Mental health technicians* may conduct interviews or even serve as teachers for patients. They may provide assistance to mentally ill or mentally retarded patients who need to adjust to the surrounding community.

Index